ADVANCED MECHANICS AND GENERAL RELATIVITY

Aimed at advanced undergraduates with background knowledge of classical mechanics and electricity and magnetism, this textbook presents both the particle dynamics relevant to general relativity, and the field dynamics necessary to understand the theory.

Focusing on action extremization, the book develops the structure and predictions of general relativity by analogy with familiar physical systems. Topics ranging from classical field theory to minimal surfaces and relativistic strings are covered in a consistent manner. Nearly 150 exercises and numerous examples throughout the textbook enable students to test their understanding of the material covered. A tensor manipulation package to help students overcome the computational challenge associated with general relativity is available on a site hosted by the author. A link to this and to a solutions manual can be found at www.cambridge.org/9780521762458.

JOEL FRANKLIN is an Assistant Professor in the physics department of Reed College. His work spans a variety of fields, including stochastic Hamiltonian systems (both numerical and mathematical), modifications of general relativity, and their observational implications.

ADVANCED MECHANICS AND GENERAL RELATIVITY

JOEL FRANKLIN

Reed College

CAMBRIDGE
UNIVERSITY PRESS

University Printing House, Cambridge CB2 8BS, United Kingdom

One Liberty Plaza, 20th Floor, New York, NY 10006, USA

477 Williamstown Road, Port Melbourne, VIC 3207, Australia

314-321, 3rd Floor, Plot 3, Splendor Forum, Jasola District Centre, New Delhi - 110025, India

79 Anson Road, #06-04/06, Singapore 079906

Cambridge University Press is part of the University of Cambridge.

It furthers the University's mission by disseminating knowledge in the pursuit of
education, learning and research at the highest international levels of excellence.

www.cambridge.org
Information on this title: www.cambridge.org/9780521762458

© J. Franklin 2010

This publication is in copyright. Subject to statutory exception
and to the provisions of relevant collective licensing agreements,
no reproduction of any part may take place without the written
permission of Cambridge University Press.

First published 2010

A catalogue record for this publication is available from the British Library

Library of Congress Cataloging in Publication data
Franklin, Joel, 1975–
Advanced mechanics : an introduction to general relativity / Joel Franklin.
p. cm.
Includes bibliographical references and index.
ISBN 978-0-521-76245-8 (hardback)
1. General relativity (Physics) 2. Mechanics. I. Title.
QC173.6.F73 2010
530.11 – dc22 2010011442

ISBN 978-0-521-76245-8 Hardback

Additional resources for this publication at www.cambridge.org/9780521762458

Cambridge University Press has no responsibility for the persistence or
accuracy of URLs for external or third-party internet websites referred to in
this publication, and does not guarantee that any content on such websites is,
or will remain, accurate or appropriate.

For Lancaster, Lewis, and Oliver

Contents

Preface

Classical mechanics, as a subject, is broadly defined. The ultimate goal of mechanics is a complete description of the motion of particles and rigid bodies. To find $\mathbf{x}(t)$ (the position of a particle, say, as a function of time), we use Newton's laws, or an updated (special) relativistic form that relates changes in momenta to forces. Of course, for most interesting problems, it is not possible to solve the resulting second-order differential equations for $\mathbf{x}(t)$. So the content of classical mechanics is a variety of techniques for describing the motion of particles and systems of particles in the absence of an explicit solution. We encounter, in a course on classical mechanics, whatever set of tools an author or teacher has determined are most useful for a partial description of motion. Because of the wide variety of such tools, and the constraints of time and space, the particular set that is presented depends highly on the type of research, and even personality of the presenter.

This book, then, represents a point of view just as much as it contains information and techniques appropriate to further study in classical mechanics. It is the culmination of a set of courses I taught at Reed College, starting in 2005, that were all meant to provide a second semester of classical mechanics, generally to physics seniors. One version of the course has the catalog title "Classical Mechanics II", the other "Classical Field Theory". I decided, in both instantiations of the course, to focus on general relativity as a target. The classical mechanical tools, when turned to focus on problems like geodesic motion, can take a student pretty far down the road toward motion in arbitrary space-times. There, the Lagrangian and Hamiltonian are used to expose various constants of the motion, and applying these to more general space-times can be done easily. In addition, most students are familiar with the ideas of coordinate transformation and (Cartesian) tensors, so much of the discussion found in a first semester of classical mechanics can be modified to introduce the geometric notions of metric and connection, even in flat space and space-time.

So my first goal was to exploit students' familiarity with classical mechanics to provide an introduction to the geometric properties of motion that we find in general relativity. We begin, in the first chapter, by reviewing Newtonian gravity, and simultaneously, the role of the Lagrangian and Hamiltonian points of view, and the variational principles that connect the two. Any topic that benefits from both approaches would be a fine vehicle for this first chapter, but given the ultimate goal, Newtonian gravity serves as a nice starting point. Because students have seen Newtonian gravity many times, this is a comfortable place to begin the shift from $L = \frac{1}{2} m v^2 - U$ to an understanding of the Lagrangian as a geometric object. The metric and its derivatives are introduced in order to make the "length-minimizing" role of the free Lagrangian clear, and to see how the metric dependence on coordinates can show up in the equations of motion (also a familiar idea).

Once we have the classical, classical mechanics reworked in a geometric fashion, we are in position to study the simplest modification to the underlying geometry – moving the study of dynamics from Euclidean flat space (in curvilinear coordinates) to Minkowski space-time. In the second chapter, we review relativistic dynamics, and its Lagrange and Hamiltonian formulation, including issues of parametrization and interpretation that will show up later on. Because of the focus on the role of forces in determining the dynamical properties of relativistic particles, an advertisement of the "problem" with the Newtonian gravitational force is included in this chapter. That problem can be seen by analogy with electrodynamics – Newtonian gravity is not in accord with special relativity, with deficiency similar in character to Maxwell's equations with no magnetic field component. So we learn that relativistic dynamics requires relativistic forces, and note that Newtonian gravity is not an example of such a force.

Going from Euclidean space in curvilinear coordinates to Minkowski space-time (in curvilinear coordinates, generally) represents a shift in geometry. In the third chapter, we return to tensors in the context of these flat spaces, introducing definitions and examples meant to motivate the covariant derivative and associated Christoffel connection. These exist in flat space(-time), so there is an opportunity to form a connection between tensor ideas and more familiar versions found in vector calculus. To understand general relativity, we need to be able to characterize space-times that are not flat. So, finally, in the fourth chapter, we leave the physical arena of most of introductory physics and discuss the idea of curvature, and the manner in which we will quantify it. This gives us our first introduction to the Riemann tensor and a bit of Riemannian geometry, just enough, I hope, to keep you interested, and provide a framework for understanding Einstein's equation. At the end of the chapter, we see the usual motivation of Einstein's equation, as an attempt to modify Newton's second law, together with Newtonian gravity, under the influence of the weak equivalence principle – we are asking: "under what conditions can the motion of classical bodies that interact gravitationally, be viewed as length-minimizing paths in a curved space-time?" This is Einstein's idea, if everything undergoes the same motion (meaning acceleration, classically), then perhaps that motion is a feature of space-time, rather than forces.

At this point in the book, an abrupt shift is made. What happened is that I was asked to teach "Classical Field Theory", a different type of second semester of classical mechanics geared toward senior physics majors. In the back of most classical mechanics texts, there is a section on field theory, generally focused on fluid dynamics as its end goal. I again chose general relativity as a target – if geodesics and geometry can provide an introduction to the motion side of GR in the context of advanced mechanics, why not use the techniques of classical field theory to present the field-theoretic (meaning Einstein's equation again) end of the same subject? This is done by many authors, notably Thirring and Landau and Lifschitz. I decided to focus on the idea that, as a point of physical model-building, if you start off with a second-rank, symmetric tensor field on a Minkowski background, and require that the resulting theory be self-consistent, you end up, almost uniquely, with general relativity. I learned this wonderful idea (along with most of the rest of GR) directly from Stanley Deser, one of its originators and early proponents. My attempt was to build up enough field theory to make sense of the statement for upper-level undergraduates with a strong background in E&M and quantum mechanics.

So there is an interlude, from one point of view, amplification, from another, that covers an alternate development of Einstein's equation. The next two chapters detail the logic of constructing relativistic field theories for scalars (massive Klein–Gordon), vectors (Maxwell and Proca), and second-rank symmetric tensors (Einstein's equation). I pay particular attention to the vector case – there, if we look for a relativistic, linear, vector field equation, we get E&M almost uniquely (modulo mass term). The coupling of E&M to other field theories also shares similarities with the coupling of field theories to GR, and we review that aspect of model-building as well. As we move, in Chapter 6, to general relativity, I make heavy use of E&M as a theory with much in common with GR, another favorite technique of Professor Deser. At the end of the chapter, we recover Einstein's equation, and indeed, the geometric interpretation of our second-rank, symmetric, relativistic field as a metric field. The digression, focused on fields, allows us to view general relativity, and its interpretation, in another light.

Once we have seen these two developments of the same theory, it is time (late in the game, from a book point of view) to look at the physical implications of solutions. In Chapter 7, we use the Weyl method to develop the Schwarzschild solution, appropriate to the exterior of spherically symmetric static sources, to Einstein's equation. This is the GR analogue of the Coulomb field from E&M,

and shares some structural similarity with that solution (as it must, in the end, since far away from sources, we have to recover Newtonian gravity), and we look at the motion of test particles moving along geodesics in this space-time. In that setting, we recover perihelion precession (massive test particles), the bending of light (massless test particles), and gravitational redshift. This first solution also provides a venue for discussing the role of coordinates in a theory that is coordinate-invariant, so we look at the various coordinate systems in which the Schwarzschild space-time can be written and its physical implications uncovered.

Given the role of gravitational waves in current experiments (like LIGO), I choose radiation as a way of looking at additional solutions to Einstein's equation in vacuum. Here, the linearized form of the equations is used, and contact is again made with radiation in E&M. There are any number of possible topics that could have gone here – cosmology would be an obvious one, as it allows us to explore non-vacuum solutions. But, given the field theory section of the book, and the view that Maxwell's equations can be used to inform our understanding of GR, gravitational waves are a natural choice.

I have taken two routes through the material found in this book, and it is the combination of these two that informs its structure. For students who are interested in classical mechanical techniques and ideas, I cover the first four chapters, and then move to the last three – so we see the development of Einstein's equation, its role in determining the physical space-time outside a spherically symmetric massive body, and the implications for particles and light. If the class is focused on field theory, I take the final six chapters to develop content. Of course, strict adherence to the chapters will not allow full coverage – for a field theory class, one must discuss geodesic and geometric notions for the punchline of Chapter 7 to make sense. Similarly, if one is thinking primarily about classical mechanics, some work on the Einstein–Hilbert action must be introduced so that the Weyl method in Chapter 8 can be exploited.

Finally, the controversial ninth chapter – here I take some relevant ideas from the program of "advanced mechanics" and present them quickly, just enough to whet the appetite. The Kerr solution for the space-time outside a spinning massive sphere can be understood, qualitatively and only up to a point, by analogy with a spinning charged sphere from E&M. The motion of test bodies can be qualitatively understood from this analogy. In order to think about more exotic motion, we spend some time discussing numerical solution to ODEs, with an eye toward the geodesic equation of motion in Kerr space-time. Then, from our work understanding metrics, and relativistic dynamics, combined with the heavy use of variational ideas throughout the book, a brief description of the physics of relativistic strings is a natural topic. We work from area-minimization in Euclidean spaces to area-minimization in Minkowski space-times, and end up with the standard equations of motion for strings.

I have made available, and refer to, a minimal `Mathematica` package that is meant to ease some of the computational issues associated with forming the fundamental tensors of Riemannian geometry. While I do believe students should compute, by hand, on a large piece of paper, the components of a nontrivial Riemann tensor, I do not want to let such computations obscure the utility of the Riemann tensor in geometry or its role for physics. So, when teaching this material, I typically introduce the package (with supporting examples, many drawn from the longer homework calculations) midway through the course. Nevertheless, I hope it proves useful for students learning geometry, and that they do not hesitate to use the package whenever appropriate.

A note on the problems in this book. There are the usual set of practice problems, exercises to help learn and work with definitions. But, in addition, I have left some relatively large areas of study in the problems themselves. For example, students develop the Weyl metric, appropriate to axially symmetric space-times, in a problem. The rationale is that the Weyl metric is an interesting solution to Einstein's equation in vacuum, and yet, few astrophysical sources exhibit this axial symmetry. It is an important solution, but exploring the detailed physics of the solution is, to a certain extent, an aside. In the end, I feel that students learn best when they develop interesting (if known) ideas on their own. That is certainly the case for research, and I think problems can provide an introduction to that process. In addition to practicing the techniques discussed in the text, working out long, involved, and physically interesting problems gives students a sense of ownership, and aids retention. Another example is the verification that the Kerr solution to Einstein's equation is in fact a vacuum solution. Here, too, a full derivation of Kerr is beyond the techniques introduced within the book, so I do not consider the derivation to be a primary goal – verification, however, is a must, and can be done relatively quickly with the tools provided. I have marked these more involved problems with a ∗ to indicate that they are important, but may require additional tools or time.

As appears to be current practice, I am proud to say that there are no new ideas in this book. General relativity is, by now, almost a century old, and the classical mechanical techniques brought to its study, much older. I make a blanket citation to all of the components of the Bibliography (found at the end), and will point readers to specific works as relevant within the text.

Acknowledgments

I would like to thank my teachers, from undergraduate to postdoctoral: Nicholas Wheeler, Stanley Deser, Sebastian Doniach, and Scott Hughes, for their thoughtful advice, gentle criticism, not-so-gentle criticism, and general effectiveness in teaching me something (not always what they intended).

I have benefitted greatly from student input,[1] and have relied almost entirely on students to read and comment on the text as it was written. In this context, I would like to thank Tom Chartrand, Zach Schultz, and Andrew Rhines. Special thanks goes to Michael Flashman who worked on the solution manual with me, and provided a careful, critical reading of the text as it was prepared for publication.

The Reed College physics department has been a wonderful place to carry out this work – my colleagues have been helpful and enthusiastic as I attempted to first teach, and then write about, general relativity. I would like to thank Johnny Powell and John Essick for their support and advice. Also within the department, Professor David Griffiths read an early draft of this book, and his comments and scathing criticism have been addressed in part – his help along the way has been indispensable.

Finally, Professor Deser introduced me to general relativity, and I thank him for sharing his ideas, and commentary on the subject in general, and for this book in particular. Much of the presentation has been informed by my contact with him – he has been a wonderful mentor and teacher, and working with him is always a learning experience, that is to say, a great pleasure.

---

[1] The *Oxford English Dictionary* defines a student to be "A person who is engaged in or addicted to study" – from that point of view, we are all students, so here I am referring to "younger" students, and specifically, younger students at Reed College.

1

Newtonian gravity

The first job, in studying general relativity, must be to establish the predictions and limitations of Newtonian gravity, its immediate predecessor. So we begin by studying the usual mechanical formulations of central potentials associated with spherically symmetric central bodies. We shall study the orbital motion of "test particles", and in general, review the language and results of classical gravity.

1.1 Classical mechanics

The program of classical mechanics is to determine the trajectory of a particle or system of particles moving under the influence of some force. The connection between force and motion is provided by Newton's second law:

$$m \ddot{\mathbf{x}}(t) = \mathbf{F}, \tag{1.1}$$

supplemented by appropriate initial or boundary conditions. The forces are provided ($\mathbf{F}$ might be made up of a number of different forces) and we solve the above for $\mathbf{x}(t)$, from which any measurable quantity can be predicted.

As an approach to solving problems, Newton's second law can be difficult to work with. Given a generic force and multiple particles, a direct and complete solution proceeding from (1.1) is often unattainable. So we content ourselves with supplying less information than the full $\mathbf{x}(t)$ (sometimes, for example, we can easily find $\dot{\mathbf{x}}(t)$, but cannot proceed to $\mathbf{x}(t)$), or, we work on special classes of forces for which alternate formulations of the second law are tractable. It is with the latter that we will begin our work on Newtonian gravity – following a short review of the Lagrangian formulation of the equations of motion given a force derivable from a potential, we will see how the Lagrange approach can be used to simplify and solve for the trajectories associated with the Newtonian central potential. From there, we will move on to the Hamiltonian formulation of the same problem.

1.2 The classical Lagrangian

Here we will define the Lagrangian formulation of the fundamental problem of classical mechanics: "Given a potential, how do particles move?" This section serves as a short review of Lagrangians in general, and the next section will specialize to focus on Keplerian orbits in the classical setting – if we are to understand the changes to the motion of particles in general relativity (GR), it behooves us to recall the motion in the "normal" case. Our ultimate goal is to shift from the specific sorts of notations used in introductory cases (for example, spherical coordinates), to a more abstract notation appropriate to the study of particle motion in general relativity.

As we go, we will introduce some basic tensor operations, but there will be more of this to come in Chapter 3. We just need to become comfortable with summation notation for now.

1.2.1 Lagrangian and equations of motion

A Lagrangian is the integrand of an action – while this is not the usual definition, it is, upon definition of action, more broadly applicable than the usual "kinetic minus potential" form. In classical mechanics, the Lagrangian leading to Newton's second law reads, in Cartesian coordinates:[1]

$$L = \underbrace{\frac{1}{2} m \left(\mathbf{v}(t) \cdot \mathbf{v}(t) \right)}_{\equiv T} - U(\mathbf{x}(t)), \tag{1.2}$$

where we view x, y and z as functions of a parameter t which we normally interpret as "time". The first term is the kinetic energy (denoted T), the second is the potential energy. Remember, the ultimate goal of classical mechanics is to find the trajectory of a particle under the influence of a force. Physically, we control the description of the system by specifying the particle mass, form of the force or potential, and boundary conditions (particle starts from rest, particle moves from point a to point b, etc.). Mathematically, we use the equations of motion derived from the Lagrangian, together with the boundary conditions, to determine the curve $\mathbf{x}(t) = x(t)\,\hat{\mathbf{x}} + y(t)\,\hat{\mathbf{y}} + z(t)\,\hat{\mathbf{z}}$ through three-dimensional space.

---

[1] I will refer to the "vector" (more appropriately, the coordinate differential is the vector) of coordinates as $\mathbf{x} = x\,\hat{\mathbf{x}} + y\,\hat{\mathbf{y}} + z\,\hat{\mathbf{z}}$, and its time-derivative (velocity) as $\mathbf{v} = \frac{d\mathbf{x}}{dt}$.

Extremization of an action

The Euler–Lagrange equations come from the extremization, in the variational calculus sense, of the action:

$$S[\mathbf{x}(t)] = \int L(\mathbf{x}(t), \mathbf{v}(t))\, dt. \tag{1.3}$$

We imagine a path connecting two points $\mathbf{x}(0)$ and $\mathbf{x}(T)$, say. Then we define the dynamical trajectory to be the unique path that extremizes S. Suppose we have an arbitrary $\mathbf{x}(t)$ with the correct endpoints, and we perturb it slightly via $\mathbf{x}(t) \longrightarrow \mathbf{x}(t) + \boldsymbol{\eta}(t)$. In order to leave the physical observation of the endpoints unchanged, we require $\boldsymbol{\eta}(0) = \boldsymbol{\eta}(T) = 0$. The action responds to this change:

$$\begin{aligned} S[\mathbf{x}(t) + \boldsymbol{\eta}(t)] &= \int_0^T L(\mathbf{x}(t) + \boldsymbol{\eta}(t), \dot{\mathbf{x}}(t) + \dot{\boldsymbol{\eta}}(t))\, dt \\ &\approx \int_0^T \left(L(\mathbf{x}(t), \dot{\mathbf{x}}(t)) + \frac{\partial L}{\partial \mathbf{x}} \cdot \boldsymbol{\eta} + \frac{\partial L}{\partial \dot{\mathbf{x}}} \cdot \dot{\boldsymbol{\eta}} \right) dt, \end{aligned} \tag{1.4}$$

where the second line is just the Taylor expansion of the integrand to first order in $\boldsymbol{\eta}$. The unperturbed value $S[\mathbf{x}(t)]$ is recognizable from the leading term in the integrand, all the rest is the change:

$$\Delta S = S[\mathbf{x}(t) + \boldsymbol{\eta}(t)] - S[\mathbf{x}(t)] = \int_0^T \left(\frac{\partial L}{\partial \mathbf{x}} \cdot \boldsymbol{\eta} + \frac{\partial L}{\partial \dot{\mathbf{x}}} \cdot \dot{\boldsymbol{\eta}} \right) dt. \tag{1.5}$$

The small change $\boldsymbol{\eta}(t)$ is arbitrary, but once chosen, its time-derivative is fixed. We would like to write the integrand of (1.5) entirely in terms of the arbitrary trajectory perturbation, $\boldsymbol{\eta}(t)$, rather than quantities derived from this. We can use integration by parts on the second term to "flip" the t-derivative onto the L-derivative:

$$\int_0^T \frac{\partial L}{\partial \dot{\mathbf{x}}} \cdot \dot{\boldsymbol{\eta}}\, dt = \int_0^T \left(\frac{d}{dt}\left(\frac{\partial L}{\partial \dot{\mathbf{x}}} \cdot \boldsymbol{\eta} \right) - \boldsymbol{\eta} \cdot \frac{d}{dt}\left(\frac{\partial L}{\partial \dot{\mathbf{x}}} \right) \right) dt. \tag{1.6}$$

The first term, as a total time-derivative, gets evaluated at $t = 0$ and $t = T$ where $\boldsymbol{\eta}$ vanishes. We can use (1.6) to make the replacement under the integral in (1.5):

$$\frac{\partial L}{\partial \dot{\mathbf{x}}} \cdot \dot{\boldsymbol{\eta}} \longrightarrow -\boldsymbol{\eta} \cdot \frac{d}{dt}\left(\frac{\partial L}{\partial \dot{\mathbf{x}}} \right), \tag{1.7}$$

and this leaves us with ΔS that depends on $\boldsymbol{\eta}(t)$ only:

$$\Delta S = \int_0^T \left(\frac{\partial L}{\partial \mathbf{x}} - \frac{d}{dt}\left(\frac{\partial L}{\partial \dot{\mathbf{x}}} \right) \right) \cdot \boldsymbol{\eta}\, dt. \tag{1.8}$$

Now extremization means $\Delta S = 0$, and the arbitrary value of $\eta(t)$ allows us to set the term in parentheses equal to zero by itself (that's the only way to get $\Delta S = 0$ for arbitrary $\eta(t)$).

As a point of notation, we use the variational derivative symbol δ to indicate that we have performed all appropriate integration by parts, so you will typically see (1.8) written as:

$$\delta S = \int_0^T \left(\frac{\partial L}{\partial \mathbf{x}} - \frac{d}{dt} \left(\frac{\partial L}{\partial \dot{\mathbf{x}}} \right) \right) \cdot \delta \mathbf{x} \, dt, \tag{1.9}$$

where $\delta \mathbf{x}$ replaces η – this tells us that it is the variation with respect to $\mathbf{x}$ that is inducing the change in S. For actions that depend on more than one variable that can be varied, the notation makes it clear which one is being varied. In addition to this δS, $\delta \mathbf{x}$ shift, we will also refer to the Euler–Lagrange equations from variation with respect to $\mathbf{x}$ as:

$$\frac{\delta S}{\delta \mathbf{x}} = \left(\frac{\partial L}{\partial \mathbf{x}} - \frac{d}{dt} \left(\frac{\partial L}{\partial \dot{\mathbf{x}}} \right) \right), \tag{1.10}$$

the "variational derivative" of S with respect to $\mathbf{x}$. Extremization is expressed by $\delta S = 0$, or equivalently in this case, $\frac{\delta S}{\delta \mathbf{x}} = 0$.

Variation provides the ordinary differential equation (ODE) structure of interest, a set of three second-order differential equations, the Euler–Lagrange equations of motion:

$$\boxed{\frac{d}{dt} \frac{\partial L}{\partial \mathbf{v}} - \frac{\partial L}{\partial \mathbf{x}} = 0.} \tag{1.11}$$

In Cartesian coordinates, with the Lagrangian from (1.2), the Euler–Lagrange equations reproduce Newton's second law given a potential U:

$$m \, \ddot{\mathbf{x}}(t) = -\nabla U. \tag{1.12}$$

The advantage of the action approach, and the Lagrangian in particular, is that the equations of motion can be obtained for any coordinate representation of the kinetic energy and potential. Although it is easy to define and verify the correctness of the Euler–Lagrange equations in Cartesian coordinates, they are not necessary to the formulation of valid equations of motion for systems in which Cartesian coordinates are less physically and mathematically useful.

The Euler–Lagrange equations, in the form (1.11), hold regardless of our association of $\mathbf{x}$ with Cartesian coordinates. Suppose we move to cylindrical coordinates $\{s, \phi, z\}$, defined by

$$x = s \cos \phi \quad y = s \sin \phi \quad z = z, \tag{1.13}$$

then the Lagrangian in Cartesian coordinates can be transformed to cylindrical coordinates by making the replacement for $\{x, y, z\}$ in terms of $\{s, \phi, z\}$ (and associated substitutions for the Cartesian velocities):

$$L(s, \phi, z) = L(\mathbf{x}(s, \phi, z)) = \frac{1}{2} m \left(\dot{s}^2 + s^2 \dot{\phi}^2 + \dot{z}^2 \right) - U(s, \phi, z). \qquad (1.14)$$

But, the Euler–Lagrange equations require no modification, the variational procedure that gave us (1.11) can be applied in the cylindrical coordinates, giving three equations of motion:

$$0 = \frac{d}{dt} \frac{\partial L}{\partial \dot{s}} - \frac{\partial L}{\partial s}$$

$$0 = \frac{d}{dt} \frac{\partial L}{\partial \dot{\phi}} - \frac{\partial L}{\partial \phi} \qquad (1.15)$$

$$0 = \frac{d}{dt} \frac{\partial L}{\partial \dot{z}} - \frac{\partial L}{\partial z}.$$

The advantage is clear: coordinate transformation occurs once and only once, in the Lagrangian. If we were to start with Newton's second law, we would have three equations with acceleration and coordinates coupled together. The decoupling of these would, in the end, return (1.15).

1.2.2 Examples

In one dimension, we can consider the Lagrangian $L = \frac{1}{2} m \dot{x}^2 - \frac{1}{2} k (x - a)^2$, appropriate to a spring potential with spring constant k and equilibrium spacing a. Then the Euler–Lagrange equations give:

$$\frac{d}{dt} \frac{\partial L}{\partial \dot{x}} - \frac{\partial L}{\partial x} = m \ddot{x} + k (x - a) = 0. \qquad (1.16)$$

Notice that for a real physical problem, the above equation of motion is not enough – we also need to specify two boundary conditions. We can phrase this choice in terms of boundaries in time at $t = t_0$ and $t = t_f$ (particle starts at 1 m from the origin at $t = 0$ and ends at 2 m from the origin at $t = 10$ s), or as an initial position and velocity (particle starts at equilibrium position with speed 5 m/s) – there are other choices as well, depending on our particular experimental setup.

In two dimensions, we can express a radial spring potential as:

$$L = \frac{1}{2} m (\dot{x}^2 + \dot{y}^2) - \frac{1}{2} k (\sqrt{x^2 + y^2} - a)^2, \qquad (1.17)$$

giving us two equations of motion:

$$0 = m\,\ddot{x} + \frac{k\,x\,(\sqrt{x^2 + y^2} - a)}{\sqrt{x^2 + y^2}}$$

$$0 = m\,\ddot{y} + \frac{k\,y\,(\sqrt{x^2 + y^2} - a)}{\sqrt{x^2 + y^2}}. \tag{1.18}$$

Suppose we want to transform to two-dimensional polar coordinates via $x = s\cos\phi$ and $y = s\sin\phi$ – we can write (1.18) in terms of the derivatives of $s(t)$ and $\phi(t)$ and solve for $\ddot{s}$ and $\ddot{\phi}$ to get:

$$\ddot{s} = -\frac{k}{m}\,(s - a) + s\,\dot{\phi}^2$$

$$\ddot{\phi} = -\frac{2\,\dot{\phi}\,\dot{s}}{s}. \tag{1.19}$$

Actually performing this decoupling is not easy, and you should try the procedure for a few different coordinate system choices to appreciate the efficiency of the Lagrangian approach.

We can take advantage of the coordinate-independence of the Lagrangian to rewrite L directly in terms of $s(t)$ and $\phi(t)$, where it becomes:

$$L = \frac{1}{2}\,m\,(\dot{s}^2 + s^2\dot{\phi}^2) - \frac{1}{2}\,k\,(s - a)^2. \tag{1.20}$$

Then the Euler–Lagrange analogues of (1.15) become:

$$\frac{d}{dt}\frac{\partial L}{\partial \dot{s}} - \frac{\partial L}{\partial s} = m\,\ddot{s} - m\,s\,\dot{\phi}^2 + k\,(s - a) = 0$$

$$\frac{d}{dt}\frac{\partial L}{\partial \dot{\phi}} - \frac{\partial L}{\partial \phi} = m\,(2\,s\,\dot{s}\,\dot{\phi} + s^2\,\ddot{\phi}) = 0, \tag{1.21}$$

precisely (1.19), and well worth the effort of rewriting just the Lagrangian.

The Lagrangian written in polar coordinates (1.20) also provides insight into the motivation for using different coordinate systems. Notice that in the cylindrical case, the ϕ coordinate does not appear in the Lagrangian at all, only $\dot{\phi}$ shows up. Then we know from the Euler–Lagrange equations of motion that:

$$\frac{d}{dt}\frac{\partial L}{\partial \dot{\phi}} = 0 \longrightarrow \frac{\partial L}{\partial \dot{\phi}} = \text{constant of the motion.} \tag{1.22}$$

When possible, this type of observation can be useful in actually solving the equations of motion. Finding (or constructing) a coordinate system in which one or more of the coordinates do not appear is one of the goals of Hamilton–Jacobi theory.

Problem 1.1

Generate the equations of motion for the one-dimensional Lagrangian:

$$L = \frac{1}{2} m \dot{x}^2 - (A x + B).\tag{1.23}$$

From the equations of motion (and the implicit definition of the potential), provide a physical interpretation for the constants A and B.

Problem 1.2

The Euler–Lagrange equations come from extremization of the action. So we expect the "true", dynamical trajectory to minimize (in this case) the value of $S = \int L\,dt$. For free particle motion, the Lagrangian is $L = \frac{1}{2} m v^2$ (in one dimension, for a particle of mass m). Suppose we start at $x(0) = 0$ and at time T, we end up at $x(T) = x_f$. The solution to the equation of motion is:

$$x(t) = \frac{x_f\,t}{T}.\tag{1.24}$$

(a) Compute $S = \int_0^T L\,dt$ for this trajectory.

(b) Any function that goes to zero at the endpoints can be represented in terms of the sine series, vanishing at $t = 0$ and $t = T$: $\sum \alpha_j \sin\left(\frac{j\pi t}{T}\right)$ by appropriate choice of the coefficients $\{\alpha_j\}_{j=1}^\infty$. So "any" trajectory for the particle could be represented by:

$$x(t) = \frac{x_f\,t}{T} + \sum_{j=1}^{\infty} \alpha_j \sin\left(\frac{j\pi t}{T}\right).\tag{1.25}$$

Find the value of the action $S = \int_0^T L\,dt$ for this arbitrary trajectory, show (assuming $\alpha_j \in \mathbb{R}$) that the value of the action for this arbitrary trajectory is greater than the value you get for the dynamical trajectory.

Problem 1.3

Take a potential in cylindrical coordinates $U(s)$ and write out the Euler–Lagrange equations (1.15) for the Lagrangian written in cylindrical coordinates (1.14). Verify that you get the same equations starting from Newton's second law in Cartesian coordinates and transforming to cylindrical coordinates, then isolating $\ddot{s}$, $\ddot{\phi}$ and $\ddot{z}$.

1.3 Lagrangian for $U(r)$

We want to find the parametrization of a curve $\mathbf{x}(t)$ corresponding to motion under the influence of a central potential. Central potentials depend only on a particle's distance from some origin, so they take the specific form: $U(x, y, z) = U(r)$ with $r^2 \equiv x^2 + y^2 + z^2$. We know, then, that the associated force will be directed, from

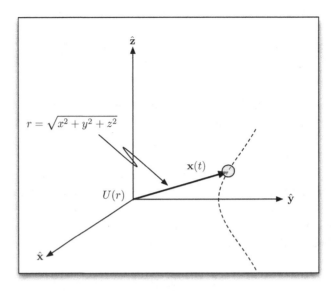

Figure 1.1   A particle trajectory. We want to find the curve $\mathbf{x}(t)$ from the equations of motion.

some origin, either toward or away from the particle (the force is $-\nabla U \sim \hat{\mathbf{r}}$, the usual result familiar from electrostatics, for example). Refer to Figure 1.1.

The Lagrangian for the problem is:

$$L = T - U = \frac{1}{2} m \left(\dot{x}^2 + \dot{y}^2 + \dot{z}^2 \right) - U(r). \tag{1.26}$$

We can move to spherical coordinates using the definition of $\{x, y, z\}$ in terms of $\{r, \theta, \phi\}$:

$$
\begin{aligned}
x &= r \, \sin\theta \, \cos\phi \\
y &= r \, \sin\theta \, \sin\phi \\
z &= r \, \cos\theta.
\end{aligned}
\tag{1.27}
$$

Then the kinetic term in the Lagrangian will be made up of the following derivatives:

$$
\begin{aligned}
\dot{x}^2 &= \left(\dot{r} \, \sin\theta \, \cos\phi + r \, \dot{\theta} \, \cos\theta \, \cos\phi - r \, \dot{\phi} \, \sin\theta \, \sin\phi \right)^2 \\
\dot{y}^2 &= \left(\dot{r} \, \sin\theta \, \sin\phi + r \, \dot{\theta} \, \cos\theta \, \sin\phi + r \, \dot{\phi} \, \sin\theta \cos\phi \right)^2 \\
\dot{z}^2 &= \left(\dot{r} \, \cos\theta - r \, \dot{\theta} \, \sin\theta \right)^2,
\end{aligned}
\tag{1.28}
$$

and inserting the coordinate and velocity forms into (1.26), we have:

$$\boxed{L = \frac{1}{2} m \left(\dot{r}^2 + r^2 \, \dot{\theta}^2 + r^2 \, \sin^2\theta \, \dot{\phi}^2 \right) - U(r).} \tag{1.29}$$

The equations of motion for this Lagrangian are the usual ones:

$$0 = \frac{d}{dt}\frac{\partial L}{\partial \dot r} - \frac{\partial L}{\partial r} = m\ddot r - m r\left(\dot\theta^2 + \sin^2\theta\,\dot\phi^2\right) + \frac{\partial U}{\partial r}$$

$$0 = \frac{d}{dt}\frac{\partial L}{\partial \dot\theta} - \frac{\partial L}{\partial \theta} = m r^2\ddot\theta + 2m r\dot r\dot\theta - m r^2\sin\theta\,\cos\theta\,\dot\phi^2$$

$$0 = \frac{d}{dt}\frac{\partial L}{\partial \dot\phi} - \frac{\partial L}{\partial \phi} = m r^2\sin^2\theta\,\ddot\phi + 2m r\sin^2\theta\,\dot r\dot\phi + 2m r^2\sin\theta\,\cos\theta\,\dot\theta\,\dot\phi.$$

$$(1.30)$$

Now none of us would have gone this far without using some simplifying assumptions (angular momentum conservation, for instance), but the form of the above is interesting. Notice that there is only one term that we would associate with the physical environment (only one term involves the potential), the rest are somehow residuals of the coordinate system we are using.

More interesting than that is the structure of the equations of motion – everything that isn't $\ddot X$ (or $\frac{\partial U}{\partial r}$) looks like $f(r, \theta)\,\dot X\,\dot Y$ (here, $X, Y \in \{r, \theta, \phi\}$). That is somewhat telling, and says more about the structure of the Lagrangian and its quadratic dependence on velocities than anything else. Setting aside the details of spherical coordinates and central potentials, we can gain insight into the classical Lagrangian by looking at it from a slightly different point of view – one that will allow us to generalize it appropriately to both special relativity and general relativity. We will return to the central potential after a short notational aside.

1.3.1 The metric

And so, innocuously, begins our journey. Let's rewrite (1.26) in matrix–vector notation (we'll take potentials that are arbitrary functions of all three coordinates), the kinetic term is the beneficiary here:

$$L = \frac{1}{2}m\begin{pmatrix}\dot x & \dot y & \dot z\end{pmatrix}\begin{pmatrix}1 & 0 & 0\\ 0 & 1 & 0\\ 0 & 0 & 1\end{pmatrix}\begin{pmatrix}\dot x\\ \dot y\\ \dot z\end{pmatrix} - U(x, y, z). \qquad (1.31)$$

We can make the move to spherical coordinates just by changing our curve coordinates from $\{x, y, z\}$ to $\{r, \theta, \phi\}$ and modifying the matrix:

$$L = \frac{1}{2}m\begin{pmatrix}\dot r & \dot\theta & \dot\phi\end{pmatrix}\begin{pmatrix}1 & 0 & 0\\ 0 & r^2 & 0\\ 0 & 0 & r^2\sin^2\theta\end{pmatrix}\begin{pmatrix}\dot r\\ \dot\theta\\ \dot\phi\end{pmatrix} - U(r, \theta, \phi). \qquad (1.32)$$

But while it takes up a lot more space on the page, this is not so trivial a statement. Think of it this way – consider two points infinitesimally close together in the two coordinate systems, the infinitesimal distance (squared) between the two points can be written in Cartesian or spherical coordinates:

$$ds^2 = dx^2 + dy^2 + dz^2$$
$$ds^2 = dr^2 + r^2(d\theta^2 + \sin^2\theta \, d\phi^2).$$

(1.33)

These are just statements of the Pythagorean theorem in two different coordinate systems. The *distance* between the two points is the same in both, that can't change, but the representation is different.

These distances can also be expressed in matrix–vector form:

$$ds^2 = \begin{pmatrix} dx & dy & dz \end{pmatrix} \begin{pmatrix} 1 & 0 & 0 \\ 0 & 1 & 0 \\ 0 & 0 & 1 \end{pmatrix} \begin{pmatrix} dx \\ dy \\ dz \end{pmatrix}$$

(1.34)

and

$$ds^2 = \begin{pmatrix} dr & d\theta & d\phi \end{pmatrix} \begin{pmatrix} 1 & 0 & 0 \\ 0 & r^2 & 0 \\ 0 & 0 & r^2\sin^2\theta \end{pmatrix} \begin{pmatrix} dr \\ d\theta \\ d\phi \end{pmatrix}.$$

(1.35)

The matrix form for infinitesimal length should look familiar (compare with (1.31) and (1.32)), and this isn't so surprising – velocity is intimately related to infinitesimal displacements. If we ask for the distance traveled along a curve parametrized by t in an infinitesimal interval dt, the answer is provided by:

$$ds = \sqrt{\left(\frac{dx}{dt}dt\right)^2 + \left(\frac{dy}{dt}dt\right)^2 + \left(\frac{dz}{dt}dt\right)^2} = \sqrt{\mathbf{v}\cdot\mathbf{v}}\,dt.$$

(1.36)

We call the matrix in (1.34) and (1.35) the *metric*. It is often represented not as a matrix but as a "second-rank tensor" and denoted $g_{\mu\nu}$. It tells us, given a coordinate system, how to measure distances. In classical mechanics, we usually go the other way around, as we have done here – we figure out how to measure distances in the new coordinates and use that to find $g_{\mu\nu}$ (actually, we rarely bother with the formal name or matrix, just transform kinetic energies and evaluate the equations of motion).

Label the vectors appearing in (1.34) and (1.35) dx^μ, so that the three components associated with $\mu = 1, 2, 3$ correspond to the three components of the vector:

$$dx^\mu \doteq \begin{pmatrix} dx \\ dy \\ dz \end{pmatrix},$$

(1.37)

for example. Then we can define the "Einstein summation notation" to express lengths. Referring to (1.34) for the actual matrix–vector form, we can write:

$$ds^2 = \sum_{\mu=1}^{3} \sum_{\nu=1}^{3} dx^\mu \, g_{\mu\nu} \, dx^\nu$$

$$\equiv dx^\mu \, g_{\mu\nu} \, dx^\nu.$$

(1.38)

The idea behind the notation is that when you have an index appearing twice, as in the top line, the explicit $\sum$ is redundant. The prescription is: take each repeated index and sum it over the dimension of the space. Rename $x = x^1, y = x^2, z = x^3$, then:

$$dx^\mu \, g_{\mu\nu} \, dx^\nu = dx^1 \, g_{11} \, dx^1 + dx^2 \, g_{22} \, dx^2 + dx^3 \, g_{33} \, dx^3$$

$$= \left(dx^1\right)^2 + \left(dx^2\right)^2 + \left(dx^3\right)^2.$$

(1.39)

In general, there would be more terms in the sum, the diagonal form of $g_{\mu\nu}$ simplified life (here). The same equation, $ds^2 = dx^\mu \, g_{\mu\nu} \, dx^\nu$, holds if we take $g_{\mu\nu}$ to be the matrix defined in (1.35) and $x^1 = r$, $x^2 = \theta$, $x^3 = \phi$.

In Einstein summation notation, we sum over repeated indices where one is up, one is down (objects like $g_{\mu\nu}dx_\mu$ are nonsense and will never appear). The repeated index, because it takes on all values $1 \longrightarrow D$ (in this case, $D = 3$ dimensions) has no role in labeling a component, and so can be renamed as we wish, leading to statements like[2] (we reintroduce the summation symbols to make the point clear):

$$ds^2 = dx^\mu \, g_{\mu\nu} \, dx^\nu = \sum_{\mu=1}^{3} \sum_{\nu=1}^{3} dx^\mu \, g_{\mu\nu} \, dx^\nu$$

$$= \sum_{\gamma=1}^{3} \sum_{\nu=1}^{3} dx^\gamma \, g_{\gamma\nu} dx^\nu = dx^\gamma \, g_{\gamma\nu} \, dx^\nu$$

$$= dx^\alpha \, g_{\alpha\beta} \, dx^\beta.$$

(1.40)

Finally, the explicit form of the metric can be recovered from the "line element" (just ds^2 written out). If we are given the line element:

$$ds^2 = dr^2 + r^2 \, d\theta^2 + r^2 \, \sin^2 \theta \, d\phi^2,$$

(1.41)

---

[2] There are a few general properties of the metric that we can assume for all metrics considered here. 1. The metric is symmetric, this is a convenient notational device, we have no reason to expect $dxdy \neq dydx$ in a line element. 2. It does not have to be diagonal. 3. It can depend (as with the spherical metric) on position.

then we know that the metric, in matrix form, is:

$$g_{\mu\nu} \doteq \begin{pmatrix} 1 & 0 & 0 \\ 0 & r^2 & 0 \\ 0 & 0 & r^2 \sin^2\theta \end{pmatrix} \tag{1.42}$$

with coordinate differential:

$$dx^\alpha \doteq \begin{pmatrix} dr \\ d\theta \\ d\phi \end{pmatrix}. \tag{1.43}$$

If, instead, we are given:

$$ds^2 = d\rho^2 + \rho^2 \, d\phi^2 + dz^2, \tag{1.44}$$

then we read off the metric:

$$g_{\mu\nu} \doteq \begin{pmatrix} 1 & 0 & 0 \\ 0 & \rho^2 & 0 \\ 0 & 0 & 1 \end{pmatrix} \tag{1.45}$$

and coordinate differential:

$$dx^\alpha \doteq \begin{pmatrix} d\rho \\ d\phi \\ dz \end{pmatrix}. \tag{1.46}$$

1.3.2 Lagrangian redux

Returning to the Lagrangian, we can write:

$$\boxed{L = \frac{1}{2} m \, \dot{x}^\mu \, g_{\mu\nu} \, \dot{x}^\nu - U(x, y, z).} \tag{1.47}$$

That's sensible, *and* I don't even have to tell you which coordinates I mean. We can vary the action associated with this Lagrangian as before to find the equations of motion (alternatively, we can appeal directly to the Euler–Lagrange equations). We have to be a little careful, it is possible (as in the spherical case) that $g_{\mu\nu}(x^\mu)$, i.e. the metric depends on the coordinates. The equations of motion, written in tensor form, are:

$$0 = \frac{d}{dt} \frac{\partial L}{\partial \dot{x}^\alpha} - \frac{\partial L}{\partial x^\alpha}, \tag{1.48}$$

where α goes from 1 to 3, covering all three coordinates.

Our goal is to obtain a general explicit form for these, written in terms of x^α, its first and second derivatives, the metric, and $U(x, y, z)$. For starters, we need to take a $\dot{x}^\alpha$ derivative of the kinetic term:

$$\frac{\partial}{\partial \dot{x}^\alpha} \left(\frac{1}{2} m \, \dot{x}^\mu \, g_{\mu\nu} \, \dot{x}^\nu \right) = \frac{1}{2} m \left(\frac{\partial \dot{x}^\mu}{\partial \dot{x}^\alpha} g_{\mu\nu} \, \dot{x}^\nu + \dot{x}^\mu \, g_{\mu\nu} \frac{\partial \dot{x}^\nu}{\partial \dot{x}^\alpha} \right),$$ (1.49)

but it is clear that:[3]

$$\frac{\partial \dot{x}^\mu}{\partial \dot{x}^\alpha} = \delta^\mu_\alpha,$$ (1.51)

so that:

$$\frac{1}{2} m \left(\frac{\partial \dot{x}^\mu}{\partial \dot{x}^\alpha} g_{\mu\nu} \, \dot{x}^\nu + \dot{x}^\mu \, g_{\mu\nu} \frac{\partial \dot{x}^\nu}{\partial \dot{x}^\alpha} \right) = \frac{1}{2} m \left(\delta^\mu_\alpha \, g_{\mu\nu} \, \dot{x}^\nu + \dot{x}^\mu \, g_{\mu\nu} \, \delta^\nu_\alpha \right)$$
$$= \frac{1}{2} m \left(g_{\alpha\nu} \, \dot{x}^\nu + \dot{x}^\mu \, g_{\mu\alpha} \right)$$ (1.52)

(using $g_{\mu\nu} \, \delta^\nu_\alpha = g_{\mu\alpha}$). Because, in the first term, ν is a "dummy index" (it is summed over) and in the second, μ is a dummy, we can relabel these however we like. In addition, we can use the symmetry property of $g_{\mu\nu}$ to write, finally:

$$\frac{1}{2} m \left(g_{\alpha\nu} \, \dot{x}^\nu + \dot{x}^\mu \, g_{\mu\alpha} \right) = \frac{1}{2} m \left(g_{\alpha\nu} \, \dot{x}^\nu + \dot{x}^\nu \, g_{\alpha\nu} \right) = m \, g_{\alpha\nu} \, \dot{x}^\nu.$$ (1.53)

That's just the flavor of the sorts of calculation we will be doing over (and over) again.

Turning to the derivative $\frac{\partial L}{\partial x^\alpha}$:

$$\frac{\partial L}{\partial x^\alpha} = \frac{1}{2} m \, \dot{x}^\mu \frac{\partial g_{\mu\nu}}{\partial x^\alpha} \dot{x}^\nu - \frac{\partial U}{\partial x^\alpha},$$ (1.54)

where the first term comes from any coordinate-dependence hidden in the metric, and the second term reflects the coordinate-dependence of the potential.

The equations of motion now read:

$$\left(\frac{d}{dt} \frac{\partial L}{\partial \dot{x}^\alpha} - \frac{\partial L}{\partial x^\alpha} \right) = m \frac{d}{dt} \left(g_{\alpha\nu} \, \dot{x}^\nu \right) - \frac{1}{2} m \frac{\partial g_{\mu\nu}}{\partial x^\alpha} \dot{x}^\mu \, \dot{x}^\nu + \frac{\partial U}{\partial x^\alpha}$$
$$= m \frac{\partial g_{\alpha\nu}}{\partial x^\gamma} \dot{x}^\gamma \, \dot{x}^\nu + m \, g_{\alpha\nu} \, \ddot{x}^\nu - \frac{1}{2} m \frac{\partial g_{\mu\nu}}{\partial x^\alpha} \dot{x}^\mu \, \dot{x}^\nu + \frac{\partial U}{\partial x^\alpha}$$
$$= m \, g_{\alpha\nu} \, \ddot{x}^\nu + m \, \dot{x}^\nu \, \dot{x}^\gamma \left(\frac{\partial g_{\alpha\nu}}{\partial x^\gamma} - \frac{1}{2} \frac{\partial g_{\gamma\nu}}{\partial x^\alpha} \right) + \frac{\partial U}{\partial x^\alpha}.$$
(1.55)

---

[3] The "Kronecker" delta is defined to be:

$$\delta^\mu_\alpha \equiv \begin{cases} 1 & \mu = \alpha \\ 0 & \mu \neq \alpha \end{cases}$$ (1.50)

Now for a little sleight-of-hand which you will prove in Problem 1.7 – notice in this last line that the second term has the factor $\dot{x}^\nu \dot{x}^\gamma$, which is symmetric in $\nu \leftrightarrow \gamma$. Using the result from Problem 1.7, we have:

$$m\,\dot{x}^\nu\,\dot{x}^\gamma \left(\frac{\partial g_{\alpha\nu}}{\partial x^\gamma} - \frac{1}{2} \frac{\partial g_{\gamma\nu}}{\partial x^\alpha} \right) = m\,\dot{x}^\nu\,\dot{x}^\gamma \left(\frac{1}{2} \frac{\partial g_{\alpha\nu}}{\partial x^\gamma} + \frac{1}{2} \frac{\partial g_{\alpha\gamma}}{\partial x^\nu} - \frac{1}{2} \frac{\partial g_{\gamma\nu}}{\partial x^\alpha} \right) \quad (1.56)$$

and we can write the equations of motion as:

$$m\,g_{\alpha\nu}\,\ddot{x}^\nu + m\,\dot{x}^\nu\,\dot{x}^\gamma \left(\frac{1}{2} \frac{\partial g_{\alpha\nu}}{\partial x^\gamma} + \frac{1}{2} \frac{\partial g_{\alpha\gamma}}{\partial x^\nu} - \frac{1}{2} \frac{\partial g_{\gamma\nu}}{\partial x^\alpha} \right) = -\frac{\partial U}{\partial x^\alpha}. \quad (1.57)$$

The term in parentheses appears a lot, and is given a special name – it is called the "connection for the metric $g_{\mu\nu}$" and is denoted $\Gamma_{\alpha\gamma\nu}$. At this point, we are just giving a name and symbol to a particular combination of derivatives of the metric:

$$\boxed{\Gamma_{\alpha\gamma\nu} = \frac{1}{2}\left(\frac{\partial g_{\alpha\nu}}{\partial x^\gamma} + \frac{\partial g_{\alpha\gamma}}{\partial x^\nu} - \frac{\partial g_{\gamma\nu}}{\partial x^\alpha} \right),} \quad (1.58)$$

and we'll see the significance of this object in Chapter 3.

We have, in their final form, the equations of motion for any coordinate choice:

$$\boxed{m\,g_{\alpha\nu}\,\ddot{x}^\nu + m\,\Gamma_{\alpha\nu\gamma}\,\dot{x}^\nu\,\dot{x}^\gamma = -\frac{\partial U}{\partial x^\alpha},} \quad (1.59)$$

where the index α appearing only once in each term on the left and right is an "open" index – there are three equations here, one for each value $\alpha = 1$, 2, and 3.

The terms in (1.59) that are not explicitly second derivatives of the coordinates, or derivatives of U, are quadratic in the first derivatives $\dot{x}^\mu$, just as we saw explicitly for spherical coordinates in Section 1.3 (in particular (1.30)).

We went slow, but we have made some progress, especially in terms of our later work. Setting $U = 0$ in (1.59), we have the so-called "geodesic" equation for a generic metric $g_{\mu\nu}$. The geodesic equation has solutions which are interpreted as "straight lines" in general. Certainly in the current setting, if we take $U = 0$ and work in Cartesian coordinates, where $\Gamma_{\alpha\nu\gamma} = 0$ (since the metric does not, in this case, depend on position), the solutions are manifestly straight lines. Later on, in special and general relativity, we will lose the familiar notion of length, but if we accept a generalized length interpretation, we can still understand solutions to the geodesic equation as length-extremizing curves (the natural generalization of straight lines).

In developing (1.59), we made no assumptions about dimension or form for the metric (beyond symmetric, invertible, and differentiable), and these equations hold for arbitrary dimension, coordinates, and potential. The geodesic form ($U = 0$) is precisely the starting point for studying particle motion in general relativity – there, the moral is that we have no forces (no potential), and the curved trajectories (orbits, for example) we see are manifestations of the curvature of the space-time (expressed via the metric). In that context, it must be the $\Gamma_{\alpha\nu\gamma}$ term that approximates the forcing we would normally associate with a Newtonian gravitational source.

Problem 1.4

Start with the "usual" Cartesian Lagrangian for a central potential:

$$L = \frac{1}{2} m \left(\dot{x}^2 + \dot{y}^2 + \dot{z}^2 \right) - U \left(\sqrt{x^2 + y^2 + z^2} \right) \tag{1.60}$$

and transform the coordinates (and velocities) directly (no metric allowed) to find the Lagrangian associated with a central potential in cylindrical coordinates, with $x^1 = s$, $x^2 = \phi$, $x^3 = z$. From the Lagrangian itself (most notably, its kinetic term), write the metric associated with cylindrical coordinates.

Problem 1.5

(a) For the metric $g_{\mu\nu}$ in spherical coordinates, with $x^1 = r$, $x^2 = \theta$, $x^3 = \phi$, find the r component of the equation of motion (i.e. $\alpha = 1$):

$$m\, g_{\alpha\nu}\, \ddot{x}^\nu + \frac{1}{2} m\, \dot{x}^\nu \dot{x}^\gamma \left(\frac{\partial g_{\alpha\nu}}{\partial x^\gamma} + \frac{\partial g_{\alpha\gamma}}{\partial x^\nu} - \frac{\partial g_{\gamma\nu}}{\partial x^\alpha} \right) = -\frac{\partial U}{\partial x^\alpha}. \tag{1.61}$$

(b) Starting from the Lagrangian in spherical coordinates, calculate the r equation of motion directly from:

$$\frac{d}{dt} \frac{\partial L}{\partial \dot{r}} - \frac{\partial L}{\partial r} = 0 \tag{1.62}$$

and verify that you get the same result.

Problem 1.6

Consider a matrix $\mathbb{A}$ with entries A^{ij}. In two dimensions, for example:

$$\mathbb{A} \doteq \begin{pmatrix} A^{11} & A^{12} \\ A^{21} & A^{22} \end{pmatrix}. \tag{1.63}$$

Take a vector $\mathbf{x}$ with entries x_k, and write the matrix–vector products $\mathbb{A}\,\mathbf{x}$ and $\mathbf{x}^T \mathbb{A}$ in indexed notation.

Problem 1.7

(a) Given a second-rank tensor $T_{\mu\nu}$, often viewed as an $N \times N$ matrix (for a space of dimension N), show by explicit construction that one can always decompose $T_{\mu\nu}$ into a symmetric ($S_{\mu\nu} = S_{\nu\mu}$) and antisymmetric part ($A_{\mu\nu} = -A_{\nu\mu}$) via $T_{\mu\nu} = S_{\mu\nu} + A_{\mu\nu}$. The symmetric portion is often denoted $T_{(\mu\nu)}$, and the antisymmetric $T_{[\mu\nu]}$.

(b) In terms of the decomposition, $T_{\mu\nu} = S_{\mu\nu} + A_{\mu\nu}$, evaluate the sums:

$$T_{\mu\nu}\, Q^{\mu\nu} \qquad T_{\mu\nu}\, P^{\mu\nu}, \tag{1.64}$$

with $Q^{\mu\nu}$ symmetric and $P^{\mu\nu}$ antisymmetric.

(c) Using the decomposition above, show that, as in (1.56):

$$\dot{x}^\nu \dot{x}^\gamma \left(\frac{\partial g_{\alpha\nu}}{\partial x^\gamma} - \frac{1}{2} \frac{\partial g_{\gamma\nu}}{\partial x^\alpha} \right) = \frac{1}{2} \dot{x}^\nu \dot{x}^\gamma \left(\frac{\partial g_{\alpha\nu}}{\partial x^\gamma} + \frac{\partial g_{\alpha\gamma}}{\partial x^\nu} - \frac{\partial g_{\gamma\nu}}{\partial x^\alpha} \right) \qquad (1.65)$$

(note that $\dot{x}^\nu \dot{x}^\gamma$ plays the role of a symmetric second-rank tensor). Remember that metrics are symmetric, $g_{\mu\nu} = g_{\nu\mu}$.

1.4 Classical orbital motion

After that indexing tour-de-force, we are still stuck solving the problem of orbital motion. The point of that introduction to the "covariant formulation" (meaning coordinate-independent) of the equations of motion will become clear as we proceed. Now we take a step back, dangling indices do not help to solve the problem in this case. It is worth noting that we have taken coordinate-independence to a new high (low?) with our metric notation – in (1.59), you *really* don't know what coordinate system you are in.

So we will solve the equations of motion for the gravitational central potential in a chosen set of coordinates – the standard spherical ones. There is a point to the whole procedure – GR is a coordinate-independent theory, we will write statements that look a lot like (1.59), *but*, in order to "solve" a problem, we will always have to introduce coordinates. That is the current plan. After we have dispensed with Keplerian orbits, we will move on and solve the exact same problem using the Hamiltonian formulation, and for that we will need to discuss vectors and tensors again.

We have the following basic mechanics problem shown in Figure 1.2: given a body of mass M generating a Newtonian gravitational potential $\phi(r) = -\frac{MG}{r}$, how does a particle of mass m move under the influence of $\phi(r)$? In particular, we are interested in the orbital motion, and we'll tailor our discussions to this form, namely, the target elliptical orbits.

Ellipse

Going back to the Lagrangian for a generic central potential (we will have $U(r) = m\,\phi(r)$ eventually), in abstract language, we have:

$$L = \frac{1}{2} m\, \dot{x}^\mu\, g_{\mu\nu}\, \dot{x}^\nu - U(r). \qquad (1.66)$$

Suppose we start off in spherical coordinates, so that we know the metric is:

$$g_{\mu\nu} \doteq \begin{pmatrix} 1 & 0 & 0 \\ 0 & r^2 & 0 \\ 0 & 0 & r^2 \sin^2\theta \end{pmatrix} \qquad (1.67)$$

with coordinate differential $dx^\alpha \doteq (dr, d\theta, d\phi)^T$.

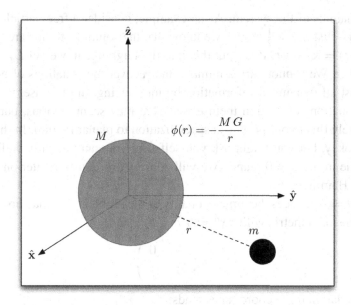

Figure 1.2 A central body of mass M generates a gravitational field, given by the potential $\phi(r)$. A "test" mass m responds to the force of gravity.

We will transform the radial coordinate: let $\rho = r^{-1}$, then the metric (specified equivalently by the associated line element) becomes:

$$ds^2 = dr^2 + r^2\,d\theta^2 + r^2\,\sin^2\theta\,d\phi^2$$

$$= \frac{1}{\rho^4}\,d\rho^2 + \frac{1}{\rho^2}\,d\theta^2 + \frac{1}{\rho^2}\,\sin^2\theta\,d\phi^2. \tag{1.68}$$

In matrix form, the metric reads:

$$g_{\mu\nu} \doteq \begin{pmatrix} \frac{1}{\rho^4} & 0 & 0 \\ 0 & \frac{1}{\rho^2} & 0 \\ 0 & 0 & \frac{1}{\rho^2}\sin^2\theta \end{pmatrix} \tag{1.69}$$

with the new coordinate differential $dx^\alpha \doteq (d\rho, d\theta, d\phi)^T$.

The potential is spherically symmetric, meaning that there are no preferred directions, or functionally, that it depends only on r (or, equivalently, ρ). We can *set* $\theta = \frac{\pi}{2}$ and $\dot\theta = 0$ to put the motion in a specific plane (the horizontal plane – for Cartesian coordinates in their standard configuration, this is the $x - y$ plane). Our choice here is motivated by the symmetry of the potential – since the potential depends only on r, the particular plane in which the motion occurs cannot matter. In fact, if we went through the equations of motion, we would find that $\theta = \frac{\pi}{2}$ and $\dot\theta = 0$ do not limit our solution. That's all well and good, but we have to be careful – when we use information about a solution prior to variation, we can lose the full dynamics of the system. As an example, consider a free particle classical Lagrangian – just $L_f = \frac{1}{2} m \dot{x}^2$ – we know that the solutions to this are vectors of the form $\mathbf{x}(t) = \mathbf{x}_0 + \mathbf{v} t$. If we put this into the Lagrangian, we get $L_f = \frac{1}{2} m v^2$, just a number. We cannot vary a number and recover the equations of motion, so we have lost all dynamical information by introducing, in this case, the solution from the equations of motion themselves. That may seem obvious, but we have done precisely this in our proposed specialization to planar motion. In this case, it works out okay, but you might ask yourself why you can't equally well take the motion to lie in the $\theta = 0$ plane? We will address this question later on when we discuss the Hamiltonian.

Putting $\theta = \frac{\pi}{2}$ reduces the dimensionality of the problem. We may now consider a two-dimensional metric with $dx^\alpha \doteq (d\rho, d\phi)^T$, and:

$$g_{\mu\nu} \doteq \begin{pmatrix} \frac{1}{\rho^4} & 0 \\ 0 & \frac{1}{\rho^2} \end{pmatrix}. \tag{1.70}$$

Our Lagrangian in these coordinates reads:

$$\boxed{L = \frac{1}{2} m \left(\frac{\dot{\rho}^2}{\rho^4} + \frac{\dot{\phi}^2}{\rho^2} \right) - U(\rho).} \tag{1.71}$$

At this point, we could find and solve the equations of motion, but it is easiest to note that there is no ϕ dependence in the above, i.e. ϕ is an ignorable coordinate. From the equation of motion, then:

$$\frac{d}{dt} \left(\frac{\partial L}{\partial \dot{\phi}} \right) = 0 \tag{1.72}$$

we learn that the momentum conjugate to ϕ (defined by $\frac{\partial L}{\partial \dot{\phi}}$) is conserved. So we substitute a constant for $\dot{\phi}$:

$$\frac{\partial L}{\partial \dot{\phi}} = J_z = \frac{m \dot{\phi}}{\rho^2} \longrightarrow \boxed{\dot{\phi} = \frac{J_z}{m} \rho^2.} \tag{1.73}$$

Now take the equation of motion for ρ, substituting the $J_z \rho^2 / m$ for $\dot{\phi}$ after varying:

$$\left(\frac{d}{dt} \left(\frac{\partial L}{\partial \dot{\rho}} \right) - \frac{\partial L}{\partial \rho} \right) \Bigg|_{\dot{\phi} = \frac{J_z}{m} \rho^2} = m \frac{d}{dt} \left(\frac{\dot{\rho}}{\rho^4} \right) + 2 m \dot{\rho}^2 \rho^{-5} + \frac{J_z^2}{m} \rho + \frac{dU}{d\rho} \tag{1.74}$$

$$= m \frac{\ddot{\rho}}{\rho^4} - 2 m \frac{\dot{\rho}^2}{\rho^5} + \frac{J_z^2 \rho}{m} + \frac{dU}{d\rho}.$$

We can reparametrize – rather than finding the time development of the $\rho(t)$ and $\phi(t)$ coordinates, the geometry of the solution can be uncovered by expressing ρ, and later r, in terms of ϕ. That is, we want to replace the functional dependence of ρ on t with $\rho(\phi)$. To that end, define $\rho' \equiv \frac{d\rho}{d\phi}$ and use change of variables to rewrite $\dot{\rho}$ and $\ddot{\rho}$:

$$\dot{\rho} = \rho' \dot{\phi} = \rho' \frac{J_z}{m} \rho^2$$

$$\ddot{\rho} = \rho'' \dot{\phi} \frac{J_z}{m} \rho^2 + \rho' \frac{J_z}{m} 2 \rho \dot{\rho} = \rho'' \left(\frac{J_z^2}{m^2} \rho^4 \right) + 2 \rho'^2 \frac{J_z^2}{m^2} \rho^3. \tag{1.75}$$

Putting these into the equation of motion gives:

$$0 = \frac{J_z^2}{m} \rho'' + \frac{2 J_z^2}{m} \rho'^2 \rho^{-1} - 2m \left(\rho'^2 \frac{J_z^2}{m^2} \right) \rho^{-1} + \frac{J_z^2 \rho}{m} + \frac{dU}{d\rho}, \tag{1.76}$$

or, finally:

$$\boxed{ -\frac{dU}{d\rho} = \frac{J_z^2}{m} \rho'' + \frac{J_z^2}{m} \rho. } \tag{1.77}$$

So far, the analysis applies to any spherically symmetric potential. We now specialize to the potential for Newtonian gravity, $U = -\frac{G m M}{r} = -G m M \rho$, and that's easy to differentiate, insert in (1.77), and solve:

$$G M \left(\frac{m}{J_z} \right)^2 = \rho'' + \rho \longrightarrow \rho(\phi) = G M \left(\frac{m}{J_z} \right)^2 + \alpha \cos \phi + \beta \sin \phi, \tag{1.78}$$

where α and β are arbitrary constants.

What type of solution is this? Keep in mind that $A \equiv G M \frac{m^2}{J_z^2}$ is just a constant, so what we really have is:

$$r(\phi) = \frac{1}{A + \alpha \cos \phi + \beta \sin \phi}. \tag{1.79}$$

Let's agree to start with $r'(\phi = 0) = 0$, this amounts to starting with no radial velocity, and tells us that $\beta = 0$:

$$r'(0) = -\frac{(-\alpha \sin \phi + \beta \cos \phi)}{(A + \alpha \cos \phi + \beta \sin \phi)^2} \bigg|_{\phi=0} = 0 \longrightarrow \beta = 0. \tag{1.80}$$

So, finally:

$$\boxed{ r(\phi) = \frac{1}{A + \alpha \cos \phi}, } \tag{1.81}$$

and this familiar solution is shown in Figure 1.3.

That's the story with elliptical orbits. We used the Lagrange approach to find a first integral of the motion (J_z), then we solved the problem using ϕ as the parameter

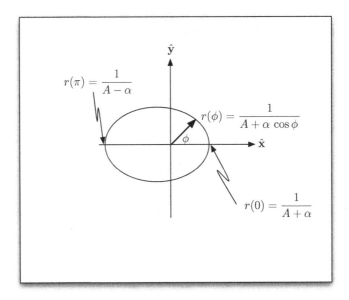

Figure 1.3 Ellipse in $r(\phi)$ parametrization.

for the curve $(r(\phi), \phi)$. There are a couple of things we will be dropping from our analysis. Set $G = 1$, which just changes how we measure masses. We can also set the test mass $m = 1$, it cannot be involved in the motion – this choice rescales J_z. We make these simplifying omissions to ease the transition to the usual choice of units $G = c = 1$.

Problem 1.8

We have been focused on bound, elliptical, orbits, but one can also approach a massive object along a straight line, this is called radial infall.

(a) From our radial equation for the ϕ-parametrized $\rho(\phi) = 1/r(\phi)$ curve, we had, for arbitrary $U(\rho)$:

$$\frac{J_z^2}{m}\left(\rho''(\phi) + \rho(\phi)\right) = -\frac{dU(\rho)}{d\rho}. \tag{1.82}$$

Can this equation be used to develop the ODE appropriate for radial infall with the Newtonian point potential? (i.e. A particle falls inward from $r(t = 0) = R$ with $\dot{r}(t = -\infty) = 0$ toward a spherically symmetric central body with mass M sitting at $r = 0$.) If not, explain why. If so, prepare to solve the relevant ODE for $r(t)$ in the next part.

(b) Solve the radial infall problem with initial conditions from part a. i.e. Find $r(t)$ appropriate for a particle of mass m falling *toward* $r = 0$ along a straight line – assume a spherically symmetric massive body is located at $r = 0$ with mass M.

Problem 1.9

Inspired by conservation of energy $H = E$ (Hamiltonian is a constant), and the Lagrangian for orbital motion in the $\theta = \frac{1}{2}\pi$ plane (setting $m = 1$):

$$L = \frac{1}{2}\left(\dot{r}^2 + r^2\dot{\phi}^2\right) - U(r) \tag{1.83}$$

with $U(r) = -\frac{M}{r}$ as usual, one can reinterpret orbital motion as one-dimensional motion in r (the Hamiltonian involves only the r-coordinate and the constant associated with angular momentum, J_z).

(a) Write the effective potential energy (i.e. the potential energy for the one-dimensional problem) in this setting. Make a sketch of the effective potential, label its behavior for $r \sim 0$, $r \sim \infty$. On your sketch, identify any zero-crossings and minima (both location r and value U_{min}).

(b) Solve for $r(t)$ when $E = U_{min}$ from part a. What does the full two-dimensional solution look like in the $x - y$ plane for this case?

Problem 1.10

(a) For an ellipse parametrized by:

$$r(\phi) = \frac{p}{1 + e\cos\phi}, \tag{1.84}$$

sketch the trajectory (in the $x - y$ plane) for $(p, e) = \left(1, \frac{1}{2}\right)$, $(p, e) = \left(\frac{1}{2}, \frac{1}{2}\right)$, and $(p, e) = \left(1, \frac{1}{4}\right)$ by considering the points defined by $\phi = 0, \frac{\pi}{2}, \pi$. What values of (p, e) correspond to a circle of radius R?

(b) Write $\{r_a, r_p\}$ (the radii for aphelion – furthest and perihelion – closest approach to the central body) in terms of $\{p, e\}$.

(c) Using the above, find the relationship between the constants of integration $\{\alpha, J_z\}$ in:

$$r(\phi) = \frac{1}{\frac{M}{J_z^2} + \alpha\cos\phi} \tag{1.85}$$

(from our solution using the Lagrangian) and $\{r_a, r_p\}$ – that is, find $\alpha(r_a, r_p)$ and $J_z(r_a, r_p)$.

1.5 Basic tensor definitions

In preparation for the Hamiltonian form, I want to be more careful with the definition of tensors. We will be covering this topic in more detail in Chapter 3, but we can discuss a few key points right now.

There are two varieties of tensor with which we will need to become familiar. A tensor, for us, is defined by its behavior under a coordinate transformation. In particular, if we have a singly indexed object $f^\alpha(x)$ that depends on a set of coordinates x,[4] and we change to a new coordinate system $x \longrightarrow \bar{x}$, then a *contravariant* tensor transforms via:

$$\bar{f}^\alpha(\bar{x}) = f^\nu(\bar{x}) \frac{\partial \bar{x}^\alpha}{\partial x^\nu},$$ (1.86)

and it is important to remember that the right-hand side should (morally speaking) be expressed entirely in terms of the new coordinate system (so that, for example, what we mean by $f^\nu(\bar{x})$ is really $f^\nu(x(\bar{x}))$, i.e. we take the original vector in x coordinates, and rewrite the x in terms of the $\bar{x}$, the same holds for the transformation factor $\frac{\partial \bar{x}^\alpha}{\partial x^\nu}$). We indicate that an indexed object has contravariant transformation character by putting the index up:

A *covariant* tensor $f_\alpha(x)$ responds to coordinate transformations according to:

$$\bar{f}_\alpha(\bar{x}) = f_\nu(\bar{x}) \frac{\partial x^\nu}{\partial \bar{x}^\alpha}.$$ (1.87)

Covariant tensors have lowered indices. The standard example of a contravariant tensor is the coordinate differential dx^α, and a familiar covariant tensor is the gradient of a scalar: $\frac{\partial \phi}{\partial x^\alpha} \equiv \phi_{,\alpha}$.[5]

The fun continues when we introduce objects with more indices – each index comes with the appropriate transformation (so lower indices pick up factors associated with the covariant transformation rule, upper indices have contravariant transformation). For doubly indexed objects:

$$\bar{f}^{\alpha\beta}(\bar{x}) = \frac{\partial \bar{x}^\alpha}{\partial x^\mu} \frac{\partial \bar{x}^\beta}{\partial x^\nu} f^{\mu\nu}$$

$$\bar{f}_{\alpha\beta}(\bar{x}) = \frac{\partial x^\mu}{\partial \bar{x}^\alpha} \frac{\partial x^\nu}{\partial \bar{x}^\beta} f_{\mu\nu}.$$ (1.89)

There is no particular reason to imagine that these two different objects (covariant and contravariant tensors) are connected, but within the confines of the geometry we will be studying, they can be related. We use precisely the metric and its matrix inverse to *define* "raising" and "lowering" operations. So, for example, we have:

$$f_\alpha \equiv g_{\alpha\nu} f^\nu$$ (1.90)

---

[4] For tensors like $f^\alpha(x)$ with coordinate-dependence, I'll omit the explicit index that should go on the coordinates, just to avoid clutter – so from now on, $f^\alpha(x)$ is shorthand for $f^\alpha(x^\mu)$ which is itself shorthand for $f^\alpha(x_1, x_2, \ldots, x_D)$.

[5] We will use the comma notation to indicate derivatives. The index appearing after the comma should be lower, since the derivative operator is itself covariant. There are a couple of other conventions that we will employ where necessary, in particular the naked partial can be handy:

$$\frac{\partial \phi(x)}{\partial x^\alpha} \equiv \partial_\alpha \phi(x) = \phi_{,\alpha}.$$ (1.88)

and if we set $g^{\mu\nu} \equiv (g_{\mu\nu})^{-1}$, the matrix inverse of $g_{\mu\nu}$,[6] the raised version of f_ν is then defined to be:

$$f^\alpha = g^{\alpha\nu} f_\nu. \tag{1.92}$$

The importance of the inverse is to recover the tautology:

$$f_\alpha = g_{\alpha\nu} f^\nu = g_{\alpha\nu} \left(g^{\nu\beta} f_\beta\right) = \delta^\beta_\alpha f_\beta = f_\alpha. \tag{1.93}$$

There is nothing special about the metric in this context – all we need is a second-rank symmetric tensor that is invertible (think of square, symmetric matrices). The metric is a good choice for us because it is the object that is governed by general relativity. As a field theory, GR connects fields to sources in a manner analogous to Maxwell's equations for E&M, and the field that is sourced is precisely a metric field. It is in this sense that GR makes a connection between mass distributions and "the structure of space-time" – the metric provides all of that structure.

Just the facts

In Chapter 3, we will cover tensors again. For now, all you need to know are the following definitions.

Under a coordinate transformation: $x \longrightarrow \bar{x}(x)$, a contravariant tensor of rank n (meaning n indices) transforms as:

$$\bar{f}^{\alpha_1\alpha_2\alpha_3...\alpha_n}(\bar{x}) = \left(\prod_{j=1}^{n} \frac{\partial \bar{x}^{\alpha_j}}{\partial x^{\mu_j}}\right) f^{\mu_1\mu_2\mu_3...\mu_n}(x(\bar{x})). \tag{1.94}$$

A covariant tensor of rank n transforms as:

$$\bar{f}_{\beta_1\beta_2\beta_3...\beta_n}(\bar{x}) = \left(\prod_{j=1}^{n} \frac{\partial x^{\nu_j}}{\partial \bar{x}^{\beta_j}}\right) f_{\nu_1\nu_2\nu_3...\nu_n}(x(\bar{x})). \tag{1.95}$$

A "mixed" tensor of rank n with m contravariant indices, $k \equiv n - m$ covariant ones, transforms according to the general rule:

$$\bar{f}^{\alpha_1\alpha_2\alpha_3...\alpha_m}_{\beta_1\beta_2\beta_3...\beta_k} = \left(\prod_{j=1}^{m} \frac{\partial \bar{x}^{\alpha_j}}{\partial x^{\mu_j}} \prod_{i=1}^{k} \frac{\partial x^{\nu_i}}{\partial \bar{x}^{\beta_i}}\right) f^{\mu_1\mu_2\mu_3...\mu_m}_{\nu_1\nu_2\nu_3...\nu_k}(x(\bar{x})). \tag{1.96}$$

The metric, as a matrix that tells us how to measure local distances, is a second-rank covariant tensor $g_{\mu\nu}$, we define the contravariant tensor $g^{\mu\nu}$ to be the

---

[6] Matrix inverse in the sense that you display $g_{\mu\nu}$ as a $D \times D$ matrix for a space of dimension D, take its inverse, and associate the resulting matrix entries with the components of $g^{\mu\nu}$. We will then have:

$$g_{\alpha\nu} g^{\nu\beta} = \delta^\beta_\alpha. \tag{1.91}$$

matrix inverse of $g_{\mu\nu}$ so that $g^{\mu\alpha} g_{\alpha\nu} = \delta^\mu_\nu$, the Kronecker delta (with value 1 if $\mu = \nu$, else 0). Then the relation between the contravariant and covariant forms of a tensor is defined to be:

$$f_\alpha = g_{\alpha\beta} f^\beta \tag{1.97}$$

and we also have:

$$f^\alpha = g^{\alpha\beta} f_\beta. \tag{1.98}$$

Problem 1.11

(a) Using a generic transformation $x^\alpha \longrightarrow \bar{x}^\alpha$, establish the contravariant character of the coordinate differential dx^μ, and the covariant character of the gradient $\frac{\partial \phi(x)}{\partial x^\mu} \equiv \phi_{,\mu}$.

(b) From the definition of polar coordinates in terms of Cartesian: $x = s \cos \phi$, $y = s \sin \phi$, construct the matrix–vector equation relating $\{dx, dy\}$ to $\{ds, d\phi\}$. If we take the polar $\{s, \phi\}$ to be the transformed coordinates: $\{\bar{x}^1 = s, \bar{x}^2 = \phi\}$, and the Cartesian to be the original set: $\{x^1 = x, x^2 = y\}$, show that your matrix–vector equation represents precisely the contravariant transformation rule.

(c) Work out the gradient for the scalar $\psi = k\,x\,y$ in both Cartesian $\{x^1 = x, x^2 = y\}$ and polar $\{\bar{x}^1 = s, \bar{x}^2 = \phi\}$ coordinates. Show, by explicit construction, that the covariant transformation law relating $\bar\psi_{,\alpha}$ to $\psi_{,\alpha}$ holds in this case (i.e. start with $\psi_{,\mu}$ in Cartesian coordinates, transform according to $\bar\psi_{,\mu} = \frac{\partial x^\alpha}{\partial \bar{x}^\mu} \psi_{,\alpha}$, and show that this is what you get when you calculate $\bar\psi_{,\mu}$ explicitly).

Problem 1.12

Using the matrix point of view, construct the matrix form of both $\frac{\partial x^\alpha}{\partial \bar{x}^\beta}$ and $\frac{\partial \bar{x}^\beta}{\partial x^\gamma}$ for the explicit transformation from Cartesian to polar coordinates: $\{x^1 = x, x^2 = y\}$, $\{\bar{x}^1 = s, \bar{x}^2 = \phi\}$ (where $x = s \cos \phi$, $y = s \sin \phi$ defines $\{s, \phi\}$). Write both transformation "matrices" in the original and new variables (so construct $\frac{\partial x^\alpha}{\partial \bar{x}^\beta}(x)$ and $\frac{\partial x^\alpha}{\partial \bar{x}^\beta}(\bar{x})$ and then similarly for $\frac{\partial \bar{x}^\alpha}{\partial x^\beta}$). Verify, in both coordinate systems, that:

$$\frac{\partial x^\alpha}{\partial \bar{x}^\beta} \frac{\partial \bar{x}^\beta}{\partial x^\gamma} = \delta^\alpha_\gamma. \tag{1.99}$$

With Leibniz notation, this relationship is clear, but work out the two matrices on the left and multiply them to see how this goes in real coordinate systems. The moral value of doing the matrix–matrix multiplication in both the original Cartesian, and the polar coordinate systems is to drive home the point that the relation is coordinate-independent, but in order to obtain it correctly you must work in one set of coordinates or the other.

Problem 1.13

By combining tensors, we can form new objects with the correct tensorial character.

(a) Take two first-rank contravariant tensors f^μ and h^ν. If we form a direct product, $T^{\mu\nu} = f^\mu h^\nu$, we get a second-rank contravariant tensor. By transforming f^μ and h^ν (for $x \longrightarrow \bar{x}$) in the product, write down the second-rank contravariant tensor transformation law ($\bar{T}^{\mu\nu} = -T^{\mu\nu}$).

(b) Do the same for the covariant second-rank tensor constructed out of f_μ and h_ν via $T_{\mu\nu} = f_\mu h_\nu$.

(c) A scalar transforms as: $\bar{\phi}(\bar{x}) = \phi(x(\bar{x}))$ (i.e. a transcription, no transformation). Show that by taking a contravariant f^α and covariant h_β, the product $\psi = f^\alpha h_\alpha$ is a scalar.

(d) If $h_{\mu\nu}$ is a covariant second-rank tensor, show that $h^{\mu\nu} \equiv (h_{\mu\nu})^{-1}$ (the matrix inverse) is a contravariant second-rank tensor.

Problem 1.14

Suppose we form the second-rank tensor: $A_{\mu\nu} = p_\mu\, q_\nu$ from two first-rank tensors p_μ and q_ν. Show that this second-rank $A_{\mu\nu}$, viewed as a matrix, is *not* invertible – work in two dimensions for concreteness (the same argument holds, by induction, for all dimensions).

1.6 Hamiltonian definition

We turn now to Hamilton's formulation of the equations of motion. This discussion parallels the Lagrangian one, but as in strict classical mechanics, new avenues of discovery are open to us using the Hamiltonian. In particular, there is no better place to discuss invariance and conservation, a beautiful correspondence that is important in general relativity.

As has already been suggested, general relativity can be viewed as a "theory without forces" (or, in the Hamiltonian setting, potentials), and there the Hamiltonian plays an even more interesting role, because it is numerically identical to the Lagrangian. So it is beneficial to use as much of one or the other approach as we find useful. In order to set the stage, we re-derive and solve the equations of motion for the Keplerian ellipse. I hope the clothing is different enough to hold your interest.

We'll start with the definition of the Hamiltonian from the Lagrangian via a short review of the implications/utility of the Legendre transform. From there, we will look at the relation between allowed (canonical) coordinate and momentum transformations and constants of the motion (Noether's theorem).

1.6.1 Legendre transformation

The Legendre transformation is used to define the Hamiltonian, the starting point for Hamilton's formulation of mechanics. While the Hamiltonian has definite physical meaning in many cases, the Legendre transformation that defines it is a general technique for replacing a function's dependent variable with a different, geometrically related, one. We'll start with the definition and some examples in one dimension.

One-dimensional Legendre transformation

Consider an arbitrary function of x: $f(x)$. We know that locally, the slope of this curve is precisely its derivative with respect to x, so the change in the function $f(x)$ at the point x for a small change in the argument, dx, is:

$$df = \frac{df}{dx}\,dx = p(x)\,dx \quad p(x) \equiv \frac{df(x)}{dx}, \tag{1.100}$$

as usual. Now suppose we want to find a function that reverses the roles of the slope and infinitesimal, i.e. a function $g(p)$ such that $dg = x\,dp$ (where we now view x as a function of p defined by the inverse of $p = f'(x)$). It is easy to see that the function:

$$\boxed{g(p) = p\,x(p) - f(p),} \tag{1.101}$$

called the Legendre transform, has the property:

$$dg = x\,dp + p\,dx - df = x\,dp, \tag{1.102}$$

as desired. Notice that since $g(p)$ is a function of p only, we must have $x = \frac{dg}{dp}$ just as we had $p = \frac{df}{dx}$ before. So we have a pair of functions $f(x)$ and $g(p)$ related in the following way:

$$f(x) \longrightarrow g(p) = x(p)\,p - f(x(p)) \quad x(p) : \frac{df(x)}{dx} = p$$

$$g(p) \longrightarrow f(x) = p(x)\,x - g(p(x)) \quad p(x) : \frac{dg(p)}{dp} = x, \tag{1.103}$$

where the pair $f(x)$ and $g(p)$ are Legendre transforms *of each other*. There is a nice symmetry here, the same transformation takes us back and forth.

Example

Let's look at an example to see how the transformation works in practice.

Consider the function $f(x) = \alpha\,x^m$ for integer m. We can define p from the derivative of $f(x)$ as prescribed above:

$$\frac{df}{dx} = \alpha\,m\,x^{m-1} = p \longrightarrow x(p) = \left(\frac{p}{\alpha\,m}\right)^{\frac{1}{m-1}}. \tag{1.104}$$

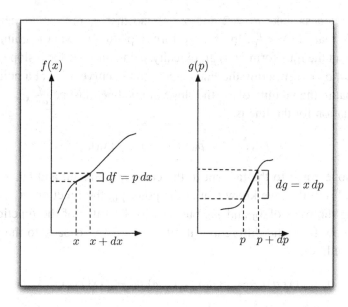

Figure 1.4   The Legendre transform constructs a function $g(p)$ from $f(x)$ by swapping the role of the local slope of the curve for its argument.

With this assignment, we can construct $f(x(p)) \equiv f(p)$ – it is:

$$f(p) = \alpha \left(\frac{p}{\alpha m} \right)^{\frac{m}{m-1}} , \tag{1.105}$$

and the Legendre transform is:

$$g(p) = p\,x(p) - f(p) = \left(\frac{1}{\alpha m} \right)^{\frac{1}{m-1}} p^{\frac{m}{m-1}} - \alpha \left(\frac{p}{\alpha m} \right)^{\frac{m}{m-1}} \tag{1.106}$$

$$= p^{\frac{m}{m-1}} \left[\left(\frac{1}{\alpha m} \right)^{\frac{1}{m-1}} - \alpha \left(\frac{1}{\alpha m} \right)^{\frac{m}{m-1}} \right].$$

Now take the reverse point of view – suppose we were given $g(p)$ as above, if we define $x = \frac{dg}{dp}$, and form the Legendre transform of $g(p)$: $h(x) = p(x)\,x - g(x)$ (with $g(x) \equiv g(p(x))$), then we would find that $h(x) = \alpha\,x^m$, precisely our starting point (you should try it).

The polynomial example is a good place to see the geometrical implications of the Legendre transformation. Remember the goal, unmotivated at this point: we took a function $f(x)$ with infinitesimal $df = p\,dx$ (for $p \equiv \frac{df}{dx}$), and found a function $g(p)$ with infinitesimal $dg = x\,dp$, reversing the roles of the derivative and argument of the original $f(x)$. The procedure is sketched in Figure 1.4.

For concreteness, take $m = 2$ and $\alpha = 1$, so that $f(x) = x^2$ – then we find from (1.106) that $g(p) = \frac{p^2}{4}$. In order to form a picture of what is going on, let's try to construct the transform $g(p)$ graphically, using only the local slope and value of $f(x)$. Start by constructing the line tangent to the curve $f(x)$ at a point x_0, say. We know that in the vicinity of x_0, the slope of this line must be $\frac{df}{dx}|_{x=x_0} \equiv p_0$, and then the equation for the line is:

$$\bar{f}(x, x_0) = p_0 (x - x_0) + f(x_0), \tag{1.107}$$

i.e. it has slope equal to the tangent to the curve $f(x)$ at x_0, and takes the value $f(x_0)$ at x_0. From this, we know that at the point p_0, the function $g(p)$ has slope x_0, reversing the roles of x_0 and p_0, but what is the value of the function $g(p_0)$? We don't know from the differential itself – but if we go back to the Legendre transform, we have:

$$g(p) = x(p)\, p - f(x(p)) \longrightarrow g(p_0) = x(p_0)\, p_0 - f(x(p_0)) = x_0\, p_0 - f(x_0), \tag{1.108}$$

and we can use this value to construct the line going through the point $g(p_0)$ having slope x_0:

$$\bar{g}(p, p_0) = x_0 (p - p_0) + g(p_0) = x_0\, p - f(x_0), \tag{1.109}$$

which means that the $\bar{g}$-intercept of this line (i.e. $\bar{g}(0, p_0)$) has value $-f(x_0)$. So we have a graphical prescription: for every point x_0, write down the slope of the line tangent to $f(x_0)$ (called p_0) – on the transform side, draw a line having g-intercept $-f(x_0)$ and slope given by x_0 – mark its value at p_0, and you will have the graph of $g(p)$.

In Figure 1.5, we see the graph of $f(x) = x^2$ with four points picked out, the lines tangent to those points are defined by (1.107), with $x_0 = 0, 1, 2, 3$. For example, for $x_0 = 3$, we have $p_0 = 6$, and $f(x_0) = 9$. Referring to Figure 1.6, the point at $p_0 = 6$ can be found by drawing a line with g-intercept $-f(x_0) = -9$, and slope $x_0 = 3$ – we mark where this line crosses $p_0 = 6$, as shown. In this manner, we can convert a graph of $f(x)$ into a graph of $g(p)$, and, of course, vice versa.

Notice, finally, that the curve defined by the three points in Figure 1.6 is precisely described by $g(p) = \frac{p^2}{4}$.

Hamiltonian in classical mechanics

We can easily extend the above one-dimensional discussion to higher dimensions. In classical mechanics, we often start with a Lagrangian, defined as a function of $x(t)$ and $\dot{x}(t)$, say. Then we have, in a sense, two variables in $L(x, \dot{x})$, and we

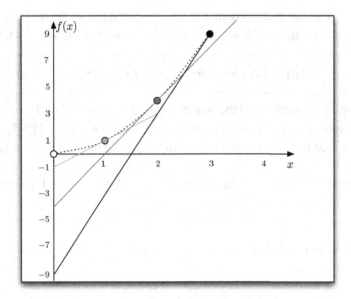

Figure 1.5   A graph of $f(x) = x^2$ with some representative points and tangent lines shown.

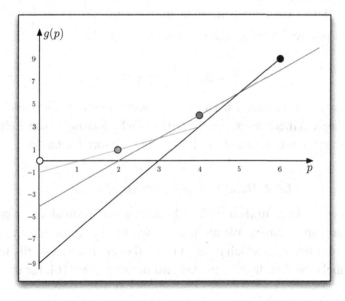

Figure 1.6   The three relevant lines coming from the data in Figure 1.5, used to define the curve $g(p)$.

can replace $\dot{x}$ with an independent variable by setting $p = \frac{\partial L}{\partial \dot{x}}$, and performing a Legendre transformation of L to eliminate $\dot{x}$ in favor of p. Define H via:

$$H(x, p) = p\,\dot{x}(p) - L(x, p) \qquad p = \frac{\partial L}{\partial \dot{x}} \qquad L(x, p) \equiv L(x, \dot{x}(p)). \qquad (1.110)$$

Notice that we have performed the transformation on only one of the two variables in the Hamiltonian. In words, we construct the Hamiltonian in (1.110) by using the definition of p to find $\dot{x}(p)$ and then writing $p\,\dot{x} - L(x, \dot{x})$ entirely in terms of p and x.

For example, if we have a simple harmonic oscillator potential, then:

$$L(x, \dot{x}) = \frac{1}{2} m\,\dot{x}^2 - \frac{1}{2} k\,x^2 \qquad (1.111)$$

and $p \equiv \frac{\partial L}{\partial \dot{x}} = m\,\dot{x}$, so that $\dot{x} = \frac{p}{m}$, $L(x, p) = \frac{1}{2}\frac{p^2}{m} - \frac{1}{2} k\,x^2$ and we can form $H(x, p)$ from (1.110):

$$H(x, p) = p\,\frac{p}{m} - \left(\frac{1}{2}\frac{p^2}{m} - \frac{1}{2} k\,x^2\right) = \frac{1}{2}\frac{p^2}{m} + \frac{1}{2} k\,x^2. \qquad (1.112)$$

We recognize this as the total energy of the system (numerically).

For the usual Lagrangian in three dimensions: $L = \frac{1}{2} m\,v^2 - U(\mathbf{x})$, with $v^2 = \dot{\mathbf{x}} \cdot \dot{\mathbf{x}}$, we can define the *canonical momentum vector* via $p_x = \frac{\partial L}{\partial \dot{x}}$, $p_y = \frac{\partial L}{\partial \dot{y}}$ and $p_z = \frac{\partial L}{\partial \dot{z}}$, and proceed to the Hamiltonian once again:

$$H(\mathbf{x}, \mathbf{p}) = \frac{p^2}{2m} + U(\mathbf{x}), \qquad (1.113)$$

at least, in Cartesian coordinates. This is the starting point for Hamiltonian considerations in classical mechanics, and we will begin by looking at some changes that must occur to bring this natural form into usable, tensorial notation.

1.6.2 Hamiltonian equations of motion

The first thing we have to deal with is Legendre transformations in our generic space (or, later, space-time). We are used to writing the Hamiltonian in terms of generalized coordinates q which have an index (but are not necessarily tensors, the index just labels the coordinate). In standard notation (see [16], for example), we have:

$$H = \sum_i p_i\,\dot{q}_i - L$$
$$p_i \equiv \frac{\partial L}{\partial \dot{q}_i}. \qquad (1.114)$$

In our new notation and with our sensitivity to tensor character and index placement, we see there is already an issue here – we must have one index up and one index down, according to our summation convention, and the sum in (1.114) does not clearly fit this prescription. Cartesian coordinates form a contravariant tensor x^α, but then the canonical momenta are given by the manifestly covariant form:

$$p_\alpha = \frac{\partial L}{\partial \dot{x}^\alpha},\qquad (1.115)$$

i.e. canonical momenta are covariant first-rank tensors. The distinction does not come up if we use Cartesian coordinates, because there is no numerical difference between the contravariant and covariant forms of a tensor (since the metric can be represented by the identity matrix). We want the Hamiltonian to be a scalar, and this can be achieved if we write our Legendre transform as:

$$H = p_\alpha \dot{x}^\alpha - L.\qquad (1.116)$$

Remember the point of the Hamiltonian approach – we want to treat p_α and x^α as independent entities. The variational principle gives us the equations of motion directly:

$$
\boxed{
\begin{aligned}
S &= \int dt\, L = \int dt\, (p_\alpha \dot{x}^\alpha - H) \\[4pt]
\frac{\delta S}{\delta p_\alpha} &= \dot{x}^\alpha - \frac{\partial H}{\partial p_\alpha} = 0 \\[4pt]
\frac{\delta S}{\delta x^\alpha} &= -\dot{p}_\alpha - \frac{\partial H}{\partial x^\alpha} = 0,
\end{aligned}
}
\qquad (1.117)
$$

where we have varied with respect to p_α and x^α *separately*.

Using our tensor notation (and setting $m = 1$), $L = \frac{1}{2} \dot{x}^\mu g_{\mu\nu} \dot{x}^\nu - U(x, y, z)$. We get, as the canonical momenta:

$$\frac{\partial L}{\partial \dot{x}^\alpha} = g_{\mu\alpha} \dot{x}^\mu = p_\alpha,\qquad (1.118)$$

so that

$$
\begin{aligned}
H = p_\alpha \dot{x}^\alpha - L &= p_\alpha p^\alpha - \left(\frac{1}{2} \dot{x}^\mu g_{\mu\nu} \dot{x}^\nu - U(x, y, z) \right) \\[4pt]
&= p_\alpha g^{\alpha\beta} p_\beta - \frac{1}{2} p_\alpha g^{\alpha\beta} p_\beta + U(x, y, z) \qquad (1.119) \\[4pt]
&= \frac{1}{2} p_\alpha g^{\alpha\beta} p_\beta + U(x, y, z).
\end{aligned}
$$

From here, we can write down the equations of motion using (1.117):

$$\dot{x}^\alpha = \frac{\partial H}{\partial p_\alpha} = g^{\alpha\beta} p_\beta$$

$$\dot{p}_\alpha = -\frac{\partial H}{\partial x^\alpha} = -\left(\frac{1}{2} p_\mu g^{\mu\nu}{}_{,\alpha} p_\nu + U_{,\alpha} \right). \tag{1.120}$$

As we shall soon see, the triply indexed object $g_{\mu\nu,\alpha}$[7] is *not* a tensor – this will be one of the highlights of our upcoming work. The local point is, we cannot raise and lower the indices of the derivative of the metric as we would like – instead note the identity (to be shown later):

$$\boxed{g^{\mu\nu}{}_{,\alpha} = -g^{\mu\gamma} g^{\nu\delta} g_{\gamma\delta,\alpha},} \tag{1.122}$$

then the equation for $\dot{p}_\alpha$ from above is:

$$\dot{p}_\alpha = \frac{1}{2} p_\mu \left(g^{\mu\gamma} g^{\nu\delta} g_{\gamma\delta,\alpha} \right) p_\nu - U_{,\alpha}$$

$$= \frac{1}{2} p^\gamma g_{\gamma\delta,\alpha} p^\delta - U_{,\alpha}. \tag{1.123}$$

Let's introduce coordinates and work on a familiar example – suppose we are already in the reduced two-dimensional space, $\{\rho, \phi\}$ with metric (inspired by our Lagrangian studies (1.70)):

$$g_{\mu\nu} \doteq \begin{pmatrix} \frac{1}{\rho^4} & 0 \\ 0 & \frac{1}{\rho^2} \end{pmatrix}. \tag{1.124}$$

We have:

$$H = \frac{1}{2} \left(\rho^4 p_\rho^2 + \rho^2 p_\phi^2 \right) + U(\rho), \tag{1.125}$$

then the equations of motion are given by:

$$\dot{\rho} = \frac{\partial H}{\partial p_\rho} = \rho^4 p_\rho$$

$$\dot{\phi} = \frac{\partial H}{\partial p_\phi} = \rho^2 p_\phi \tag{1.126}$$

---

[7] Remember, the comma denotes partial derivatives with respect to the coordinates:

$$g_{\mu\nu,\alpha} \equiv \frac{\partial g_{\mu\nu}}{\partial x^\alpha}. \tag{1.121}$$

and we recognize the second equation from the definition of J_z (i.e. $\dot{\phi} = \frac{J_z}{r^2}$ from (1.73)). For the rest:

$$\dot{p}_\rho = -\frac{\partial H}{\partial \rho} = -2\rho^3 \, p_\rho^2 - \rho \, p_\phi^2 - \frac{\partial U}{\partial \rho}$$

$$\dot{p}_\phi = -\frac{\partial H}{\partial \phi} = 0. \tag{1.127}$$

Here again, the second equation is familiar – "ignorable coordinates have conserved momenta", and the canonical p_ϕ is a constant of the motion.

Putting the derivatives together, we can write the equation of motion in terms of $\ddot{\rho}$:

$$\dot{p}_\rho = \frac{d}{dt}\left(\frac{\dot{\rho}}{\rho^4}\right) = \frac{\ddot{\rho}}{\rho^4} - 4\frac{\dot{\rho}^2}{\rho^5} = -2\rho^3\left(\frac{\dot{\rho}^2}{\rho^8}\right) - \rho \, p_\phi^2 - \frac{\partial U}{\partial \rho} \tag{1.128}$$

or

$$\frac{\ddot{\rho}}{\rho^4} - 2\frac{\dot{\rho}^2}{\rho^5} + \rho \, p_\phi^2 = -\frac{\partial U}{\partial \rho}, \tag{1.129}$$

which is the same as our Lagrange equation of motion (with $p_\phi = J_z$).

We haven't yet got to the point of using H rather than L – it's coming, and to set the stage we must discuss transformations. So far, all we have done is verify that we get the same equation of motion content from both approaches.

1.6.3 Canonical transformations

When we work within the Lagrangian formulation of mechanics, our basic object is $L(x, \dot{x}, t)$. If we were to transform coordinates from x to $\bar{x}$, say, the velocities associated with $\bar{x}$ would be determined. That is, $\bar{x}$ and $\dot{\bar{x}}$ are not independent. For example, if we have a Lagrangian written in terms of x, and we decide to rewrite it in terms of $\bar{x} \equiv 3x$, then we know that $\dot{\bar{x}} = 3\dot{x}$, you don't get to pick $\dot{\bar{x}}$ separately.

From a Hamiltonian point of view, though, the position and momentum variables are independent. A transformation takes a set of coordinates and momenta $\{x^\alpha, p_\alpha\}$ and changes each (in theory, independently) into a new set $\{\bar{x}^\alpha, \bar{p}_\alpha\}$. But we have to be careful here, not all transformations will preserve the equations of motion. A transformation is called "canonical" if the new system is a Hamiltonian system with equations of motion:

$$\frac{\partial \bar{H}}{\partial \bar{p}_\alpha} = \dot{\bar{x}}^\alpha$$

$$\frac{\partial \bar{H}}{\partial \bar{x}^\alpha} = -\dot{\bar{p}}_\alpha, \tag{1.130}$$

corresponding directly to the transformation of the original equations of motion: $\frac{\partial \bar{H}}{\partial p_\alpha} = \dot{x}^\alpha$ and $\frac{\partial \bar{H}}{\partial x^\alpha} = -\dot{p}_\alpha$. If not, then simply taking $\bar{H}(\bar{x}, \bar{p}) = H(x(\bar{x}, \bar{p}), p(\bar{x}, \bar{p}))$ will not describe the same physical system. We have no such restriction for the Lagrangian, there we can transform coordinates all we want, the equations of motion are always consistent because the transformation of the derivative, $\dot{x}$, goes along for the ride. We will develop a sure-fire method for generating transformations that are canonical.

Example

To see the problem with an arbitrary transformation of coordinates and momenta to some new set, take a simple harmonic oscillator, we have $H = \frac{1}{2} p^2 + \frac{1}{2} x^2$. Now if we "transform" using $\bar{x} = \frac{1}{2} x$ and $\bar{p} = p$, then $\bar{H}$ "$=$" $\frac{1}{2} \bar{p}^2 + 2 \bar{x}^2$ and the equations of motion, if we just viewed this as a Hamiltonian, give the wrong frequency of oscillation (when compared with the frequency given by the original Hamiltonian).

If instead we went back to the Lagrangian $L = \frac{1}{2} \dot{x}^2 - \frac{1}{2} x^2$, then we carry the transformation through the $\dot{x}$ term, and $L = 2 \dot{\bar{x}}^2 - 2 \bar{x}^2$ and we find out we should have had $\frac{\partial L}{\partial \dot{\bar{x}}} = 4 \dot{\bar{x}} = \bar{p}$ as the canonical momentum so that:

$$H = (4 \dot{\bar{x}}) \dot{\bar{x}} - L = 4 \dot{\bar{x}}^2 - (2 \dot{\bar{x}}^2 - 2 \bar{x}^2) = 2 \dot{\bar{x}}^2 + 2 \bar{x}^2 = \frac{\bar{p}^2}{8} + 2 \bar{x}^2, \quad (1.131)$$

which again gives the correct equation of motion.

While we are free to make any transformation we like, somehow the $\bar{x}$ and $\bar{p}$ variables are coupled if we want to make the same Hamiltonian system out of the resulting set. What constraints can we place on the transformation taking $\{x, p\} \longrightarrow \{\bar{x}, \bar{p}\}$ to ensure this happens?

1.6.4 Generating functions

There is a nice way to ensure that the transformation taking us from $\{x, p\}$ to $\{\bar{x}, \bar{p}\}$ is canonical – we had a variational principle that generated the equations of motion, so equivalent to asking that the *form* of the equations of motion be retained is the requirement that the variations be identical:

$$\delta \left(\int dt \, (p_\alpha \dot{x}^\alpha - H(x, p)) \right) = \delta \left(\int dt \, (\bar{p}_\alpha \dot{\bar{x}}^\alpha - \bar{H}(\bar{x}, \bar{p})) \right). \quad (1.132)$$

This is one of those cases where it is not enough to just set the integrands equal (although that would certainly be *a* solution). What if we had a total time-derivative on the right? That would look something like:

$$\int_{t_0}^{t_f} dt \, \dot{K} = K(t_f) - K(t_0), \quad (1.133)$$

i.e. it would contribute a constant – the variation of a constant is zero, so evidently, we can add any total time-derivative to the integrand on the right-hand side without changing the equations of motion.

Our expanded requirement is that:

$$p_\alpha \dot{x}^\alpha - H(x, p) = \bar{p}_\alpha \dot{\bar{x}}^\alpha - \bar{H}(\bar{x}, \bar{p}) + \dot{K}. \qquad (1.134)$$

$\dot{K}$ is fun to write down – but what is it? Hamiltonians are generally functions of position and momentum, so K must be some function of these, possibly with explicit time-dependence thrown in there as well.

Remember the target – we want to connect two sets of data: $\{x, p\}$ and $\{\bar{x}, \bar{p}\}$. We can use K to *generate* a valid transformation (hence its name "generating function"), that is where we are headed, but then we must make K a function of at least one variable from the original $\{x, p\}$ set, one variable from the $\{\bar{x}, \bar{p}\}$ set. There are four ways to do this, and it doesn't much matter which one we pick. For now, let $K = K(x, \bar{x}, t)$, then:

$$\frac{d}{dt}K = \frac{\partial K}{\partial x}\dot{x} + \frac{\partial K}{\partial \bar{x}}\dot{\bar{x}} + \frac{\partial K}{\partial t}. \qquad (1.135)$$

Inputting this into (1.134) gives us:

$$\left(p_\alpha - \frac{\partial K}{\partial x^\alpha}\right)\dot{x}^\alpha - \left(\bar{p}_\alpha + \frac{\partial K}{\partial \bar{x}^\alpha}\right)\dot{\bar{x}}^\alpha = H - \bar{H} + \frac{\partial K}{\partial t}. \qquad (1.136)$$

We can satisfy the equality in (1.136) by setting:

$$\boxed{p_\alpha = \frac{\partial K}{\partial x^\alpha} \qquad \bar{p}_\alpha = -\frac{\partial K}{\partial \bar{x}^\alpha} \qquad \frac{\partial K}{\partial t} = 0 \qquad H = \bar{H}.} \qquad (1.137)$$

While this does have the correct counting so that, in theory, we can invert the above to find $\bar{x}(x, p)$ and $\bar{p}(x, p)$, $K(x, \bar{x})$ itself is not the most useful generating function. For reasons that will become clear later, we would prefer to use a generator $\hat{K}(x, \bar{p})$ (the hat is just to distinguish this new generator from K), a function of the old coordinates and new momenta. How can we find such a function? Well, from our discussion of Legendre transformations and the second equation in (1.137), it is apparent that we could replace $\bar{x}$ with $\bar{p}$ via a Legendre transformation taking $K(x, \bar{x}) \longrightarrow \hat{K}(x, \bar{p})$. In order for these to be Legendre duals, we must have $\frac{\partial K}{\partial \bar{x}} = \bar{p}$ – that is precisely (modulo a minus sign) the requirement in (1.137). Suppose, then, that we construct $\hat{K}(x, \bar{p})$ as the Legendre transform of $K(x, \bar{x})$ (again, the signs are different but don't let that bother you):

$$\hat{K}(x, \bar{p}) = \bar{p}_\alpha \bar{x}^\alpha + K(x, \bar{x}) \qquad \frac{\partial K}{\partial \bar{x}^\alpha} = -\bar{p}_\alpha. \qquad (1.138)$$

As an explicit check, we can show that our $\hat{K}(x, \bar{p})$ is independent of $\bar{x}$:

$$\frac{\partial \hat{K}}{\partial \bar{x}^\alpha} = \bar{p}_\alpha + \frac{\partial K}{\partial \bar{x}^\alpha} = 0. \tag{1.139}$$

Now (1.134) reads:

$$p_\alpha \dot{x}^\alpha - H = \bar{p}_\alpha \dot{\bar{x}}^\alpha - \bar{H} + \frac{d}{dt}\left(\hat{K} - \bar{p}_\alpha \bar{x}^\alpha\right), \tag{1.140}$$

and the total time-derivative of our new generating function $\hat{K}$ is:

$$\frac{d}{dt}\hat{K} = \frac{\partial \hat{K}}{\partial x^\alpha} \dot{x}^\alpha + \frac{\partial \hat{K}}{\partial \bar{p}_\alpha} \dot{\bar{p}}_\alpha. \tag{1.141}$$

Inserting (1.141) into (1.140), we can write the requirement:

$$\left(p_\alpha - \frac{\partial \hat{K}}{\partial x^\alpha}\right) \dot{x}^\alpha + \left(\bar{x}^\alpha - \frac{\partial \hat{K}}{\partial \bar{p}_\alpha}\right) \dot{\bar{p}}_\alpha = H - \bar{H}, \tag{1.142}$$

and the relevant transformation connection is:

$$\boxed{p_\alpha = \frac{\partial \hat{K}}{\partial x^\alpha} \qquad \bar{x}^\alpha = \frac{\partial \hat{K}}{\partial \bar{p}_\alpha} \qquad H = \bar{H}.} \tag{1.143}$$

For our harmonic oscillator example, we want $\bar{x} = \frac{1}{2}x$ as the coordinate transformation, so from the second equation in (1.143) we have $\hat{K} = \frac{1}{2}x\,\bar{p}$, then the first tells us $p = \frac{1}{2}\bar{p}$, and we can transform the Hamiltonian by replacing $x = 2\bar{x}$, $p = \frac{1}{2}\bar{p}$:

$$H = \frac{1}{2}p^2 + \frac{1}{2}x^2 \longrightarrow \bar{H} = \frac{1}{8}\bar{p}^2 + 2\bar{x}^2, \tag{1.144}$$

which is what we got from Lagrangian considerations (see (1.131)). The role of the generating function is to ensure that transforming a Hamiltonian does not transform the physical problem.

Problem 1.15
For the function $f(x) = \sin(x)$:
(a) Find the Legendre transform, $g(p)$ with $p = \frac{df}{dx}$.
(b) Take your $g(p)$ from above, and find the Legendre transform of it, $h(x)$ with $x = \frac{dg}{dp}$. Show that $h(x)$ is precisely $\sin(x)$, your starting point.

Problem 1.16
Find $K(x, \bar{x})$ (see (1.137)) generating $\bar{x} = p$, $\bar{p} = -x$ (so that the roles of x and p are interchanged). Starting from $H = \frac{p^2}{2m} + \frac{1}{2}kx^2$, find the transformed Hamiltonian,

$\bar{H}(\bar{x}, \bar{p}) = H(x(\bar{x}, \bar{p}), p(\bar{x}, \bar{p}))$ associated with this transformation (this particular Hamiltonian has a special form for this transformation).

Problem 1.17
For Newtonian gravity in one dimension, the Lagrangian is:

$$L = \frac{1}{2} m \dot{x}^2 + \frac{G M m}{x}. \tag{1.145}$$

(a) Find the equation of motion for x given this Lagrangian.
(b) Find the associated Hamiltonian, and generate the equations of motion from the Hamiltonian – show that your pair of first-order ODEs is equivalent to the second-order ODE you got in part a.

*Problem 1.18
Write the Hamiltonian for the electromagnetic potential energy: $U = q\,\phi - q\,\mathbf{v} \cdot \mathbf{A}$ starting from the Lagrangian. You should be able to express the Hamiltonian as $H = \mathbf{Q} \cdot \mathbf{Q} + q\,\phi$, where $\mathbf{Q}$ is a vector constructed from $\mathbf{p}$ and $\mathbf{A}$.

1.7 Hamiltonian and transformation

Continuing with the Hamiltonian formulation of the central body problem – we will uncover the real power of the approach by considering transformations, finding conserved quantities, and using them to reduce the number (and degree) of ODEs we get in the equations of motion.

Our first goal is to prove Noether's theorem on the Hamiltonian side, and we are poised to do this. Then we will develop constants of the motion for Euclidean space written in spherical coordinates. These correspond to angular momentum conservation and total energy conservation.

1.7.1 Canonical infinitesimal transformations

We have the generic form for a canonical transformation, one that leads to a new Hamiltonian system. The transformation is generated by a function which we call $\hat{K}(x, \bar{p})$, and connects the "old" coordinates and momenta $\{x^\alpha, p_\alpha\}$ to the new set $\{\bar{x}^\alpha, \bar{p}_\alpha\}$ via:

$$\boxed{p_\alpha = \frac{\partial \hat{K}}{\partial x^\alpha} \qquad \bar{x}^\alpha = \frac{\partial \hat{K}}{\partial \bar{p}_\alpha},} \tag{1.146}$$

and then we set $\bar{H} = H$, meaning that we take:

$$\bar{H}(\bar{x}, \bar{p}) = H(x(\bar{x}, \bar{p}), p(\bar{x}, \bar{p})). \tag{1.147}$$

The advantage of the generator $\hat{K}(x, \bar{p})$, and the transformation it generates, over $K(x, \bar{x})$ form is clear from the identity transformation:

$$\hat{K} = x^\alpha \, \bar{p}_\alpha. \tag{1.148}$$

This generates $\bar{x}^\alpha = x^\alpha$ and $\bar{p}_\alpha = p_\alpha$ trivially. We will be looking at small perturbations from identity, and the $\hat{K}(x, \bar{p})$ generator is well-suited to this task because it is easy to express the identity transformation with it.

We want to consider canonical *infinitesimal* transformations – the question will eventually be: what are the transformations that leave the entire physical problem invariant? So we must first be able to talk about generic transformations.

Of course, most transformations are complicated, so in order to discuss a generic one, we make a small (read "linear") transformation. This turns out (for deep and not so deep reasons) to suffice – we can put together multiple infinitesimal transformations to make big complicated ones.

To start off, then, we make a small change to the coordinates and momenta:

$$\bar{x}^\alpha = x^\alpha + \epsilon \, f^\alpha(x, p)$$
$$\bar{p}_\alpha = p_\alpha + \epsilon \, h_\alpha(x, p). \tag{1.149}$$

That is: our new coordinates and momenta differ from the old by a small amount, labeled with ϵ, and depend on functions of the old coordinates and momenta. This is our target transformation, and we will now find the form of the generating function that supports it.

We want to express (1.149) in terms of $\hat{K}$ – we want $\hat{K}$ to generate the small transformations above, so we add a little piece to the identity:

$$\hat{K} = x^\alpha \, \bar{p}_\alpha + \epsilon \, J(x, \bar{p}). \tag{1.150}$$

We can view the $\bar{p}$-dependence in $J(x, \bar{p})$ as, itself, a function of x and p, so think of:

$$J(x, \bar{p}) = J(x, \bar{p}(x, p)), \tag{1.151}$$

and this will allow us to expand the transformation generated by $\hat{K}$ in powers of ϵ for ultimate comparison with (1.149).

From the generator $\hat{K}$, with its identity plus correction form, we have, to first order in ϵ:

$$p_\alpha = \frac{\partial \bar{K}}{\partial x^\alpha} = \bar{p}_\alpha + \epsilon \, \frac{\partial J}{\partial x^\alpha}$$

$$\bar{x}^\alpha = \frac{\partial \bar{K}}{\partial \bar{p}_\alpha} = x^\alpha + \epsilon \, \frac{\partial J}{\partial \bar{p}_\alpha} = x^\alpha + \epsilon \left(\frac{\partial J}{\partial p_\beta} \frac{\partial p_\beta}{\partial \bar{p}_\alpha} \right) = x^\alpha + \epsilon \left(\frac{\partial J}{\partial p_\beta} \delta_\beta^\alpha + O(\epsilon) \right). \tag{1.152}$$

The second equation, relating $\bar{x}^\alpha$ to x^α, required Taylor expansion (an application of the chain rule here) – we can drop the $O(\epsilon)$ term inside the parentheses because it will combine with the ϵ term out front to give an overall ϵ^2 dependence in which we are uninterested (that is zero, to first order in ϵ). When we drop the offending term, our linear relation for $\bar{x}^\alpha$ is:

$$\bar{x}^\alpha = x^\alpha + \epsilon \frac{\partial J}{\partial p_\alpha}. \tag{1.153}$$

Comparing (1.152) and (1.153) with (1.149), we can relate the functions f^α and h_α to derivatives of J:

$$\boxed{\begin{aligned} f^\alpha &= \frac{\partial J}{\partial p_\alpha} \\ h_\alpha &= -\frac{\partial J}{\partial x^\alpha}. \end{aligned}} \tag{1.154}$$

1.7.2 Rewriting H

Excellent – now what? Well, we have a transformation, generated by $J(x, p)$ (an infinitesimal generator) that is defined entirely in terms of the original coordinates and momenta. We can now ask the separate question: how does H change under this transformation? To answer this question, we form $H(\bar{x}, \bar{p})$, expand in the small parameter ϵ, and subtract $H(x, p)$:

$$\begin{aligned} H(x + \epsilon f, p + \epsilon h) - H(x, p) &\cong \epsilon \left(\frac{\partial H}{\partial x^\alpha} f^\alpha + \frac{\partial H}{\partial p_\alpha} h_\alpha \right) \\ &= \epsilon \left[\frac{\partial H}{\partial x^\alpha} \frac{\partial J}{\partial p_\alpha} - \frac{\partial H}{\partial p_\alpha} \frac{\partial J}{\partial x^\alpha} \right] \\ &\equiv \epsilon [H, J], \end{aligned} \tag{1.155}$$

where the final line serves to define the usual "Poisson bracket". That's nice, because consider the flip side of the coin – the change in J as a particle moves along its trajectory is given by:

$$\begin{aligned} \frac{dJ}{dt} &= \frac{\partial J}{\partial x^\alpha} \dot{x}^\alpha + \frac{\partial J}{\partial p_\alpha} \dot{p}_\alpha \\ &= \frac{\partial J}{\partial x^\alpha} \frac{\partial H}{\partial p_\alpha} - \frac{\partial J}{\partial p_\alpha} \frac{\partial H}{\partial x^\alpha} = -[H, J]. \end{aligned} \tag{1.156}$$

We have used the Hamiltonian equations of motion to rewrite $\dot{x}^\alpha$ and $\dot{p}_\alpha$ – so we are assuming that $x^\alpha(t)$ and $p_\alpha(t)$ take their dynamical form (i.e. that they satisfy the equations of motion). This gives us the time-derivative of J along the trajectory.

Now, the point. If we have a function J such that $[H, J] = 0$, then we know that:

- The Hamiltonian remains unchanged under the coordinate transformation implied by J:
$\Delta H = [H, J] = 0$.
- The quantity J is a constant of the motion $\dot{J} = -[H, J] = 0$.

But keep in mind, J is intimately tied to a transformation. This is the Noetherian sentiment.

Noether's theorem itself, here, reads: if a transformation generated by J leaves the Hamiltonian unchanged, then the generator J is conserved along the dynamical trajectory. We have to be careful, we are talking about an "isometry" of the Hamiltonian, an additional feature not present in a generic canonical transformation. Isometry means that the Hamiltonian in the new coordinates is obtained by literal replacement of the old coordinates, so that it is "form-invariant". For example, if we have the Hamiltonian:

$$H(x, p) = \frac{p^2}{2m},$$ (1.157)

and we let $J = a\,p$ so that $\bar{x} = x + a$ and $\bar{p} = p$, then:

$$H(x(\bar{x}), p(\bar{p})) = \frac{\bar{p}^2}{2m}$$ (1.158)

and we see that we can replace p in (1.157) with $\bar{p}$ directly, so the transformation generated by J leaves the Hamiltonian "form-invariant", and Noether's theorem applies. The generator $J = a\,p$ is obviously a constant of the motion (since p itself is for uniform motion).

We have expanded our notion of transformation from its Lagrangian roots – now we can talk about transformations to coordinates and momenta as long as they can be generated (at least infinitesimally) by J.

Example

As a simple example to see how the program of symmetry implies time-independence is applied, let's take one-dimensional motion under the potential $U(x)$ – the Hamiltonian is:

$$H = \frac{p^2}{2m} + U(x).$$ (1.159)

We will attempt to solve the partial differential equation (PDE) $[H, J] = 0$ for $J(x, p)$ a function of x and p. The vanishing Poisson bracket is easy to write:

$$U'(x)\frac{\partial J}{\partial p} - \frac{p}{m}\frac{\partial J}{\partial x} = 0.$$ (1.160)

Now, how can we solve for J? Let's use separation of variables for starters (and finishers). Take an additive separation $J = J_x(x) + J_p(p)$, then we have:

$$U'(x) J'_p(p) - \frac{p}{m} J'_x(x) = 0, \tag{1.161}$$

and to make the usual separation argument (separability relies on an identification of terms that depend only on one of a set of variables – so that we have, for example: $f(x) + g(y) = 0$, which cannot be true for all values of x and y unless $f(x)$ and $g(y)$ are equal to constants), we must divide by the product $(U'(x) \, p)$, and then we can simply set:

$$\frac{J'_p(p)}{p} = \alpha = \frac{J'_x(x)}{mU'(x)} \tag{1.162}$$

in order to solve (1.161). The constant α is the "separation constant" and, we will assume, is some real number. Solving (1.162) gives $J'_p(p) = \frac{1}{2} \alpha \, p^2$ and $J_x(x) = \alpha \, m \, U(x)$ (plus constants of integration, which just add overall constants to J – since J generates transformations via its derivatives, additive constants do not play an interesting role). Then we have:

$$J = \alpha \, m \left(\frac{p^2}{2m} + U(x) \right) = \alpha \, m \, H. \tag{1.163}$$

This tells us that any generator proportional to H is a constant of the motion (can you see what transformation is generated by this choice?) – no surprise, $[H, H] = 0$ automatically. As for the utility, once we have learned that, for example, $J = H$ has $\frac{dJ}{dt} = 0$ along the trajectory, then our work is highly simplified – let $J = E$, a number (which we recognize as the total energy) – then:

$$E = \frac{p^2}{2m} + \frac{1}{2} k x^2, \tag{1.164}$$

and we can generate a geometrical relation between p and x. Figure 1.7 shows the constant-energy contours for $\frac{1}{2} p^2 + \frac{1}{2} x^2 = E$ – it is no surprise here that isosurfaces of constant energy E form circles (for $m = k = 1$).

If we write H in terms of $\dot{x}$ by replacing p, we have simplified our problem – now we have a single first-order ODE for $\dot{x}$ to solve. This is the functional utility of the constants of the motion we discover through the Poisson bracket PDE.

Suppose we instead consider multiplicative separation (familiar from, for example, E&M) – take $J(x, p) = J_x(x) J_p(p)$, then the Poisson bracket relation reads:

$$J_x J'_p U' - \frac{p}{m} \left(J_p J'_x \right) = 0 \tag{1.165}$$

and dividing by $(J \, U' \, p)$ allows separation with constant α:

$$\frac{J'_p}{p J_p} = \alpha = \frac{J'_x}{m J_x U'} \tag{1.166}$$

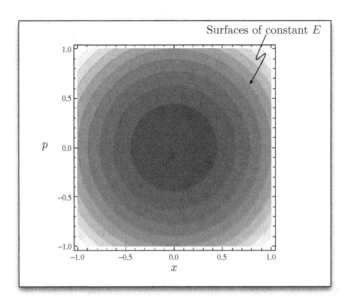

Figure 1.7   The relation between x and p for a constant E.

and this has solution:

$$J_x = \gamma\, e^{\alpha\, m\, U} \qquad J_p = \kappa\, e^{\frac{\alpha\, p^2}{2}}, \tag{1.167}$$

for constants γ and κ. Putting it together with an overall factor out front:

$$J = J_0\, e^{\alpha\, m\left(\frac{p^2}{2m}+U\right)} = J_0\, e^{\alpha\, m\, H}. \tag{1.168}$$

We can now define the infinitesimal transformation for this J:

$$\bar{x} = x + \epsilon\, \frac{\partial J}{\partial p} = x + \underbrace{\epsilon\, J_0\, \alpha\, m}_{\equiv\beta}\, \frac{\partial H}{\partial p}\, e^{\alpha\, m\, H} = x + \beta\, \frac{p}{m}\, e^{\alpha\, m\, H}$$

$$\bar{p} = p - \epsilon\, \frac{\partial J}{\partial x} = p - \beta\, \frac{\partial H}{\partial x}\, e^{\alpha\, m\, H} = p - \beta\, U'\, e^{\alpha\, m\, H}, \tag{1.169}$$

and, as advertised, if we rewrite the Hamiltonian in terms of these variables:

$$\begin{aligned}
H &= \frac{1}{2\,m}\left(\bar{p} + \beta\, U'\, e^{\alpha\, m\, H}\right)^2 + U\!\left(\bar{x} - \beta\, \frac{p}{m}\, e^{\alpha\, m\, H}\right) \\
&= \frac{1}{2\,m}\left(\bar{p}^2 + 2\,\bar{p}\,\beta\, U'\, e^{\alpha\, m\, H}\right) + U(\bar{x}) - U'(\bar{x})\,\beta\, \frac{p}{m}\, e^{\alpha\, m\, H} + O(\beta^2) \\
&= \frac{\bar{p}^2}{2\,m} + U(\bar{x}) + O(\beta^2),
\end{aligned} \tag{1.170}$$

so that literal replacement of $x \to \bar{x}$, $p \to \bar{p}$ in $H(x, p)$ gives $H(\bar{x}, \bar{p})$, which was the point of the Poisson brackets in the first place.

1.7.3 A special type of transformation

Let's see how all of this machinery applies to our generic form for H for a specific type of transformation (coordinates-to-coordinates). Our goal is to find transformations that leave our Hamiltonian unchanged, and the associated constants of the motion. The utility of such constants should be clear: the more constants of the motion we have, the less integration we have to do. In particular, as with the above example, it should be possible to find direct, functional, relations between the first derivatives of the coordinates and the coordinates themselves.

The Hamiltonian written in generic-coordinate form, is:

$$H = \frac{1}{2} p_\alpha g^{\alpha\beta} p_\beta + U(x). \tag{1.171}$$

Now all we have to do is pick a J, a generator. Suppose we ask: what coordinate transformations can I have that set the new coordinates to a function only of the old? In other words, we want $\bar{x}^\alpha = x^\alpha + \epsilon f^\alpha(x)$ where $f^\alpha(x)$ has no dependence on p. That tells us the form for J immediately – from $f^\alpha = \frac{\partial J}{\partial p_\alpha}$, we set:

$$J = p_\alpha f^\alpha(x) \tag{1.172}$$

by quadrature, and then we also obtain the transformation for momenta: $h_\alpha = -p_\beta \frac{\partial f^\beta}{\partial x^\alpha}$.

We have satisfied the "canonical infinitesimal transformation" requirement, now we want to deal with $\Delta H = 0$. We are concerned with finding specific constraints on f^α that will set $[H, J] = 0$, then we know that J is conserved. The Poisson bracket PDEs are:

$$[H, J] = \frac{\partial H}{\partial x^\alpha} f^\alpha - \frac{\partial H}{\partial p_\alpha} p_\beta \frac{\partial f^\beta}{\partial x^\alpha} = 0 \tag{1.173}$$

and from the form of the Hamiltonian in (1.171):

$$\frac{\partial H}{\partial x^\alpha} = \frac{1}{2} p_\beta p_\gamma g^{\gamma\beta}{}_{,\alpha} + U_{,\alpha}$$

$$\frac{\partial H}{\partial p_\alpha} = p^\alpha. \tag{1.174}$$

Then we have:[8]

$$[H, J] = \left(-\frac{1}{2} p^\gamma p^\beta g_{\gamma\beta,\alpha} + U_{,\alpha} \right) f^\alpha - g^{\gamma\delta} p_\delta p_\alpha f^\alpha{}_{,\gamma}$$

$$= -p^\alpha p^\beta \left(\frac{1}{2} g_{\alpha\beta,\gamma} f^\gamma + g_{\alpha\gamma} f^\gamma{}_{,\beta} \right) + U_{,\alpha} f^\alpha = 0. \tag{1.176}$$

---

[8] We use the result

$$g_{\mu\alpha} g^{\alpha\beta}{}_{,\gamma} g_{\beta\nu} = -g_{\mu\nu,\gamma}, \tag{1.175}$$

obtainable via the product rule for the ordinary derivative.

Notice the two separate pieces to the Poisson bracket: we must have *both* the term multiplying $p^\alpha p^\beta$ equal to zero, and $U_{,\alpha} f^\alpha = 0$ (which says that the coordinate transformation must be orthogonal to the force). The first term is a geometric statement (meaning that it is defined once a metric is given), the second is physical (it depends on the environment through U). They do not mix, since the orthogonality of transformation and force does not involve momenta, and so cannot cancel the first term everywhere.

In general relativity, where there is no potential, it is only the first term that counts, and vectors f^α satisfying this equation are called "Killing vectors". The side-constraint imposed by U can be dealt with after the general form implied by the PDE:

$$\frac{1}{2} g_{\alpha\beta,\gamma} f^\gamma + g_{\alpha\gamma} f^\gamma{}_{,\beta} = 0 \tag{1.177}$$

is satisfied.

The above equation, with some massaging, is "Killing's equation", and more typically written as:

$$\boxed{f_{\mu;\nu} + f_{\nu;\mu} = 0 \qquad f_{\mu;\nu} \equiv f_{\mu,\nu} - \Gamma^\alpha{}_{\mu\nu} f_\alpha,} \tag{1.178}$$

where the "connection" with one index raised, from (1.58), is:

$$\Gamma^\alpha{}_{\mu\nu} = \frac{1}{2} g^{\alpha\rho} \left(g_{\rho\mu,\nu} + g_{\rho\nu,\mu} - g_{\mu\nu,\rho} \right). \tag{1.179}$$

Killing's equation is a *tensor* statement – if true in one coordinate system, true in all. Keep in mind the point of these vectors – if we can solve the above PDE, we have a transformation that leaves the Hamiltonian unchanged, and is generated by $J = f^\alpha p_\alpha$. Since the Hamiltonian is invariant under the transformation, J is itself conserved. There are a few notational shifts attached to (1.178) – we are defining the "semicolon" form of the derivative, and this does end up being significant. For now, it is just a handy way to keep track of the baggage we are attaching to the partial derivatives $f_{\mu,\nu}$. That baggage is familiar from Section 1.3.2 where it was originally defined, an unenlightening combination of partial derivatives of the metric. We will have more to say about both the form and interpretation of these objects in Chapter 3. For now it is interesting to note that the semicolon derivative (called the "covariant" derivative) is coming to us from a fairly straightforward application of Poisson brackets in familiar classical mechanics, requiring no fancy geometry.

Example

Let's calculate a Killing vector in spherical coordinates. For a central potential, the extra portion of (1.176) reads: $\frac{\partial U}{\partial r} f^r = 0$, which we can accomplish simply by setting $f^r = 0$. For the other two components, take:

$$f^\alpha \doteq \begin{pmatrix} 0 \\ f^\theta(r, \theta, \phi) \\ f^\phi(r, \theta, \phi) \end{pmatrix} \longrightarrow f_\alpha \doteq \begin{pmatrix} 0 \\ r^2 f^\theta \\ r^2 \sin^2\theta \ f^\phi \end{pmatrix}, \tag{1.180}$$

where we have used the metric appropriate to spherical coordinates (1.35) to lower the index and make the covariant form f_α. Now we need to construct the *a priori* nine equations (in three dimensions) implied by (1.178). Since the left-hand side of (1.178) is manifestly symmetric, there are really only six equations, and we will look at the relevant, independent ones.

We'll start with the equations for f_2:

$$2 f_{2;2} = 2 \left(f_{2,2} - \Gamma^\sigma_{22} f_\sigma \right) \tag{1.181}$$

and

$$\Gamma^\sigma_{\alpha\beta} = \frac{1}{2} g^{\sigma\rho} \left(g_{\alpha\rho,\beta} + g_{\beta\rho,\alpha} - g_{\alpha\beta,\rho} \right), \tag{1.182}$$

so that $\Gamma^\sigma_{22} f_\sigma = \frac{1}{2} (f^2 g_{22,2}) = 0$, and we have:

$$2 f_{2;2} = 2 f_{2,2} = 2 r^2 \frac{\partial f^\theta}{\partial \theta}, \tag{1.183}$$

from which we learn that $f^\theta = f^\theta(r, \phi)$. By similar arguments involving the $f_{1;2}$, $f_{1;3}$ equations, there is no r-dependence in either f^θ or f^ϕ. We are left with:

$$f_{2;3} + f_{3;2} = f_{2,3} + f_{3,2} - \Gamma_{\sigma 23} f^\sigma - \Gamma_{\sigma 32} f^\sigma$$
$$= r^2 \frac{\partial f^\theta}{\partial \phi} + r^2 \sin^2\theta \frac{\partial f^\phi}{\partial \theta}. \tag{1.184}$$

Finally, we need:

$$f_{3;3} = f_{3,3} - \Gamma_{\sigma 33} f^\sigma = \frac{\partial f_3}{\partial \phi} - (-g_{33,2} f^2)$$

$$= r^2 \sin^2\theta \frac{\partial f^\phi}{\partial \phi} + \left(2 r^2 \sin\theta \cos\theta \ f^\theta \right). \tag{1.185}$$

So our two (remaining) PDEs are:

$$0 = r^2 \left(\frac{\partial f^\theta}{\partial \phi} + \sin^2\theta \frac{\partial f^\phi}{\partial \theta} \right)$$

$$0 = r^2 \sin\theta \left(\cos\theta \ f^\theta + \sin\theta \frac{\partial f^\phi}{\partial \phi} \right). \tag{1.186}$$

The bottom equation can be integrated:

$$f^\phi(\theta, \phi) = -\cot\theta \int f^\theta \, d\phi + f^\phi(\theta),$$ (1.187)

and then the top equation is:

$$\left(\int f^\theta \, d\phi + \frac{\partial f^\theta}{\partial \phi}\right) + \left(\sin^2\theta \, \frac{\partial f^\phi(\theta)}{\partial \theta}\right) = 0.$$ (1.188)

Because of the functional dependence here (the first term in parentheses depends only on ϕ, the second only on θ), we have a separation constant. Take that constant to be zero, and solve: we find $f^\phi = F$, a constant, and $f^\theta = A\sin\phi - B\cos\phi$ for constants A and B.

Our final form for the Killing vector f^α is:

$$f^\alpha \doteq \begin{pmatrix} 0 \\ A\sin\phi - B\cos\phi \\ F + \cot\theta\,(A\cos\phi + B\sin\phi) \end{pmatrix}.$$ (1.189)

There are three constants of integration here, and these can be used to separate the three, independent, transformations implied by f^α. We will look at the implications and physical meaning of these three in Section 1.8.

1.8 Hamiltonian solution for Newtonian orbits

Using the Killing vector(s) (1.189) appropriate to spherically symmetric potentials, and associated constants, we will finish with the solution of orbital motion in the Hamiltonian framework.

The lesson of the Hamiltonian equations of motion, together with the constants we can get from the Killing vectors, is: if possible, never form the equations of motion. In more complicated settings, of course, one must write them down, but by the time you've done that, it's almost always a computational (meaning "with a real computer") task to solve them. The solution procedure here looks quite a bit different from the equivalent Lagrange approach (which *must* involve an actual computation of the equations of motion provided $U \neq 0$). As I've mentioned before, and will probably point out a few more times: one of the interesting things about GR as a topic is its lack of forces, which makes $L = H$, and we can blend techniques between the two points of view.

1.8.1 Interpreting the Killing vector

Remember where we are. We have solved the PDE that comes from Killing's equation:[9]

$$p^\alpha p^\beta \left(\frac{1}{2} g_{\alpha\beta,\gamma} f^\gamma + g_{\alpha\gamma} f^\gamma_{,\beta}\right) = 0 \leftrightarrow \boxed{f_{(\alpha;\beta)} = 0.}$$ (1.191)

---

[9] We use the compact form for explicit symmetrization:

$$f_{(\alpha;\beta)} \equiv \frac{1}{2}\left(f_{\alpha;\beta} + f_{\beta;\alpha}\right).$$ (1.190)

To do this, we made a vector ansatz for f^α. Most importantly, we took away the f^r component. The motivation there came from the second piece of the $[H, J] = 0$ requirement in (1.176):

$$U_{,\alpha} f^\alpha = 0 \longrightarrow \frac{\partial U}{\partial r} f^r = 0 \qquad (1.192)$$

and for central potentials, this requires $f^r = 0$ (since the derivative of U is the force, which is not generally zero). Again, on the relativistic side, this is not an issue, and Killing vectors come along for the ride once a metric is specified.

With the ansatz in place, we solved the PDE to get a vector of the form:

$$f^\alpha \doteq \begin{pmatrix} 0 \\ A \sin\phi - B \cos\phi \\ F + \cot\theta \, (A \cos\phi + B \sin\phi) \end{pmatrix}. \qquad (1.193)$$

Now the question is – what is this? Well, the first question is, how *many* is this? We have three constants of integration floating around. These actually correspond to three separate Killing vectors:

$$f_1^\alpha \doteq \begin{pmatrix} 0 \\ 0 \\ F \end{pmatrix} \quad f_2^\alpha \doteq \begin{pmatrix} 0 \\ A \sin\phi \\ A \cot\theta \cos\phi \end{pmatrix} \quad f_3^\alpha \doteq \begin{pmatrix} 0 \\ -B \cos\phi \\ B \cot\theta \sin\phi \end{pmatrix}. \qquad (1.194)$$

Each of these vectors is involved in a transformation, $\bar{x}^\alpha = x^\alpha + \epsilon \, f^\alpha$, so the second vector, for example, induces:

$$\bar{r} = r \quad \bar{\theta} = \theta + \epsilon \, (A \sin\phi) \quad \bar{\phi} = \phi + \epsilon \, (A \cot\theta \cos\phi). \qquad (1.195)$$

This is an infinitesimal transformation, it is also, by virtue of its derivation, canonical.

What are the conserved quantities? For each of these transformations, we should have a constant of the motion – precisely the generator J. Remember the form of J:

$$J = p_\alpha \, f^\alpha, \qquad (1.196)$$

so suppose we take the first of the three transformations, then:

$$J = p_\phi \, F. \qquad (1.197)$$

The constant F is really just a normalization, taken care of by ϵ – here we learn that $J = p_\phi$ is conserved and this comes from the transformation $\bar{\phi} = \phi + \epsilon$. That's something we already knew, but gives a clue about the rest of the Killing vectors – angular momentum conservation comes from the rotational invariance of the equations of motion. If we took the Cartesian form of the transformation corresponding to infinitesimal rotation about a vector $\Omega \equiv \omega_x \hat{\mathbf{x}} + \omega_y \hat{\mathbf{y}} + \omega_z \hat{\mathbf{z}}$:[10]

$$\boxed{\bar{\mathbf{x}} = \mathbf{x} + \epsilon \, \Omega \times \mathbf{x}} \tag{1.198}$$

then, upon transformation to spherical coordinates, we would find (to first order in ϵ):

$$\bar{\theta} = \theta + \epsilon \, (\omega_y \cos \phi - \omega_x \sin \phi)$$
$$\bar{\phi} = \phi + \epsilon \, (\omega_z - \cot \theta \, (\omega_x \cos \phi + \omega_y \sin \phi)). \tag{1.199}$$

Evidently, for our choice of constants $\{A, B, F\}$, we are describing the vector $\Omega = -A \, \hat{\mathbf{x}} - B \, \hat{\mathbf{y}} + F \, \hat{\mathbf{z}}$.

Consider the generators J that come from all three vectors, together now:

$$\boxed{\begin{aligned} J_x &= p_\theta \, \sin \phi + p_\phi \, \cot \theta \, \cos \phi \\ J_y &= p_\theta \, \cos \phi - p_\phi \, \cot \theta \, \sin \phi \\ J_z &= p_\phi. \end{aligned}} \tag{1.200}$$

This gives us the full complement of angular momenta, agreeing with the spherical form for $\mathbf{J} = \mathbf{x} \times \mathbf{p}$.

Let's count: we have three constants of the motion so far, these can be used to eliminate momenta or coordinates. In addition to these three, we have H itself – after all, it's hard to imagine a zero more compelling than $[H, H]$, so that's four constants of the motion.

The idea is to use the three constants in (1.200) to eliminate three coordinates, momenta, or a combination. We are not free to do anything we like, there is no p_r

---

[10] To see (1.198), consider $\Omega = \phi \, \hat{\mathbf{z}}$, and set a generic vector $\mathbf{v}$ in the $x - z$ plane. Then from Figure 1.8, we rotate about the z-axis, expressible as $\bar{\mathbf{v}} = \mathbf{v} + \Omega \times \mathbf{v}$.

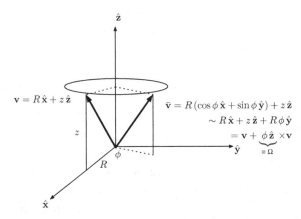

Figure 1.8 An infinitesimal rotation about the $\Omega \sim \hat{\mathbf{z}}$ axis for an arbitrary vector $\mathbf{v}$.

in the angular momentum constants, for example. Looking specifically at those, setting $\{J_x, J_y, J_z\}$ must lead to fixing two components of momentum and θ. If we take the J_z equation at face value, then the other two become:

$$J_x = p_\theta \sin\phi + J_z \cot\theta \cos\phi$$
$$J_y = p_\theta \cos\phi - J_z \cot\theta \sin\phi. \tag{1.201}$$

Suppose we want $J_x = J_y = 0$, then we get two equations for p_θ:

$$p_\theta = -J_z \cot\theta \cot\phi \qquad p_\theta = J_z \cot\theta \tan\phi \tag{1.202}$$

and this can only be true for $\cot\theta = 0 \longrightarrow \theta = \pm\frac{1}{2}\pi$.

This enforced value of θ is the source of all the commentary we had on the Lagrange side in Section 1.4. By setting the motion in the equatorial plane, we were effectively setting $J_x = J_y = 0$, leaving us with J_z. We do not have another choice of plane with $J_x = J_y = 0$.

With that in place, though, we automatically get $p_\theta = 0$ along with $\theta = \frac{1}{2}\pi$, and we can use this information in the Hamiltonian:

$$
\begin{aligned}
H &= \frac{1}{2} p_\alpha g^{\alpha\beta} p_\beta + U(r) \\
&= \frac{1}{2}\left(p_r^2 + \frac{1}{r^2} p_\theta^2 + \frac{1}{r^2 \sin^2\theta} p_\phi^2\right) + U(r) \\
&= \frac{1}{2}\left(p_r^2 + \frac{1}{r^2} J_z^2\right) + U(r).
\end{aligned} \tag{1.203}
$$

Now we can use the constancy of H to solve for p_r^2 – let's call $H = E$, the numerical constant, then solving for the p_r component of momentum gives:

$$p_r^2 = 2E - 2U(r) - \frac{1}{r^2} J_z^2. \tag{1.204}$$

To connect this to the r-velocity, we can use the equation of motion (or definition of canonical momentum from the Lagrangian) – we have $p_r = \dot{r}$, and we can specialize to the Newtonian gravitational potential to get:

$$(\dot{r})^2 = \frac{1}{r^2}\left(2Er^2 + 2Mr - J_z^2\right). \tag{1.205}$$

It is difficult to imagine solving this if we took the square root; there is a sign issue there, of course, but also the ODE gets more complicated. Let's think about what we can get out of the above with no work. The first thing we notice is that the term in parentheses is quadratic in r. The roots of this quadratic are points where $\dot{r} = 0$, which we associate with turning points of the motion. Suppose we explicitly factor the quadratic into its roots:

$$\dot{r}^2 = \frac{2E}{r^2}(r - r_+)(r - r_-)$$

$$r_+ = \frac{1}{2}\left(-\frac{M}{E} + \sqrt{\left(\frac{M}{E}\right)^2 + 2\frac{J_z^2}{E}}\right) \tag{1.206}$$

$$r_- = \frac{1}{2}\left(-\frac{M}{E} - \sqrt{\left(\frac{M}{E}\right)^2 + 2\frac{J_z^2}{E}}\right).$$

In a situation like this, a change of variables is in order – we know that whatever r is, it has two turning points, with r_- further away than r_+, so we are tempted to set:[11]

$$r(\psi) = \frac{p}{1 + e\cos\psi}, \tag{1.207}$$

with p and e arbitrary constants (although the naming here will allow us to connect to our earlier solution). The idea is to match the form of a curve (two turning points, constant angular momentum) with its functional expression. We have done this, and the freedom is granted by our use of the parameter ψ – this will eventually be related to ϕ, but for now, we are making an astute choice of curve parametrization, and ψ itself has no meaning. Continuing with our identification, the two turning points of an ellipse are:

$$r(0) \equiv r_p = \frac{p}{1 + e} \qquad r(\pi) \equiv r_a = \frac{p}{1 - e}, \tag{1.208}$$

and we can associate these with the turning points of our $\dot{r}$ equation. This tells us that:

$$\frac{p}{1 + e} = r_+ \qquad \frac{p}{1 - e} = r_- \longrightarrow \boxed{J_z = \sqrt{M p} \qquad E = \frac{M(e^2 - 1)}{2p}.} \tag{1.209}$$

Meanwhile, we also have, taking the time-derivative of (1.207):

$$\dot{r} = \frac{p\, e\sin\psi\,\dot{\psi}}{(1 + e\cos\psi)^2}, \tag{1.210}$$

and then squaring both sides and setting the whole thing equal to the right-hand side of (1.205) gives:

$$(\dot{r})^2 = \frac{p^2\, e^2\sin^2\psi\,\dot{\psi}^2}{(1 + e\cos\psi)^4} = \frac{1}{p}\left(M\, e^2\sin^2\psi\right). \tag{1.211}$$

---

[11] The $r(\psi)$ form in (1.207) is a particularly nice parametrization, and the starting point for many computational investigations when the geometry becomes more complicated.

Finally, we can solve for $\dot{\psi}^2$:

$$(\dot{\psi})^2 = \frac{M(1 + e\cos\psi)^4}{p^3}.$$ (1.212)

But what is the relationship of ψ to, for example, ϕ? We'll perform a final change of variables – let $\psi' \equiv \frac{d\psi}{d\phi}$, then by the chain rule:

$$\dot{\psi} = \frac{d\psi}{d\phi}\frac{d\phi}{dt} = \psi'\dot{\phi} = \psi'\left(\frac{J_z}{r^2}\right)$$

$$= \psi'\sqrt{pM}\left(\frac{1 + e\cos\psi}{p}\right)^2$$ (1.213)

and then:

$$(\dot{\psi})^2 = \psi'^2\frac{M}{p^3}(1 + e\cos\psi)^4 = \frac{M}{p^3}(1 + e\cos\psi)^4.$$ (1.214)

Conclusion: $\psi' = 1$, or $\psi = \phi$ (plus an arbitrary phase that we aren't interested in).

So we discover:

$$r(\psi) = r(\phi) = \frac{p}{1 + e\cos\phi}$$ (1.215)

as before, with $\{p, e\}$ related to the energy and angular momentum of the orbit through:

$$J_z = \sqrt{Mp} \qquad E = \frac{M(e^2 - 1)}{2p}.$$ (1.216)

Incidentally, $\{p, e\}$ are intrinsically greater than zero, but bound orbits have $E < 0$ (they're caught in a well), so for bound orbits, we want $e^2 < 1$. J_z can be positive or negative, $J_z = \pm\sqrt{Mp}$ corresponding to the direction (counterclockwise or clockwise) of the orbit.

We are, finally, done with orbital motion. I hope that I have convinced you that L and H lead to different ways of viewing the solutions, but they are entirely equivalent (as is evidenced by the fact that the solutions are identical).

Notice the role played by the metric in the determination of the orbits (both for L and H). In general relativity, there is no potential – all motion comes directly from the metric (and evidently, the metrics of GR are different from the ones we have been considering). It is not uncommon to characterize the more involved space-times (i.e. $g_{\mu\nu}$) that arise in general relativity in terms of the particle orbits (geodesics) they produce. This somewhat long introduction will, I hope, prove useful when you are handed some non-obvious metric and forced to discuss it in a reasonable way.

1.8.2 Temporal evolution

Starting from (1.203), we can analyze the motion of any polynomial central potential we wish using the techniques discussed above. The move to $\psi = \phi$ parametrization is a particularly useful choice, and can tell us a lot about the motion of the body, but we lose all temporal evolution information. We can recover the time-dependence *a posteriori* by explicit differentiation – for our gravitational example:

$$\frac{dr}{dt} = \frac{dr}{d\phi}\frac{d\phi}{dt},$$

(1.217)

and we know $\dot\phi = \frac{J_z}{r^2}$. We can write a small change in t in terms of a small change in ϕ from the above:

$$dt = \frac{dr}{\frac{dr}{d\phi}\dot\phi} = \frac{d\phi}{\dot\phi} = \frac{p^2\,d\phi}{J_z\,(1 + e\,\cos\phi)^2},$$

(1.218)

and we can integrate this from $\phi : 0 \longrightarrow 2\pi$ to find the total orbital period:

$$T = \int_0^{2\pi} \frac{p^2\,d\phi}{J_z\,(1 + e\,\cos\phi)^2} = \frac{J_z^3}{M^2}\frac{2\pi}{(1 - e^2)^{3/2}},$$

(1.219)

where we have used $p^2 = \frac{J_z^4}{M^2}$. Note that the semi-major axis of the orbit is just:

$$a \equiv \frac{1}{2}\left(\frac{p}{1 + e} + \frac{p}{1 - e}\right) = \frac{p}{1 - e^2},$$

(1.220)

so the period can be written in terms of this geometric parameter:

$$T = \frac{2\pi\,a^{3/2}}{\sqrt{M}},$$

(1.221)

which is Kepler's third law.

The fact that equal areas are swept out in equal times (Kepler's second law) comes directly from the infinitesimal area of an ellipse and angular momentum conservation.

For the small triangle shown (enlarged) in Figure 1.9, we have approximate area $dA = \frac{1}{2}r^2\,d\phi$, and if this is swept out in time dt, we can write:

$$\frac{dA}{dt} = \frac{1}{2}r^2\dot\phi = \frac{1}{2}J_z,$$

(1.222)

a constant.

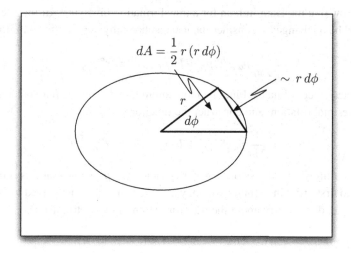

$$dA = \frac{1}{2} r \, (r \, d\phi)$$

$\sim r \, d\phi$

r

$d\phi$

Figure 1.9   An ellipse in $r(\phi)$ parametrization – we make a small triangle and use "one half base times height" with appropriate approximations to find dA, the infinitesimal area.

Problem 1.19

It is clear that $[H, H] = 0$, and so the Hamiltonian is itself a constant of any motion governed by the Hamiltonian. But what is the associated infinitesimal transformation? By treating H as an infinitesimal generator, find the transformation relating $\{\bar{x}(t), \bar{p}(t)\}$ to $\{x(t), p(t)\}$ and interpret it physically.

Problem 1.20

Write out all six equations associated with Killing's equation $f_{\mu;\nu} + f_{\nu;\mu} = 0$ for the starting point (1.180).

Problem 1.21

Killing's equation is a covariant statement (meaning that it is true in all coordinate systems):

$$f_{\mu;\nu} + f_{\nu;\mu} = 0. \tag{1.223}$$

We looked at the form of infinitesimal rotations in spherical coordinates, but the Cartesian form is even simpler – rotation through a small angle ω about the unit axis $\hat{\Omega}$ can be expressed as:

$$\bar{\mathbf{x}} = \mathbf{x} + \omega \, \hat{\Omega} \times \mathbf{x}. \tag{1.224}$$

(a) Show that the associated infinitesimal transformation satisfies Killing's equation in Cartesian coordinates.

(b) We know that the Hamiltonian for a transformation involving Killing vectors
should be unchanged – consider the usual spherically symmetric Hamiltonian:

$$H = \frac{1}{2m} p_\alpha \, g^{\alpha\beta} \, p_\beta + U(r) \qquad r^2 \equiv x^\alpha \, g_{\alpha\beta} \, x^\beta, \tag{1.225}$$

in Cartesian coordinates. Find the momentum transformation $\bar{\mathbf{p}}$ associated with
infinitesimal rotation, and construct the function:

$$\bar{H} = \frac{1}{2m} \bar{p}_\alpha g^{\alpha\beta} \, \bar{p}_\beta + U(\bar{r}) \qquad \bar{r}^2 \equiv \bar{x}^\alpha \, g_{\alpha\beta} \, \bar{x}^\beta. \tag{1.226}$$

By inputting your expressions for $\bar{p}(x, p)$ and $\bar{x}(x, p)$, show that this is the same
as H to first order in ω (note: you may assume that $g_{\mu\nu}$ is unchanged to first order
here – i.e. don't worry about the transformation of the metric itself).

Problem 1.22
We started, in (1.172), from the assumption that "coordinates go to coordinates".
Suppose we attempt to go the other direction in one dimension, and make
$\bar{x} = x + \epsilon \, F(p)$ so that the new coordinate is the old one plus a new piece that
depends only on the "old" momenta – we also demand that $\bar{p} = p$, no change to the
momenta. Construct J, the infinitesimal generator, for this type of transformation, and
constrain $F(p)$ so that your transformation satisfies $[H, J] = 0$, then H is unchanged.
What are you left with?

*Problem 1.23
Find the equations of motion for a particle moving under the influence of an electric
and magnetic field – i.e. find the Euler–Lagrange equations for:

$$L = \frac{1}{2} m \, v_j \, v^j - \left(q \, \phi - q \, v_j \, A^j\right). \tag{1.227}$$

Work in Cartesian coordinates (easiest using indexed notation), and verify
that you get the correct $\mathbf{F} = q \, \mathbf{E} + q \, \mathbf{v} \times \mathbf{B}$ at the end, with $\mathbf{E} = -\nabla\phi$ and
$\mathbf{B} = \nabla \times \mathbf{A}$.

Problem 1.24
Suppose that the Lagrangian appropriate for classical dynamical study read:

$$L = \epsilon \, a^2 + \frac{1}{2} m \, v^2 - U(x), \tag{1.228}$$

for a particle of mass m moving under the influence of a potential $U(x)$ in one
dimension. In this case, a is the acceleration, and it is multiplied by a small constant ϵ
(what dimensions must then be associated with ϵ?). Find the equations of motion
governing the particle in this setting (note, you will have to go back to the action to
develop the appropriate Euler–Lagrange equations) – assume that we can specify both
the initial/final position and initial/final velocity of the particle.

Problem 1.25

Given the Hamiltonian:

$$H = \frac{p^2}{2m} + \alpha x p, \qquad (1.229)$$

find the Lagrangian, and associated equations of motion from the Euler–Lagrange equations.

2

Relativistic mechanics

We have developed some tensor language to describe familiar physics – we reviewed orbital motion from the Lagrangian and Hamiltonian points of view, and learned how to write the equations of motion generically in terms of a metric and its derivatives (in the special combination $\Gamma^{\alpha}{}_{\mu\nu}$). All of the examples involved were in flat, three-dimensional space, and the most we have done is introduce curvilinear coordinates, i.e. there has been no fundamental alteration of Euclidean geometry.

Once a generic metric and arbitrary coordinates are introduced, we are able to discuss dynamics in more exotic geometrical settings. The field equations of general relativity provide a connection between distributions of mass (energy) and the metric determining the underlying geometry of space-time. So the target of GR is precisely a metric, and these metrics will provide examples of "exotic geometries" in which we can compute dynamical trajectories of test particles. Our notational shift from Chapter 1, in particular (1.47), (1.59), and (1.171), represents a physically significant idea when used with the metric solutions from general relativity.

Prior to any general relativistic considerations, there is a familiar metric that we can insert into our metricized dynamics that is non-Euclidean and presents fundamentally new physics. This is the Minkowski metric of special relativity, the metric describing the geometry of a four-dimensional space-*time*. When we use the Minkowski metric as our $g_{\mu\nu}$ in the equations of motion from the last chapter, we recover relativistic dynamics, and force a shift of focus from Newton's second law to the geometric notions of length extremization associated with the variation of an action as the fundamental statement of dynamics. These are important modifications when we move to full GR.[1]

---

[1] Indeed, the Minkowski metric is a solution from GR, associated with no mass, a universal vacuum.

In the current chapter, we will look at how the shift from a Euclidean metric to the Minkowski metric changes dynamical considerations. There are a few new features of interest – for example, the use of time as a coordinate makes its availability as a parameter of motion suspect. What, then, should we use to parametrize curves representing trajectories in the expanded space-time framework? How should we think about "straight-line" motion in a setting that has a metric with lengths that are naturally divided into three classes? The differences between special relativistic dynamics and classical trajectories will occupy us for much of the chapter. In the end, we will return to Newtonian gravity, and use basic ideas of length contraction and force transformation to establish that gravity, as it stands, is not in accord with the principles of special relativity.

2.1 Minkowski metric

We start by defining the four-dimensional coordinates of space-time (so-called $D = 3 + 1$ for three spatial and one temporal coordinate) as:

$$x^\mu \doteq \begin{pmatrix} c\,t \\ x \\ y \\ z \end{pmatrix}. \tag{2.1}$$

This is a contravariant four-vector, and we leave in the constant c (with units of velocity) to establish homogenous dimensions for this object. The speed c must be a physical constant that is available to all coordinate systems. A small change can be represented as:

$$dx^\mu \doteq \begin{pmatrix} c\,dt \\ dx \\ dy \\ dz \end{pmatrix}. \tag{2.2}$$

A defining feature of special relativity, that there is a universal speed measured to have the same value in any frame (any "laboratory", moving w.r.t. another or not), is expressed mathematically by the invariance of a particular combination of the coordinate differential elements. This constant speed is a fundamental input in the theory, and is not available from classical considerations – there, the measurement of the speed of an object in two different frames will differ based on the relative speed of the frames. Not so for this special velocity c.[2] For an object traveling at this

---

[2] We have just set the constant c, used to set units in (2.1), to be the special constant speed of special relativity. Why not?

speed, the measurement of c in two different frames must be numerically identical. If one frame measures a displacement dx^μ and another measures a displacement $d\bar{x}^\mu$, the requirement that c be constant is enforced by:[3]

$$-c^2\,dt^2 + dx^2 + dy^2 + dz^2 = -c^2\,d\bar{t}^2 + d\bar{x}^2 + d\bar{y}^2 + d\bar{z}^2. \qquad (2.4)$$

The scalar line element, identical in all frames:

$$ds^2 = -c^2\,dt^2 + dx^2 + dy^2 + dz^2, \qquad (2.5)$$

is the structural successor to the Pythagorean lengths of Euclidean geometry. There is an interesting new feature to the interval of special relativity, though – it is *indefinite*, meaning only that it can be greater than, less than, or equal to zero. These three options correspond to physically distinct motion. If a measurement of displacement is less than zero, $ds^2 < 0$, then the local speed is less than c, and we call such an interval "time-like" (all massive objects travel at speeds less than c, and have $ds^2 < 0$). If $ds^2 = 0$, then the motion is called "light-like" – since then the speed of a particle is equal to c, and although it is not required by special relativity, light happens to travel at precisely this universal speed. Finally, intervals with $ds^2 > 0$ are referred to as "space-like", these are associated with motion at speeds greater than c.

Just as we pulled the Euclidean metric out of the length interval in (1.34) of Section 1.3.1, we can define the Minkowski metric by writing (2.5) in matrix–vector form:

$$ds^2 = \begin{pmatrix} c\,dt & dx & dy & dz \end{pmatrix} \underbrace{\begin{pmatrix} -1 & 0 & 0 & 0 \\ 0 & 1 & 0 & 0 \\ 0 & 0 & 1 & 0 \\ 0 & 0 & 0 & 1 \end{pmatrix}}_{\equiv\, \eta_{\mu\nu}} \begin{pmatrix} c\,dt \\ dx \\ dy \\ dz \end{pmatrix}, \qquad (2.6)$$

where the matrix in the middle defines the covariant form of the metric.

The coordinate differentials as measured in two frames obey (2.4) and are related by a Lorentz transformation. The Lorentz transformation is a linear coordinate relation that can be represented with a matrix $\Lambda^\alpha{}_\beta$. The connection between $d\bar{x}^\mu$ and dx^μ is provided by $d\bar{x}^\mu = \Lambda^\mu{}_\nu\,dx^\nu$. In fact, we can view the Lorentz transformation as defined by the requirement of an invariant interval of the form (2.5) – it is the

---

[3] If the object is moving in the x-direction only, then the speed in each frame is:

$$\frac{dx}{dt} = c = \frac{d\bar{x}}{d\bar{t}}, \qquad (2.3)$$

and then (2.4) follows.

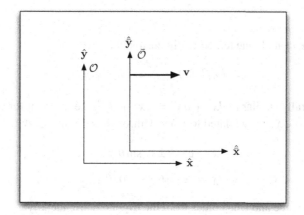

Figure 2.1   Coordinate systems related by a Lorentz boost – a frame $\mathcal{O}$ sees the $\bar{\mathcal{O}}$ frame moving to the right with speed v.

linear transformation that respects (2.4). If we take $d\bar{x}^\mu = \Lambda^\mu_{\ \nu} dx^\nu$, then:

$$d\bar{x}^\mu \eta_{\mu\nu} d\bar{x}^\nu = dx^\alpha \Lambda^\mu_{\ \alpha} \eta_{\mu\nu} \Lambda^\nu_{\ \beta} dx^\beta. \tag{2.7}$$

The requirement of (2.4), written in matrix form, is $\Lambda^T \eta \Lambda = \eta$. This is very much analogous to $\mathbb{R}^T \mathbb{I} \mathbb{R} = \mathbb{I}$ for rotations (where the metric "matrix" is just the identity).

Notice that the purely spatial Lorentz transformations, ones that set $d\bar{t} = dt$, are precisely rotations. The interesting new element of the Lorentz transformations is the "boost", a mixing of temporal and spatial coordinates. We typically describe a frame $\bar{\mathcal{O}}$ moving at speed v relative to $\mathcal{O}$ along a shared $\hat{\mathbf{x}}$-axis as in Figure 2.1. Then the Lorentz boost takes the form:

$$\Lambda^\mu_{\ \nu} \doteq \begin{pmatrix} \gamma & -\gamma\beta & 0 & 0 \\ -\gamma\beta & \gamma & 0 & 0 \\ 0 & 0 & 1 & 0 \\ 0 & 0 & 0 & 1 \end{pmatrix} \qquad \gamma \equiv \frac{1}{\sqrt{1-\beta^2}} \qquad \beta \equiv \frac{v}{c}. \tag{2.8}$$

We also have all the rotation structure from Euclidean space built in – consider a transformation defined by $\Lambda^\mu_{\ \nu}$ expressed in block form:

$$\Lambda^\mu_{\ \nu} \doteq \begin{pmatrix} 1 & 0 \\ 0 & \mathbb{R} \end{pmatrix} \tag{2.9}$$

(the upper entry is 1×1, while the lower right is a 3×3 matrix) for any matrix $\mathbb{R}$ with $\mathbb{R}^T \mathbb{R} = \mathbb{I}$, then we again preserve $dx^\alpha \eta_{\alpha\beta} dx^\beta = d\bar{x}^\alpha \eta_{\alpha\beta} d\bar{x}^\beta$.

Problem 2.1

Working in one spatial, one temporal dimension, we have:

$$-c^2 \, d\bar{t}^2 + d\bar{x}^2 = -c^2 \, dt^2 + dx^2. \tag{2.10}$$

This is structurally similar to $d\bar{x}^2 + d\bar{y}^2 = dx^2 + dy^2$, the invariant for rotation. There, we can use $\bar{x}$ and $\bar{y}$ related to x and y through $\cos\theta$ and $\sin\theta$:

$$\bar{x} = \cos\theta \, x + \sin\theta \, y$$
$$\bar{y} = -\sin\theta \, x + \cos\theta \, y. \tag{2.11}$$

The utility of cosine and sine comes from the trigonometric identity:
$\cos^2\theta + \sin^2\theta = 1$.

(a) Modify (2.11) to define $c\,\bar{t}$ and $\bar{x}$ using the hyperbolic cosine and sine, $\cosh\eta$ and $\sinh\eta$, with:

$$\cosh^2\eta - \sinh^2\eta = 1, \tag{2.12}$$

so that (2.10) holds. This provides another parametrization for the Lorentz transformation, in terms of η, the "rapidity".

(b) In the standard setup, we have the origin of $\bar{\mathcal{O}}$ aligned with the origin of $\mathcal{O}$ at time $t = \bar{t} = 0$. For $\bar{\mathcal{O}}$ traveling to the right with speed v in $\mathcal{O}$, find $\sinh\eta$ and $\cosh\eta$ in terms of v. Rewrite your transformation from part a in terms of v. You should end up with the usual expression:

$$\bar{x} = \gamma\,(x - v\,t)$$
$$c\,\bar{t} = \gamma\left(c\,t - \frac{x\,v}{c}\right). \tag{2.13}$$

Problem 2.2

Write the Minkowski metric in spherical (spatial) coordinates.

Problem 2.3

Show that $\Lambda^{\alpha}{}_{\mu}\, \eta_{\alpha\beta}\, \Lambda^{\beta}{}_{\nu} = \eta_{\mu\nu}$ for the boost in (2.8) and $\eta_{\mu\nu}$ the Minkowski metric in Cartesian coordinates.

Problem 2.4

In one dimension, the free motion of a particle is given by: $x(t) = v\,t$. Sketch this motion on a set of axes with $c\,t$ along the vertical, x along the horizontal – a sketch of motion on these axes is called a "Minkowski diagram", the resulting curve is called the "world-line" of the particle. What is the slope of your line? Draw the line associated with light (constant speed c), and indicate the regions on your graph that are inaccessible to massive particles (which must travel at speeds $< c$).

Problem 2.5

For motion under the influence of a constant force (like gravity near the surface of the earth), $F = F_0$, acting on a particle of mass m (starting from rest at the origin):

(a) Sketch the curve of $x(t)$ using axes for which ct is along the vertical, x is horizontal (i.e. make a Minkowski diagram). Indicate on your sketch the first point at which a particle travels faster than c.

(b) Sketch the speed $v(t)$ of the particle as a function of time, mark the time at which the particle travels faster than c. Sketch a plausible curve that must replace $v(t)$ in relativistic dynamics.

Problem 2.6

A mass M is attached to a spring (constant k, equilibrium location set to zero) and released from rest at an initial extension a. For what value of a does the maximum speed of the mass equal c? Use Newtonian dynamics here – the point is to see that for extensions greater than a, non-relativistic dynamics predicts maximum speeds greater than c.

2.2 Lagrangian

We have two basic problems to address in generating the (a) Lagrangian appropriate to dynamics in special relativity. The first is the form of the free particle Lagrangian, and the second is the parametrization of free particle motion. We'll introduce the free particle Lagrangian from a natural, geometrically inspired action. Remember, the Lagrangian is the integrand of an action.

2.2.1 Euclidean length extremization

To get the form of the relativistic Lagrangian, we can motivate from classical mechanics. For our usual free particle Lagrangian in three spatial dimensions with time playing the role of curve parameter, we have $L = \frac{1}{2}mv^2$. If we think of this in terms of the infinitesimal motion of a particle along a curve, we can write:

$$L = \frac{1}{2} m \frac{dx^\mu}{dt} g_{\mu\nu} \frac{dx^\nu}{dt}, \qquad (2.14)$$

with $g_{\mu\nu}$ the metric appropriate to the coordinates we are using. For concreteness, take $dx^\mu \doteq (dx, dy, dz)^T$, i.e. Cartesian coordinates – then $g_{\mu\nu}$ can be represented by the identity matrix. It is clear that the above is directly related to:

$$dr^2 = dx^\mu g_{\mu\nu} dx^\nu, \qquad (2.15)$$

or what we would call the square of an infinitesimal distance, implicitly, "along the curve parametrized by t". If we make this explicit, associating $x^\mu(t)$ with a curve, then $dx^\mu = \frac{dx^\mu}{dt} dt$, and $dr^2 = \frac{dx^\mu}{dt} g_{\mu\nu} \frac{dx^\nu}{dt} dt^2$. Think of the action, $S = \int L\, dt$ with L proportional to the length (squared) of the curve so that S is a measure of the total length (squared) that a free particle travels in moving from a to b. We are, roughly speaking, minimizing the length of the curve itself (a.k.a. making a "line") when we extremize this action. Let's start with a Lagrangian proportional to dr and see how this procedure works. The action is:

$$S = \int dr = \int \sqrt{\dot{x}^\mu g_{\mu\nu} \dot{x}^\nu}\, dt. \qquad (2.16)$$

In two dimensions, which suffices, we have the integrand $L = \sqrt{\dot{x}^2 + \dot{y}^2}$, and the variation gives us:

$$0 = \frac{\dot{x}\,(\dot{x}\,\ddot{y} - \dot{y}\,\ddot{x})}{(\dot{x}^2 + \dot{y}^2)^{3/2}}$$

$$0 = \frac{\dot{y}\,(\dot{y}\,\ddot{x} - \dot{x}\,\ddot{y})}{(\dot{x}^2 + \dot{y}^2)^{3/2}}. \qquad (2.17)$$

These are degenerate equations – the solution is, for example:

$$y(t) = A + B\,x(t), \qquad (2.18)$$

an infinite family of extremal paths. Or is it? For any A and B, what we are really doing is writing y as a linear function of x, that defines a line as the curve in the $x - y$ plane, regardless of $x(t)$. We have, in effect, parametrized a line with x itself. If you draw the curve associated with (2.18) in the $x - y$ plane, you get a straight line. So it does not matter, classically, whether we use the length along the curve or the length along the curve squared, the solutions are identical, if somewhat disguised in the latter case.

The advantage of extremizing the length rather than the length squared, when we form the free particle action, is reparametrization invariance. For classical mechanics, our parameter is t, time. We have no particular need for a different parameter, so this is an unimportant distinction for straight-line motion. But in special relativity (and general relativity), the choice of parameter for dynamical trajectories is not so straightforward. Reparametrization invariance just means that the free particle action for length (and not length-squared) extremization:

$$S_t \equiv \int \sqrt{\frac{dx^\mu}{dt} g_{\mu\nu} \frac{dx^\nu}{dt}}\, dt \qquad (2.19)$$

with "parameter" t, is unchanged when we pick a new parameter $\lambda(t)$. This is clear if we use the chain rule $\frac{dx^\mu}{dt} = \frac{dx^\mu}{d\lambda}\frac{d\lambda}{dt}$ in (2.19), and then:

$$S_\lambda = \int \sqrt{\frac{dx^\mu}{d\lambda} g_{\mu\nu} \frac{dx^\nu}{d\lambda} \frac{d\lambda}{dt} \frac{dt}{d\lambda}} \, d\lambda = \int \sqrt{\frac{dx^\mu}{d\lambda} g_{\mu\nu} \frac{dx^\nu}{d\lambda}} \, d\lambda. \qquad (2.20)$$

So we see that t really is just a parameter. It has no *a priori* physical significance from the point of view of the action. That freedom releases t from its special role as a parameter of motion from classical mechanics, and allows us to proceed to generate a dynamics framework with the roles of space and time homogenized.

2.2.2 Relativistic length extremization

On the relativistic side, we have a line element that is indefinite:

$$ds^2 = -c^2 \, dt^2 + dx^2 + dy^2 + dz^2 = dx^\mu \, \eta_{\mu\nu} \, dx^\nu. \qquad (2.21)$$

Now suppose we parametrize our curve in $D = 3 + 1$ via some ρ, to which we attach no physical significance. The goal of (special) relativistic mechanics is to find $x^\alpha(\rho)$, that is: $t(\rho)$, $x(\rho)$, etc. So we can once again write the line element, ds^2, along the ρ-parametrized curve as:

$$ds^2 = \dot{x}^\mu \, \eta_{\mu\nu} \, \dot{x}^\nu \, d\rho^2 \qquad \dot{x}^\mu \equiv \frac{dx^\mu(\rho)}{d\rho}. \qquad (2.22)$$

Motivated by the above action (2.16) for a purely spatial geometry, take:[4]

$$\boxed{S = \alpha \int ds = \alpha \int \sqrt{-\dot{x}^\mu \, \eta_{\mu\nu} \, \dot{x}^\nu} \, d\rho,} \qquad (2.23)$$

and call this the relativistic, free particle action. There is a built-in constant $\alpha \in \mathbb{R}$ which we will set later by comparing with the classical limit (particle velocity much less than c). We again expect straight lines as the free particle motion, and we have gained reparametrization invariance as before. Explicitly, suppose we have in mind a function of ρ, call it $\tau(\rho)$, and we want to rewrite this action in terms of τ. By the chain rule, we have:

$$\dot{x}^\mu = \frac{dx^\mu(\tau(\rho))}{d\rho} = \frac{dx^\mu}{d\tau}\frac{d\tau}{d\rho}, \qquad (2.24)$$

---

[4] The minus sign under the square root comes from our desire to focus on the motion of massive particles and light. Remember that a massive particle travels at speeds less than c and has interval $ds^2 < 0$, and light travels at c with interval $ds^2 = 0$. Since the term under the square root is always negative or zero for particles or light, we introduce a compensating minus sign. We could equivalently make α complex.

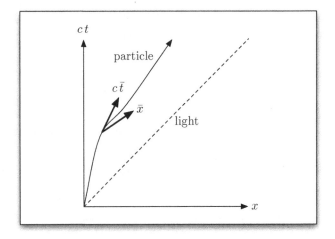

Figure 2.2   For a material particle trajectory in space-time ($D = 1 + 1$ here),
we can define a Lorentz boost to the frame ($c\,\bar{t}, \bar{x}$) shown such that the particle is
instantaneously at rest (the $c\,\bar{t}$ axis is tangent to the particle trajectory) – physically,
this is the frame in which the particle moves only in time, and defines the "proper
time" of the particle.

so that the action is:

$$
\begin{aligned}
S &= \alpha \int \frac{d\tau}{d\rho} \sqrt{-\frac{dx^\mu}{d\tau} \eta_{\mu\nu} \frac{dx^\nu}{d\tau} \frac{d\rho}{d\tau}} d\tau \\
&= \alpha \int \sqrt{-\frac{dx^\mu}{d\tau} \eta_{\mu\nu} \frac{dx^\nu}{d\tau}} d\tau,
\end{aligned}
\tag{2.25}
$$

and we have no way of establishing the difference between two different curve
parametrizations from the action integrand alone. This gives us the freedom to
define parametrizations of interest. In particular, we know that for any instantaneous
velocity of a particle (with $v < c$), it is possible to develop a Lorentz transformation
to the particle's local rest frame. That is, we can always generate a boost matrix
that takes us from a "lab" in which a particle moves in time and space to the frame
in which the particle moves only in time (use the instantaneous velocity to define
the boost parameter γ – this is shown diagramatically in Figure 2.2).

This local rest frame, in which the particle moves only in a time-like direction,
serves to define the proper time τ of the particle. Since a generic trajectory will not
have uniform velocity, τ changes along the trajectory, and we can define it only
in an infinitesimal sense – we know there exists a "barred" frame (the local rest
frame) in which the motion of the particle is purely temporal. This frame is related
to a measurement of $\{dt, dx, dy, dz\}$ in a different "lab" frame, via:

$$
-c^2\, dt^2 + dx^2 + dy^2 + dz^2 = -c^2\, d\bar{t}^2 \equiv -c^2\, d\tau^2.
\tag{2.26}
$$

Using this defining relation, we can write the derivative of τ w.r.t. t and vice versa:

$$c^2 dt^2 \left[1 - \frac{1}{c^2} \left(\left(\frac{dx}{dt}\right)^2 + \left(\frac{dy}{dt}\right)^2 + \left(\frac{dz}{dt}\right)^2 \right) \right] = c^2 d\tau^2, \qquad (2.27)$$

where the term on the left involving the squares of coordinate time-derivatives is what we would call the instantaneous lab speed of the particle (squared). This form provides a physical connection between the parameter τ and measurements made in the lab. We can write the above differential relation as:

$$\boxed{\dot{t}^2 = \frac{1}{1 - \frac{v^2}{c^2}}.} \qquad (2.28)$$

We started by suggesting that a natural Lagrangian for special relativity would be one that was proportional to the generalized length along the dynamical free particle curve. In order to see that this is a consistent description, we can take the relativistic Lagrangian (the integrand of the action) in coordinate time parametrization and see what it reduces to in the low-velocity limit where we should recover the classical free particle Lagrangian (the kinetic energy).

Since S is reparametrization-invariant, it takes the same form for any parameter. In particular, for t-parametrization:

$$S = \alpha \int \sqrt{-\frac{dx^\mu}{dt} \eta_{\mu\nu} \frac{dx^\nu}{dt}} \, dt, \qquad (2.29)$$

so that we can write, in Cartesian coordinates:

$$S = \alpha \int \sqrt{c^2 - v^2} \, dt \qquad (2.30)$$

since $\frac{d(c\,t)}{dt} = c$ and:

$$\left(\frac{dx}{dt}\right)^2 + \left(\frac{dy}{dt}\right)^2 + \left(\frac{dz}{dt}\right)^2 = v^2, \qquad (2.31)$$

the magnitude of velocity (squared) as measured in the lab with coordinate time t. The associated Lagrangian is the integrand of (2.30):

$$L = \alpha \sqrt{c^2 - v^2} = \alpha \, c \sqrt{1 - \frac{v^2}{c^2}} \approx \alpha \, c - \frac{1}{2} \alpha \frac{v^2}{c}. \qquad (2.32)$$

If we demand that this Lagrangian reduce to the classical kinetic energy in the low-speed limit, then we must set $\alpha = -m\,c$. We then have an approximate Lagrangian that differs from $T = \frac{1}{2} m\,v^2$ only by a constant, and that will not change the equations of motion, it is ignorable under variation.

Our procedure of generalized length extremization leaves us with the final form
for a free particle relativistic action and associated Lagrangian:

$$S = -m\,c \int \sqrt{-\frac{dx^\mu}{d\tau}\,\eta_{\mu\nu}\,\frac{dx^\nu}{d\tau}}\,d\tau, \tag{2.33}$$

where we use the proper time as the parameter, and we can, at any point, connect
this to coordinate time via (2.28). The advantage of proper time as a parameter is
that the action is *manifestly* a scalar (identical value) under Lorentz transformations.
We know that the four-velocity $\frac{dx^\mu}{d\tau}$ is a contravariant four-vector (since dx^μ is, and
$d\tau$ is clearly a scalar), the metric $\eta_{\mu\nu}$ is a covariant second-rank tensor, so the term
inside the square root is a scalar.

Now we can begin the same sorts of analysis we did for the classical Lagrangian,
using the above action as our starting point. In particular, it will be interesting to
find the canonical infinitesimal generators associated with constants of the motion,
although we already know basically what these are (Lorentz transformations, after
all, have $\Lambda^T \eta \Lambda = \eta$).

Problem 2.7

Show that, in proper time parametrization, the equations of motion obtained from the
Lagrangians $\dot{x}^\mu\,g_{\mu\nu}\,\dot{x}^\nu$ and $\sqrt{-\dot{x}^\mu\,g_{\mu\nu}\,\dot{x}^\nu}$ are identical (assume $g_{\mu\nu}$ is the Minkowski
metric in Cartesian coordinates).

Problem 2.8

Using only the invariance:

$$-c^2\,d\bar{t}^2 + d\bar{x}^2 = -c^2\,dt^2 + dx^2 \tag{2.34}$$

for two frames related by a Lorentz transformation, show that a particle moving with
$v \geq c$ has no (well-defined) rest frame (that is, we cannot define proper time for
motion at these speeds).

Problem 2.9

A clock is at rest at the origin of $\bar{\mathcal{O}}$ which is moving to the right with speed v in $\mathcal{O}$
along a shared x-axis. Assuming that the usual setup holds (origins of $\mathcal{O}$ and $\bar{\mathcal{O}}$
coincide at $t = \bar{t} = 0$):
(a) Write $\bar{x}^\mu(t)$, the four-vector (just the $c\,\bar{t}$ and $\bar{x}$ entries, for this problem) location
    of the clock in $\bar{\mathcal{O}}$ (its rest frame), and x^μ, the four-vector location of the clock in
    $\mathcal{O}$ (the "lab").
(b) Relate the time measured by the clock in $\bar{\mathcal{O}}$ to time elapsed in $\mathcal{O}$.
(c) Make a Minkowski diagram of the clock's motion in $\mathcal{O}$. In addition, plot the
    world-line of a clock that is at rest at the origin of $\mathcal{O}$.

2.3 Lorentz transformations

We have the relativistic Lagrangian, or at least one form of the relativistic Lagrangian. The question now becomes, what are the conserved quantities? Or, which is equivalent, what are the transformations that leave the associated Hamiltonian unchanged? We already know the answer to this, as the Lorentz transformations were developed to leave the metric unchanged, we know that the Lagrangian is insensitive to frame. But, it is of theoretical utility, at the very least, to carefully develop the connection between a general transformation like Lorentz boosts or spatial rotations, and their infinitesimal counterparts – these linearized transformations are the ones that are relevant to generators, and hence to constants of the motion. So we start by establishing, for rotations and Lorentz boosts, that it is possible to build up a general rotation (boost) out of infinitesimal ones.

We can then sensibly discuss the generators of infinitesimal transformations as a stand-in for the full transformation. But in order to check that the Poisson bracket of a generator with the Hamiltonian vanishes, we must also have the Hamiltonian – this is not so easy to develop, so we spend some time discussing the role of the Hamiltonian in free particle relativistic mechanics.

2.3.1 Infinitesimal transformations

We will look at how infinitesimal transformations can be used to build up larger ones – this is an important idea in physics, since it allows us to focus on simple, linearized forms of more general transformations to understand the action of a transformation on a physical system.

Rotations

Going back to three-dimensional space with Cartesian coordinates, we saw in Section 1.8 that an infinitesimal rotation through an angle θ about an axis $\hat{\Omega}$ as shown in Figure 2.3 could be written as (rotating the coordinate system, now, rather than a vector – hence the change in sign relative to (1.198)):

$$\bar{\mathbf{x}} = \mathbf{x} - \theta\,\hat{\Omega} \times \mathbf{x}. \tag{2.35}$$

Suppose we take $\hat{\Omega} = \hat{\mathbf{z}}$, so the rotation occurs in the $x - y$ plane. In order to keep track of the number of successive applications of this transformation, we will use a superscript on the vector $\mathbf{x}$, so one rotation is:

$$\mathbf{x}^1 = \mathbf{x} - \theta\,\hat{\mathbf{z}} \times \mathbf{x} \tag{2.36}$$

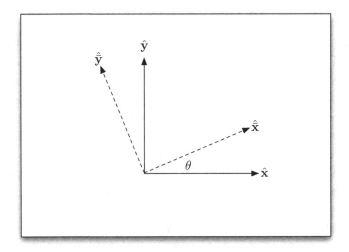

Figure 2.3   Two coordinate systems related by a rotation.

with $\mathbf{x} \equiv x\,\hat{\mathbf{x}} + y\,\hat{\mathbf{y}}$. This operation can be written using matrix–vector multiplication:

$$\begin{pmatrix} x^1 \\ y^1 \end{pmatrix} = \begin{pmatrix} x \\ y \end{pmatrix} + \begin{pmatrix} 0 & \theta \\ -\theta & 0 \end{pmatrix}\begin{pmatrix} x \\ y \end{pmatrix}. \tag{2.37}$$

Suppose we act on the $\mathbf{x}^1$ coordinates with another infinitesimal rotation:

$$\begin{aligned} \begin{pmatrix} x^2 \\ y^2 \end{pmatrix} &= \begin{pmatrix} x^1 \\ y^1 \end{pmatrix} + \begin{pmatrix} 0 & \theta \\ -\theta & 0 \end{pmatrix}\begin{pmatrix} x^1 \\ y^1 \end{pmatrix} \\ &= \left[\begin{pmatrix} 1 & 0 \\ 0 & 1 \end{pmatrix} + \begin{pmatrix} 0 & \theta \\ -\theta & 0 \end{pmatrix}\right]^2 \begin{pmatrix} x \\ y \end{pmatrix}, \end{aligned} \tag{2.38}$$

then the pattern is clear – for the nth iterate, we have:

$$\begin{pmatrix} x^n \\ y^n \end{pmatrix} = \left[\begin{pmatrix} 1 & 0 \\ 0 & 1 \end{pmatrix} + \theta\begin{pmatrix} 0 & 1 \\ -1 & 0 \end{pmatrix}\right]^n \begin{pmatrix} x \\ y \end{pmatrix}. \tag{2.39}$$

Suppose we want to build up a rotation through an angle Θ using an infinite number of the infinitesimal rotations. If we define each infinitesimal transformation to take us through an angle $\theta = \frac{\Theta}{n}$, then we can write the infinite product in (2.39) as we take $n \longrightarrow \infty$ – the result is a transformation to $\bar{x}$, $\bar{y}$ coordinates that we associate with rotation.

The point, in the end, is that to build up a net rotation through an angle Θ from the infinite application of infinitesimal θ, we have:

$$\begin{pmatrix} \bar{x} \\ \bar{y} \end{pmatrix} = \lim_{n \to \infty} \left[\begin{pmatrix} 1 & 0 \\ 0 & 1 \end{pmatrix} + \frac{\Theta}{n} \begin{pmatrix} 0 & 1 \\ -1 & 0 \end{pmatrix} \right]^n \begin{pmatrix} x \\ y \end{pmatrix}$$

$$= \begin{pmatrix} \cos\Theta & \sin\Theta \\ -\sin\Theta & \cos\Theta \end{pmatrix} \begin{pmatrix} x \\ y \end{pmatrix},$$

(2.40)

as always.

That is heartening, but what is so great about this procedure? Going back to our discussion of infinitesimal generators and Hamiltonians in Section 1.7.1, we know that the infinitesimal coordinate transformation:

$$\bar{\mathbf{x}} = \mathbf{x} - \Omega \times \mathbf{x}$$

(2.41)

leaves the Hamiltonian unchanged, and hence is associated with a conserved quantity. Using indexed notation for the cross product:[5]

$$\bar{x}^\alpha = x^\alpha - \epsilon^\alpha{}_{\beta\gamma} \, \Omega^\beta \, x^\gamma,$$

(2.43)

and in the language of generators from Section 1.7, we have coordinate transformation generated by $J = p_\alpha \, f^\alpha$ where f^α is just the second term of (2.43):

$$J = -p^\alpha \, \epsilon_{\alpha\beta\gamma} \, \Omega^\beta \, x^\gamma = -\Omega^\beta \, \epsilon_{\beta\gamma\alpha} \, x^\gamma \, p^\alpha$$

$$= -\frac{1}{2} \Omega^\beta \, \epsilon_{\beta\gamma\alpha} \, (x^\gamma \, p^\alpha - p^\gamma \, x^\alpha).$$

(2.44)

For this reason, we often say that $M^{\alpha\beta} = x^\alpha \, p^\beta - p^\alpha \, x^\beta$ is the generator of infinitesimal rotations (there are three of them, selected by the axis $\hat{\Omega}$ defined by Ω^β). As a doubly indexed object, it appears there are nine components to the matrix $\mathbb{M}$, but it is antisymmetric, so only has three independent components. Those three are, of course, the three components of angular momentum (we can write $J = -\Omega \cdot (\mathbf{x} \times \mathbf{p})$).

We could have started with (2.40) and linearized it in θ to develop the infinitesimal form:

$$\begin{pmatrix} \cos\Theta & \sin\Theta \\ -\sin\Theta & \cos\Theta \end{pmatrix} = \begin{pmatrix} 1 & 0 \\ 0 & 1 \end{pmatrix} + \begin{pmatrix} 0 & \Theta \\ -\Theta & 0 \end{pmatrix} + O(\Theta^2),$$

(2.45)

and worked from there – this is the approach we will take for the Lorentz case.

---

[5] The $\epsilon_{\alpha\beta\gamma}$ is the usual Levi–Civita symbol in three dimensions, defined as:

$$\epsilon_{\alpha\beta\gamma} = \begin{cases} 1 & (\alpha, \beta, \gamma) \text{ an even permutation of } \{1, 2, 3\} \\ -1 & (\alpha,, \beta, \gamma) \text{ an odd permutation of } \{1, 2, 3\} \\ 0 & \text{else} \end{cases}.$$

(2.42)

We put one index up in (2.43) just to match with the rest of the terms, but in this Cartesian setting, that does not change the value.

Lorentz transformations

If we consider a $D = 1 + 1$-dimensional Lorentz "boost" along a shared $\hat{\mathbf{x}}$-axis, then the matrix representing the transformation is:

$$\begin{pmatrix} c\bar{t} \\ \bar{x} \end{pmatrix} = \begin{pmatrix} \cosh\eta & \sinh\eta \\ \sinh\eta & \cosh\eta \end{pmatrix} \begin{pmatrix} c\,t \\ x \end{pmatrix}, \tag{2.46}$$

where η is the rapidity,[6] and $\cosh\eta = \gamma$, $\sinh\eta = -\gamma\beta$ for $\beta \equiv v/c$. The boost (2.46) is exact, and we want to linearize the transformation in order to make contact with an analogue of (2.43). If we take the infinitesimal parameter of the boost to be precisely η (a boost to a frame moving "slowly"), then we can develop the expansion to linear order in η (compare with (2.45)):

$$\begin{pmatrix} c\bar{t} \\ \bar{x} \end{pmatrix} \sim \left[\begin{pmatrix} 1 & 0 \\ 0 & 1 \end{pmatrix} + \begin{pmatrix} 0 & \eta \\ \eta & 0 \end{pmatrix} \right] \begin{pmatrix} c\,t \\ x \end{pmatrix}. \tag{2.47}$$

This gives us the infinitesimal transformation associated with Lorentz boosts – but what is its indexed form analogous to (2.43)? Suppose we try to represent the linearized Lorentz boost as a generalized cross product. In $D = 3 + 1$, we of course have Levi–Civita with four indices: $\epsilon^{\alpha\beta\gamma\delta}$ defined as usual. The linear transformation should be . . . linear in x^μ, but then we need a doubly indexed object $Q_{\beta\gamma}$ to soak up the additional indices. Our "reasoning" gives, thus far:

$$\bar{x}^\alpha = x^\alpha + \eta\,\epsilon^{\alpha\beta\gamma\delta}\,Q_{\beta\gamma}\,x_\delta. \tag{2.48}$$

Now suppose we exploit the analogy with rotation further – in order to rotate the $x - y$ plane, we needed a vector $\hat{\Omega} = \hat{\mathbf{z}}$ for (2.35). In the full four-dimensional space-time, there are two directions orthogonal to the $t - x$ plane, both y and z, so why not set $Q_{\beta\gamma} = \delta_{\beta y}\delta_{\gamma z}$? In this specific case, we recover (setting $c = 1$):

$$\bar{t} = t + \eta\,x \quad \bar{x} = x + \eta\,t \quad \bar{y} = y \quad \bar{z} = z, \tag{2.49}$$

as desired. In general, $Q_{\beta\gamma}$ plays the same role as Ω^α did for rotations. Only the antisymmetric part of $Q_{\beta\gamma}$ contributes, so we expect to have six independent quantities here – three rotations, and three boosts. As icing on the cake, notice that if we set $Q_{\beta\gamma} = \delta_{\beta 0}\,\delta_{\gamma 3}$, we get:

$$\bar{t} = t \quad \bar{x} = x + \eta\,y \quad \bar{y} = y - \eta\,x \quad \bar{z} = z, \tag{2.50}$$

precisely the infinitesimal rotation in (2.37). So both boosts and rotations are included here (as they should be).

---

[6] Rapidity is the natural hyperbolic analogue of an angle of rotation. This is clearly appropriate if we want to combine our notions of rotation with Lorentz boosts. The advantage of rapidity is that velocity addition takes a simple "angular addition" form.

Turning to the generator of the above transformation, we have:

$$J = \epsilon^{\alpha\beta\gamma\delta} \, Q_{\beta\gamma} \, x_\delta \, p_\alpha$$
$$= Q_{\beta\gamma} \, \epsilon^{\beta\gamma\delta\alpha} \, x_\delta \, p_\alpha. \tag{2.51}$$

We see that there are six independent generators, depending on our choice of $Q_{\beta\gamma}$ (this tensor can be taken to be antisymmetric as any symmetric portion will not contribute to the sum when contracted with the Levi–Civita symbol). Finally, we associate the tensor:

$$\boxed{M^{\alpha\beta} = x^\alpha \, p^\beta - p^\alpha \, x^\beta} \tag{2.52}$$

with the six conserved quantities. The antisymmetric $M^{\alpha\beta}$, when applied to three-dimensional, purely spatial, coordinates and momenta, has three independent elements, the components of angular momentum. Evidently, the structure is identical, and we have naturally expanded the number of conserved quantities just by introducing a fourth dimension.

We've strayed relatively far from our starting point, and the preceding argument is meant to motivate only – we will show that this set of quantities is indeed conserved, but first we have to develop the Hamiltonian associated with the relativistic Lagrangian.

2.4 Relativistic Hamiltonian

In order to show that the generator of Lorentz transformations is conserved, we need the Hamiltonian. Finding the Hamiltonian for our relativistic Lagrangian is not trivial, and the process has physical implications for proper time, motion, and the energy interpretation of the Hamiltonian, as we shall see by moving back and forth between proper time and coordinate time parametrization. There is ambiguity in the definition of the Hamiltonian when we use length and not length squared as the integrand of an action. This ambiguity comes precisely from the reparametrization invariance in actions that are proportional to length.

Remember, we are only dealing with the free particle Lagrangian (and Hamiltonian) – we have yet to introduce any potentials. So, start with:

$$L = -m \, c \, \sqrt{-\dot{x}^\mu(\tau) \, \eta_{\mu\nu} \, \dot{x}^\nu(\tau)}, \tag{2.53}$$

and we want the Legendre transform of this. In preparation, define the canonical momenta as usual:

$$p_\alpha = \frac{\partial L}{\partial \dot{x}^\alpha} = \frac{m \, c \, \dot{x}^\mu \, \eta_{\mu\alpha}}{\sqrt{-\dot{x}^\gamma \, \eta_{\gamma\nu} \, \dot{x}^\nu}}. \tag{2.54}$$

Now, in proper time parametrization, we have, by definition:

$$-c^2\,dt^2 + dx^2 + dy^2 + dz^2 = -c^2\,d\tau^2 \longrightarrow c^2\,\dot{t}^2 - \dot{x}^2 - \dot{y}^2 - \dot{z}^2 = c^2, \quad (2.55)$$

so that $\sqrt{-\dot{x}^\mu\,\eta_{\mu\nu}\,\dot{x}^\nu} = c$, and we can use this to simplify the momenta:

$$p_\alpha = m\,\dot{x}^\mu\,\eta_{\mu\alpha} = m\,\dot{x}_\alpha. \quad (2.56)$$

In order to understand the physics of this new four-vector, we can go back to temporal parametrization, then p_0 and p_1 give a sense for the whole vector – remember that $\dot{t} = \sqrt{\dfrac{1}{1-v^2/c^2}}$:[7]

$$p_0 = -m\,c\,\frac{dt}{d\tau} = -\frac{m\,c}{\sqrt{1-\frac{v^2}{c^2}}}$$

$$\qquad\qquad\qquad\qquad\qquad\qquad\qquad\qquad\qquad\qquad\qquad (2.58)$$

$$p_1 = m\,\frac{dx}{d\tau} = m\,\frac{dx}{dt}\dot{t} = \frac{m\,v^x}{\sqrt{1-\frac{v^2}{c^2}}},$$

where v^x is the velocity in coordinate time $\frac{dx(t)}{dt}$ and v is the coordinate-time velocity magnitude. So we get the usual sort of contravariant form (extending to the full, four dimensions):

$$p^\alpha \doteq \begin{pmatrix} \dfrac{m\,c}{\sqrt{1-\frac{v^2}{c^2}}} \\[2ex] \dfrac{m\,\mathbf{v}}{\sqrt{1-\frac{v^2}{c^2}}} \end{pmatrix}. \quad (2.59)$$

The physics of p^0

Before moving on to the full relativistic Hamiltonian, we will make contact with the "relativistic classical" result by considering the Lagrangian back in temporal parametrization, and forming the Hamiltonian as we normally would.

$$L^* = -m\,c^2\,\sqrt{1 - \frac{v^2}{c^2}} \quad (2.60)$$

has conjugate momenta:

$$\mathbf{p} = \frac{\partial L^*}{\partial \mathbf{v}} = \frac{m\,\mathbf{v}}{\sqrt{1-\frac{v^2}{c^2}}}, \quad (2.61)$$

---

[7] This relation comes from the definition of proper time:

$$-c^2 + v^2 = -c^2\left(\frac{d\tau}{dt}\right)^2 \longrightarrow \boxed{\dot{t}^2 = \left(1 - \frac{v^2}{c^2}\right)^{-1}}, \quad (2.57)$$

familiar from (2.28).

so that the Hamiltonian, representing the total energy is:

$$H^* = \mathbf{p} \cdot \mathbf{v} - L = \frac{m\,v^2}{\sqrt{1 - \frac{v^2}{c^2}}} + m\,c^2 \sqrt{1 - \frac{v^2}{c^2}} = \frac{m\,c^2}{\sqrt{1 - \frac{v^2}{c^2}}}. \tag{2.62}$$

This total energy can be broken down into the rest energy of a particle with $v = 0$, namely $m\,c^2$ and the "kinetic" portion, $H^* - m\,c^2$. Regardless, we see that p^0 in (2.59) is precisely the total energy (divided by c), so:

$$p^\alpha \doteq \begin{pmatrix} \frac{E}{c} \\ \frac{m\,\mathbf{v}}{\sqrt{1 - \frac{v^2}{c^2}}} \end{pmatrix}. \tag{2.63}$$

You have guessed, I hope, the reason for the delay – this is supposed to be a section on the relativistic Hamiltonian, after all. But, if we take the canonical momentum defined by (2.54), and attempt to form the relativistic Hamiltonian, we get:

$$p_\alpha \dot{x}^\alpha = -m\,c\,\sqrt{-\dot{x}^\mu\,\eta_{\mu\nu}\,\dot{x}^\nu}, \tag{2.64}$$

and this is precisely L, so subtracting L from it will give $H = p_\alpha\,\dot{x}^\alpha - L = 0$, not a very useful result for dynamics. The problem, if one can call it that, is in the form of the Lagrangian. Basically, its reparametrization invariance – so useful for the interpretation of equations of motion (via proper time) – is in a sense, too large to define H. To put it another way, there are a wide variety of Lagrangians that describe free particle special relativity and reduce to the classical one in some limit.

For now, we can focus on generating a functionally efficient Hamiltonian. Consider, for example, the equations of motion that come from our $L = -m\,c\,\sqrt{-\dot{x}^\mu\,\eta_{\mu\nu}\,\dot{x}^\nu}$. In proper time parametrization, they reduce to:

$$m\,\ddot{x}^\mu = 0. \tag{2.65}$$

So we are, once again, describing a "straight line", the general solution is $x^\mu = A^\mu\,\tau + B^\mu$ with constants used to set initial conditions and impose the proper time constraint.

From (2.65), we see that there are other Lagrangians that would lead to the same equation of motion – for example, $L^+ = \frac{1}{2}\,m\,\dot{x}^\mu\,\eta_{\mu\nu}\,\dot{x}^\nu$. And this Lagrangian has the nice property that the Hamiltonian is nontrivial. We now have canonical momenta: $p_\alpha = \frac{\partial L^+}{\partial \dot{x}^\alpha} = m\,\dot{x}^\mu\,\eta_{\mu\alpha}$. The Legendre transform of the Lagrangian is:

$$H^+ = \frac{1}{2\,m}\,p_\alpha\,\eta^{\alpha\beta}\,p_\beta \tag{2.66}$$

and the equations of motion are precisely $m \, \ddot{x}^{\mu} = 0$. Operationally, then, this Hamiltonian does what we want. In particular, it provides a way to check the constancy of the infinitesimal Lorentz generators, our original point.

We will check them directly from (2.52), without the baggage of additional constants:

$$
\begin{aligned}
[H^+, M^{\alpha\beta}] &= [H^+, x^{\alpha} \, p^{\beta}] - [H^+, x^{\beta} \, p^{\alpha}] \\
&= \frac{\partial H^+}{\partial x^{\gamma}} \frac{\partial x^{\alpha}}{\partial p_{\gamma}} p^{\beta} - \frac{\partial H^+}{\partial p_{\gamma}} \frac{\partial x^{\alpha}}{\partial x^{\gamma}} p^{\beta} - (\alpha \leftrightarrow \beta) \\
&= -\frac{1}{m} \, p^{\gamma} \, \delta^{\alpha}_{\gamma} \, p^{\beta} + \frac{1}{m} \, p^{\gamma} \, \delta^{\beta}_{\gamma} \, p^{\alpha} \\
&= 0,
\end{aligned}
\tag{2.67}
$$

so indeed, we have $M^{\alpha\beta}$ constant (all six of them).

Problem 2.10
Write the components of $M^{\alpha\beta}$ for the general solution to $\ddot{x}^{\mu} = 0$ in proper time parametrization. Do you recognize them?

*Problem 2.11
We can write a quadratic Lagrangian for free particles in special relativity without using proper time. In addition to providing a manageable Lagrangian, it allows us to sensibly define motion for particles with $m = 0$. The idea is to use a nondynamical (meaning no time derivatives appear in the Lagrangian) Lagrange multiplier to enforce the correct form. Take:

$$
L^* = \frac{1}{2} \frac{1}{Q} \dot{x}^{\mu} \, \eta_{\mu\nu} \, \dot{x}^{\nu} - Q \, m^2.
\tag{2.68}
$$

(a) Vary L^* with respect to Q and set this equal to zero to find Q. Vary with respect to x^{μ}, and use your value for Q to show you recover the correct equations of motion (2.65).
(b) If we have a massless particle, $m = 0$, we still recover the correct equations of motion. Vary with respect to Q and x^{μ} and show that a reasonable equation governing x^{μ}, and the requirement that $\dot{x}^{\mu} \, \dot{x}_{\mu} = 0$ are both enforced.

2.5 Relativistic solutions

With our relativistic equations of motion, we can study the solutions for $x(t)$ under a variety of different forces. The hallmark of a relativistic solution, as compared with a classical one, is the bound on velocity for massive particles. We shall see this in the context of a constant force, a spring force, a one-dimensional Coulomb force, and orbital motion in two dimensions.

This tour of potential interactions leads us to the question of what types of force are even allowed – we will see changes in the dynamics of a particle, but what about the relativistic viability of the mechanism causing the forces? A real spring, for example, would break long before a mass on the end of it was accelerated to speeds nearing c. So we are thinking of the spring force, for example, as an approximation to a well in a more realistic potential. To the extent that effective potentials are uninteresting (or even allowed in general), we really only have one classical, relativistic force – the Lorentz force of electrodynamics.[8]

2.5.1 Free particle motion

For the equations of motion in proper time parametrization, we have:

$$\ddot{x}^{\mu} = 0 \longrightarrow x^{\mu} = A^{\mu}\,\tau + B^{\mu}. \tag{2.69}$$

Suppose we rewrite this solution in terms of t – inverting the $x^0 = ct$ equation with the initial conditions that $t(\tau = 0) = 0$, we have:

$$\tau = \frac{ct}{A^0}, \tag{2.70}$$

and then the rest of the equations read:

$$x = A^1\frac{ct}{A^0} + B^1 \qquad y = A^2\frac{ct}{A^0} + B^2 \qquad z = A^3\frac{ct}{A^0} + B^3. \tag{2.71}$$

We can identify the components of velocity:

$$v^x = \frac{A^1 c}{A^0} \qquad v^y = \frac{A^2 c}{A^0} \qquad v^z = \frac{A^3 c}{A^0}. \tag{2.72}$$

Remember that we always have the constraint that defines proper time:

$$\frac{dt}{d\tau} = \frac{1}{\sqrt{1 - \frac{v^2}{c^2}}}, \tag{2.73}$$

and this constraint can be used with (2.70) to relate A^0 to the rest of the constants (through v^2):

$$A^0 = \frac{c}{\sqrt{1 - \frac{v^2}{c^2}}}. \tag{2.74}$$

---

[8] I guess you might argue that zero, the force associated with general relativity, is also an allowed relativistic force.

The "lab" measurements, which proceed with a clock reading coordinate time, give back constant-velocity motion with velocity $\mathbf{v}$ and we pick up a relation between the coordinate time and the proper time of the particle (in its own rest frame).

Solving the above for A^0 (implicitly defined in terms of A^1, A^2 and A^3 in (2.74)), we get $A_0 = \sqrt{A_1^2 + A_2^2 + A_3^2 + c^2}$ so that:

$$\mathbf{v} = \frac{\mathbf{A}\,c}{\sqrt{A_1^2 + A_2^2 + A_3^2 + c^2}} = \frac{\mathbf{A}}{\sqrt{1 + \frac{A^2}{c^2}}}, \tag{2.75}$$

and this has magnitude less than c for any choice of $\mathbf{A}$.

2.5.2 Motion under a constant force

A constant force, according to Newton's second law, is one for which $\frac{dp}{dt} = F$ with constant F. In one spatial dimension, we can describe relativistic constant forcing by interpreting p as the spatial, *relativistic* momentum, i.e. $p = \frac{m\,v}{\sqrt{1-v^2/c^2}}$. From this point of view, the solution is straightforward – to avoid confusion, let $v(t) = x'(t) \equiv \frac{dx(t)}{dt}$, then:

$$\frac{d}{dt}\left(\frac{m\,x'(t)}{\sqrt{1 - x'(t)^2/c^2}}\right) = F \longrightarrow x'(t) = \frac{F\,t}{m\sqrt{1 + \frac{F^2\,t^2}{m^2\,c^2}}} \tag{2.76}$$

(setting the constant of integration that would normally appear above to zero, so that our particle starts from rest), which can be integrated again. If we set $x(t = 0) = 0$, then:

$$x(t) = \frac{m\,c^2}{F}\left(\sqrt{1 + \left(\frac{F\,t}{m\,c}\right)^2} - 1\right), \tag{2.77}$$

the usual hyperbolic motion.

If we start with the relativistic free particle Lagrangian in proper time parametrization, $L = -m\,c\,\sqrt{-\dot{x}^\mu\,\eta_{\mu\nu}\,\dot{x}^\nu}$ (dots now refer to derivatives with respect to proper time, the parameter of x^μ), then we'd like to add a "potential" that gives us the constant force case. We know such a potential will be . . . strange – after all, it must be a scalar, meaning here that it will have to be constructed out of vectors (specifically, combinations of $x^\mu(\tau)$ and its τ derivatives). In this two-dimensional setting, about the only thing that suggests itself is the cross product of x^μ and $\dot{x}^\mu$. As we shall see, the Lagrangian:

$$L = -m\,c\,\sqrt{-\dot{x}^\mu\,\eta_{\mu\nu}\,\dot{x}^\nu} + \frac{1}{2}\frac{F}{c}\,\epsilon_{\mu\nu}\,\dot{x}^\mu\,x^\nu \tag{2.78}$$

corresponds to the constant force as defined above. Let's work out the equations of motion – written out, the Lagrangian is:

$$L = -mc\sqrt{c^2 \dot{t}^2 - \dot{x}^2} + F(x\dot{t} - t\dot{x}).$$ (2.79)

Remember, from the definition of proper time, we have $c^2 = c^2 \dot{t}^2 - \dot{x}^2$, which we can solve for $\dot{x}$, for example. Then:

$$\dot{x} = c\sqrt{\dot{t}^2 - 1} \qquad \ddot{x} = \frac{c\dot{t}\ddot{t}}{\sqrt{\dot{t}^2 - 1}}.$$ (2.80)

Using the proper time relation, we get equations of motion as follows:

$$\frac{d}{d\tau}\frac{\partial L}{\partial \dot{t}} - \frac{\partial L}{\partial t} = 0 \longrightarrow mc\ddot{t} = \frac{F}{c}\dot{x}$$

$$\frac{d}{d\tau}\frac{\partial L}{\partial \dot{x}} - \frac{\partial L}{\partial x} = 0 \longrightarrow m\ddot{x} = F\dot{t}.$$ (2.81)

Upon introducing the equalities (2.80), these degenerate to the single equation:

$$\ddot{t} = \frac{F}{mc}\sqrt{\dot{t}^2 - 1},$$ (2.82)

with solution:

$$t(\tau) = \alpha + \frac{mc}{F}\left(\cosh(\beta)\sinh\left(\frac{F\tau}{mc}\right) + \sinh(\beta)\cosh\left(\frac{F\tau}{mc}\right)\right).$$ (2.83)

Setting $\alpha = \beta = 0$ (our choice) to get $t(\tau = 0) = 0$, we have:

$$t(\tau) = \frac{mc}{F}\sinh\left(\frac{F\tau}{mc}\right) \longrightarrow \tau(t) = \frac{mc}{F}\sinh^{-1}\left(\frac{Ft}{mc}\right).$$ (2.84)

Meanwhile, we can solve $\dot{x}$ from (2.80):

$$x(\tau) = A + \frac{mc^2}{F}\cosh\left(\frac{F\tau}{mc}\right),$$ (2.85)

this is the spatial coordinate in proper time parametrization. To find $x(t)$, we can use the inverse relation $\tau(t)$ and just replace:

$$x(t) = A + \frac{mc^2}{F}\sqrt{1 + \frac{F^2 t^2}{m^2 c^2}}$$ (2.86)

and our lone integration constant A can be used to set $x(t = 0) = 0$, the physical boundary condition. When all is said and done, we recover (2.77).

2.5.3 Motion under the spring potential

We can also consider a relativistic spring – here, again, the difficulty is that for an arbitrary displacement, the maximum speed can be arbitrarily large classically. We expect our relativistic mechanics to take care of this, providing a maximum speed $< c$. This time, we will start from the relativistic Lagrangian written in coordinate time parametrization:[9]

$$L = -m\,c^2 \sqrt{1 - \frac{x'(t)^2}{c^2}} - \frac{1}{2} k\, x(t)^2 \tag{2.87}$$

and the potential is the usual one with spring constant k. As a check, this gives the correct Newton's second law form upon variation:

$$\frac{d}{dt} \frac{\partial L}{\partial x'(t)} - \frac{\partial L}{\partial x(t)} = 0 \longrightarrow \frac{d}{dt}\left(\frac{m\,x'(t)}{\sqrt{1 - \frac{x'(t)^2}{c^2}}} \right) = -k\,x(t). \tag{2.88}$$

We could integrate this directly (not easy). Instead, we can use the Hamiltonian associated with L:

$$H = \frac{\partial L}{\partial x'(t)} x'(t) - L = \frac{m\,x'(t)^2}{\sqrt{1 - \frac{x'(t)^2}{c^2}}} - \left(-m\,c^2 \sqrt{1 - \frac{x'(t)^2}{c^2}} - \frac{1}{2} k\, x(t)^2 \right)$$

$$= \frac{m\,c^2}{\sqrt{1 - \frac{x'(t)^2}{c^2}}} + \frac{1}{2} k\, x(t)^2, \tag{2.89}$$

and the form here comes as no surprise. But it does tell us that numerically, this particular combination is a constant, we'll call it the total energy (notice that it contains the rest energy as well as the kinetic and potential energies). Setting $H = E$ gives us a first integral of the motion, and we could solve it – instead, we can take the total time derivative to recover a second-order ODE analogous to the equation of motion itself:

$$\frac{m\,x''(t)}{\left(1 - \frac{x'(t)^2}{c^2}\right)^{3/2}} = -k\,x(t). \tag{2.90}$$

From here, it is easiest to find the solution numerically – we'll use initial conditions $x(t = 0) = x_0$, $x'(t = 0) = 0$, so we're starting at maximum amplitude, then the position and velocity for a few different starting locations are shown in Figures 2.4 and 2.5. Notice that the "low-velocity" case, corresponding to the

---

[9] Although you should ask yourself what happens if you retain proper time parametrization – in particular, what Lorentz scalar gives back the spring potential when you return to t-parametrization?

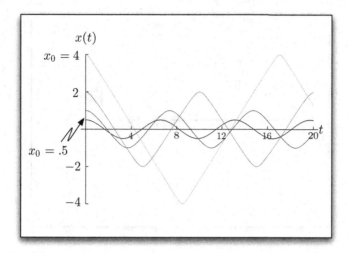

Figure 2.4   Position for a relativistic spring (this is the result of numerical integration, with $k = 1$, $c = 1$ and x_0 shown).

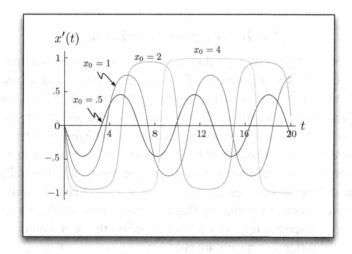

Figure 2.5   Velocity for the relativistic spring potential (here, $k = 1$, $c = 1$, and we start at a variety of maximum amplitudes as shown).

smallest starting position, looks effectively just like a classical spring, both in position and velocity. As the initial amplitude is increased, the solution for $x(t)$ begins to look more like a sawtooth as the velocity turns into a step function close to c.[10]

---

[10] To get the asymptotic behavior, solve $H = E$ for v in terms of x and send $E \longrightarrow \infty$ – the relation you will recover is $v = \pm c$, i.e. the sawtooth motion is predictable from the Hamiltonian.

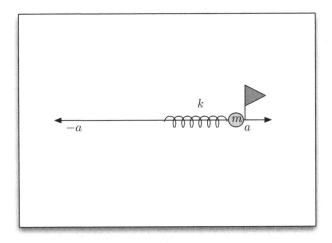

Figure 2.6   A mass m is at the full extension of a spring with constant k at a distance a. It is released from rest, and oscillates back and forth between $-a$ and a.

2.5.4 The twin paradox with springs

The spring potential gives us a good model for discussing the twin "paradox". In classical mechanics, if you put a ball of mass m on a spring, and release it at full extension from rest, we know that the mass will oscillate back and forth. If we release the mass from a location a, it will start from rest and move (speeding up and then slowing down) to $-a$, where it stops and then begins its return.

If the release point is marked with a stationary flag, as shown in Figure 2.6, the mass sees the flag move away and then come back, and classically, there is no particular difference between the two points of view. Special relativity breaks the symmetry between the stationary flag and the moving mass on a spring. Our geometric view of special relativity allows us to see the distinction, and we will also describe the difference quantitatively.

The flag, as a stationary point, moves along a straight line (geodesic) in Minkowski space-time with zero velocity, making its proper time everywhere equivalent to the coordinate time. The mass, on the other hand, is not traveling along a geodesic (i.e. $\ddot{x}^{\alpha} \neq 0$ for the mass, it experiences forcing), and moves through both space and time. Its overall length, then, is not the same as the flag's. Our goal is to calculate the time it takes to complete one full cycle in both the rest frame of the flag and the mass. The advantage to the spring force here is that we can easily and unambiguously compare temporal observations at a, since both the flag and the mass are at rest at that point.

We'll develop the solution perturbatively, starting from the equation of motion (2.90), solved for x'':

$$x'' = -\frac{kx}{m}\left[1 - \left(\frac{x'}{c}\right)^2\right]^{3/2}. \tag{2.91}$$

Suppose that $\frac{x'}{c} \sim \epsilon$, so that the speed of the ball is much less than the speed of light. Then, to first order in $\frac{x'}{c}$, the usual solution holds, set:

$$x(t) = a\,\cos(\omega t) \quad \omega^2 \equiv \frac{k}{m}. \tag{2.92}$$

The proper time requirement is already small – transforming to the rest frame of the ball is defined locally by finding a $d\tau$ such that:

$$-c^2\,d\tau^2 = -c^2\,dt^2 + dx^2, \tag{2.93}$$

and this, with $x(t)$ (so that $dx = x'\,dt$), is:

$$-c^2\,d\tau^2 = -c^2\,dt^2 + \left(x'\,dt\right)^2. \tag{2.94}$$

Taking the square root of both sides:

$$\frac{d\tau}{dt} = \sqrt{1 - \left(\frac{x'}{c}\right)^2} \sim 1 - \frac{1}{2}\left(\frac{x'}{c}\right)^2. \tag{2.95}$$

Using our first-order $x(t)$, we have:

$$\tau = t - \frac{(a\,\omega)^2\,t}{c^2} + \frac{a^2\,\omega\,\sin(2\,\omega t)}{8\,c^2}. \tag{2.96}$$

There are corrective terms of higher order, but even here we can see some interesting behavior. The fundamental inequality governing the perturbation is $a\,\omega \ll c$, and we see that τ differs from t by a linear term, just associated with the usual boost for a particle traveling at speed $a\,\omega$. The additional portion varies sinusoidally, and is associated with the acceleration – notice that it has half the period of the spatial oscillation – accelerating or decelerating both contribute to the proper time.

When the ball has returned to a after one full cycle, the flag's clock reads $t^* = \frac{2\pi}{\omega}$, while the ball's clock (proper time) reads:

$$\tau^* = t^* - \frac{a^2\,\omega\,\pi}{2\,c^2}. \tag{2.97}$$

This is less than the flag's clock. Evidently, it is possible in special relativity to distinguish between the stationary flag and the accelerating ball – just check their clocks. That's a new feature of special relativity, and focuses our attention on the underlying geometry of Minkowski space-time. The Euclidean distance traveled by the ball is $2\,a$, and the Euclidean distance the ball sees the flag travel is $2\,a$. But the Minkowski distance associated with the flag is $s = -c\,t^*$, while the Minkowski distance traveled by the ball is $s \approx -c\left(t^* - \frac{a^2\,\omega\,\pi}{2\,c^2}\right)$. It is no longer enough for two observers to see the same thing from a purely spatial point of view, the two-dimensional space-time paths are now the objects of comparison.

2.5.5 Electromagnetic infall

Consider the one-dimensional infall problem for a test charge moving under the influence of the Coulomb field coming from a point charge of magnitude Q sitting at the origin. For a test charge carrying charge $-q$, the total energy is (classically, absorbing factors of $\frac{1}{4\pi\epsilon_0}$ into the units of charge):

$$E = \frac{1}{2}m\,x'^2 - \frac{q\,Q}{x}. \tag{2.98}$$

If we start a particle from rest at spatial infinity at time $t = -\infty$, then the energy is $E = 0$, and this is conserved along the trajectory. For $E = 0$, we can solve the above for x':

$$x' = -\sqrt{\frac{2q\,Q}{m\,x}}. \tag{2.99}$$

This can be integrated once to obtain $x(t)$ – writing the integration constant in terms of t^*, the time it takes to reach the origin (starting from $t = 0$):

$$x(t) = \left(\frac{3}{2}\right)^{2/3}\left(\sqrt{\frac{2q\,Q}{m}}\,(t^* - t)\right)^{2/3}. \tag{2.100}$$

The problem, as always, with this classical solution is that the velocity grows arbitrarily large as $t \longrightarrow t^*$:

$$x'(t) = v(t) = \frac{dx}{dt} \sim (t^* - t)^{-2/3}, \tag{2.101}$$

allowing the test particle to move faster than light (already clear from (2.99)).

The relativistic Lagrangian, written in temporal parametrization, provides dynamics with a cut-off velocity. The relativistic Hamiltonian (total energy) reads:

$$E = \frac{m\,c^2}{\sqrt{1 - \frac{x'(t)^2}{c^2}}} - \frac{q\,Q}{x}. \tag{2.102}$$

Now for a particle starting at rest at spatial infinity, we have $E = m\,c^2$, not the classical zero. Once again, we can solve the above with $E = m\,c^2$ to get the

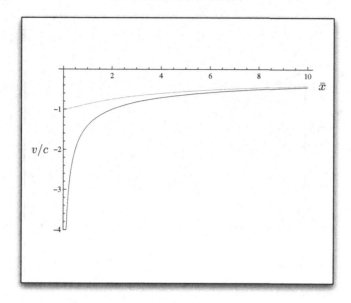

Figure 2.7   The speed as a function of (scaled) position for classical (lower curve) and relativistic (upper curve) radial infall.

dynamics. The velocity, in terms of position, is:

$$v = -\frac{c\sqrt{1+2\bar{x}}}{1+\bar{x}} \qquad \bar{x} \equiv \frac{m\,c^2 x}{q\,Q}. \tag{2.103}$$

The speed limit on v is clear even in this scaled form ($\bar{x} \geq 0$ has $v < c$ for all values). If we put the classical solution in this form, then we have $v = -c\sqrt{\frac{2}{\bar{x}}}$, and we plot both of these in Figure 2.7.

The full form of the relativistic solution is not particularly enlightening – we provide a plot in Figure 2.8 for comparison, using the same t^* ($\bar{x}(t^*) = 0$) values.

2.5.6 Electromagnetic circular orbits

We can use the time-parametrized Lagrangian to calculate the circular orbits associated with a static Coulomb field. The problem, of course, is that using the classical Lagrangian, we get the usual centripetal acceleration relation for circular orbits of radius R:

$$\frac{m\,v^2}{R} = \frac{q\,Q}{R^2} \tag{2.104}$$

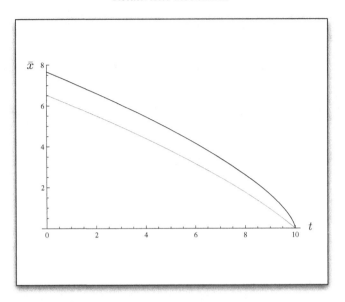

Figure 2.8   The relativistic (gray) and classical (black) infall solutions for $\frac{m\,c^2}{q\,Q} = 1$, and the same value of t^*.

for a central charge Q and test charge q (in obvious units), leading to a linear velocity given by:

$$v = \sqrt{\frac{q\,Q}{m\,R}} \qquad (2.105)$$

which can be greater than c for various ("small") R.

Let's see how the relativistic approach leads to circular orbits with a finite cut-off speed for all radii R. We now have a two-dimensional, relativistic, Lagrangian:

$$L = -m\,c^2 \sqrt{1 - \frac{r'^2 + r^2\phi'^2}{c^2}} - \frac{q\,Q}{m\,r}, \qquad (2.106)$$

where we have put the motion into the $\theta = \frac{1}{2}\pi$ plane, and we are assuming $Q > 0$, $q < 0$. This L does not depend on ϕ, so we have the immediate first-integral setting $\frac{\partial L}{\partial \phi'} = J_z''$:

$$\phi' = \frac{J_z\,c\,\sqrt{1 - \left(\frac{r'}{c}\right)^2}}{r\,\sqrt{J_z^2 + m^2\,c^2\,r^2}}. \qquad (2.107)$$

If we take the equation of motion for r, obtained as usual via $\frac{d}{dt}\left(\frac{\partial L}{\partial r'}\right) - \frac{\partial L}{\partial r} = 0$, and input (2.107) for ϕ', then we can solve for r'':

$$r'' = \frac{c}{r}\sqrt{\frac{1-\left(\frac{r'}{c}\right)^2}{J_z^2 + m^2 c^2 r^2}}\left[qQ\left(1-\left(\frac{r'}{c}\right)^2\right) + J_z^2 c\sqrt{\frac{1-\left(\frac{r'}{c}\right)^2}{J_z^2 + m^2 c^2 r^2}}\right].$$

(2.108)

To get a circular orbit, we need $r(t) = R$, so both $r' = 0$, and $r'' = 0$. If we assume $r' = 0$, then the above can be made zero by setting:

$$J_z = \frac{1}{\sqrt{2}c}\sqrt{qQ\left(qQ + \sqrt{q^2 Q^2 + 4m^2 c^4 R^2}\right)}.$$

(2.109)

Finally, we can find the linear velocity in the usual way: $v = R\phi'$, with ϕ' from (2.107), and J_z given by the above. The result is:

$$v = \frac{c}{\sqrt{1 + \frac{2m^2 c^4 R^2}{(qQ)^2\left(1+\sqrt{1+\frac{4m^2 c^4 R^2}{(qQ)^2}}\right)}}}$$

(2.110)

(using $q \longrightarrow -|q| \equiv -q$ just to include the signs), which is pretty clearly less than c (and equal to c when $R = 0$). To compare, we rewrite the result in terms of $\bar{R} \equiv \frac{mc^2 R}{qQ}$ and measure v in units of c. The classical and relativistic results are shown together in Figure 2.9.

2.5.7 *General speed limit*

All of the above are fun examples of the generic statement that relativistic Hamiltonians imply a velocity limit of some sort. We can establish this immediately, without reference to any particular potential, by inverting the Hamiltonian with constant value E in one dimension:

$$E = \frac{mc^2}{\sqrt{1 - \frac{v^2}{c^2}}} + U(x),$$

(2.111)

for any potential U. The magnitude of the velocity of a particle is:

$$v = c\sqrt{1 - \left(\frac{mc^2}{E-U}\right)^2}.$$

(2.112)

This expression clearly gives $v < c$, and we must have $mc^2 \leq |E - U|$ to get a real speed (from (2.111), we have $E = mc^2 + U$ when $v = 0$).

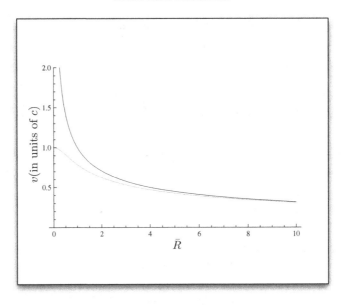

Figure 2.9   Classical (top curve) and relativistic (bottom curve) circular orbit speeds as a function of distance to the central body.

2.5.8 From whence, the force?

We have just covered some of the canonical forces (see, for example, [2, 16, 24, 32]) that can be studied in relatively closed form relativistically. But more than that, there is the question of what forces are even *consistent* with special relativity, let alone nicely solvable. The only force entirely in accord with the demands of special relativity is the Lorentz force of electrodynamics. As the theory which, to a certain extent, inspired Einstein toward a space-time view, it is not surprising that the Lorentz force is the one (and only, on the classical side) that provides a consistent description.

There are two immediate requirements for special relativity when it comes to a force associated with a field theory. The first is that the force is "tensorial enough" to offer the same prediction in two different inertial frames. The second (related) is that the mediator of the force (to wit: the field) respond to changes in its source with finite propagation speed (a causality issue). These are just rewrites of the usual two-postulate approach to special relativity – equivalence of inertial frames and finite information propagation.

General relativity is a theory of gravity that replaces Newtonian gravity. What's wrong with the Newtonian theory? As we shall see, Newtonian gravity fails to satisfy both postulates of special relativity. Given the success of electrodynamics (passes both), it is tempting to construct an analogous gravitational theory, and such a linear vector theory is successful up to a point, but cannot address the unique qualities of gravity as an observational phenomenon: everything is influenced by gravity, and the gravitational "force" between two objects is attractive (only one flavor of "charge").

2.6 Newtonian gravity and special relativity

It is interesting to note that under Lorentz transformation, while electric and magnetic fields get mixed together, the net force on a particle is identical in magnitude and direction in the two frames related by the transformation (when the force itself is appropriately transformed). Maxwell's equations and the Lorentz force law are automatically consistent with the notion that observations made in inertial frames are physically equivalent, even though observers may disagree on the names of these forces (electric or magnetic).

We will look at a force (Newtonian gravity) that does not have the property that different inertial frames agree on the physics. That will lead us to an obvious correction that is, qualitatively, a prediction of (linearized) general relativity.

The quantitative details are not currently our primary concern,[11] and will be left for later (linearized general relativity forms the bulk of Chapters 7 and 8), for now, we just want to focus on the statement: "Newtonian gravity, minimally modified so as to satisfy the demands of special relativity, yields a theory similar in form to electricity and magnetism". Indeed, to the extent that E&M is basically unique as a relativistic vector theory, it is difficult for any theory sharing its fundamental properties to look much different. Still, it is informative, and interesting to carry out the parallels – in particular, our special relativistic demands already predict the existence of something like gravitational radiation.

2.6.1 Newtonian gravity

We start with the experimental observation that for a particle of mass M and another of mass m, the force of gravitational attraction on m due to M, according to Newton, is (see Figure 2.10):

$$\mathbf{F} = -\frac{G M m}{\imath^2} \hat{\imath} \qquad \imath \equiv \mathbf{x} - \mathbf{x}'. \tag{2.113}$$

---

[11] This is good, since there are profound differences between the naive replacement we are about to make and the predictions of the full theory.

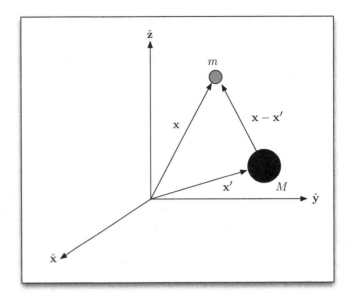

Figure 2.10   Two particles interacting via the Newtonian gravitational force.

From the force, we can, by analogy with electrostatics, construct the Newtonian gravitational field and its associated point potential:[12]

$$\mathbf{G} = -\frac{G\,M}{\imath^2}\,\hat{\boldsymbol{\imath}} = -\nabla\underbrace{\left(-\frac{G\,M}{\imath}\right)}_{\equiv\phi}. \tag{2.114}$$

By applying the Laplacian (or using our eyes), we see that the potential field ϕ satisfies:

$$\nabla^2\,\phi = 4\,\pi\,G\,\rho_m(\mathbf{x}), \tag{2.115}$$

where $\rho_m(\mathbf{x})$ is the density of a distribution of *mass*.

Comparing this to the electrostatic potential:

$$\nabla^2\,V = -\frac{\rho_e(\mathbf{x})}{\epsilon_0}, \tag{2.116}$$

we can map electrostatic results to gravitational results by the replacement:

$$\boxed{\rho_e \longrightarrow -\epsilon_0\,\rho_m\,(4\,\pi\,G).} \tag{2.117}$$

Notice that we already sense a clash with special relativity – the Newtonian gravitational potential, like the pure electrostatic one, responds instantaneously to

---

[12] The vector $\mathbf{G}$ is the Newtonian gravitational field, the force per unit mass associated with gravity. The G appearing on the right of (2.114) is the gravitational constant.

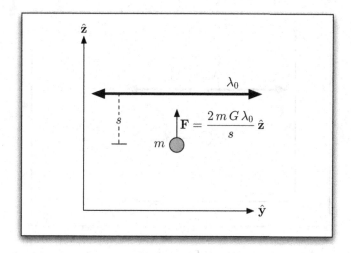

Figure 2.11   A particle of mass m feels a force in the $\hat{\mathbf{z}}$-direction due to the infinite line of uniform mass.

changes in the source distribution – that is not the sort of causal behavior we expect in a true relativistic theory. In E&M, we know that the "electrostatic potential" V gets augmented when we move to the full set of Maxwell's equations, to include a second-order temporal derivative. That is what leads to the retarded potentials that *do* have a fixed propagation speed. On the Newtonian side, we have no reason (yet) to expect such a mechanism.

2.6.2 Lines of mass

Consider an infinite rod with uniform mass density λ_0 (mass per unit length, here). What is the gravitational field associated with this configuration? If this were an electrostatics problem with charge replacing mass, the electric field would be:

$$\mathbf{E} = \frac{\lambda_0}{2\pi\,\epsilon_0\,s}\,\hat{\mathbf{s}} \tag{2.118}$$

(in cylindrical coordinates) and we apply our map (2.117) to obtain the relevant gravitational field:

$$\mathbf{G} = -\frac{2\,G\,\lambda_0}{s}\,\hat{\mathbf{s}}. \tag{2.119}$$

The force on a particle of mass m a distance s away from the line is, then, $\mathbf{F} = -\frac{2m\,G\,\lambda_0}{s}\,\hat{\mathbf{s}}$. Referring to Figure 2.11, we can write the force in terms of the local Cartesian axes for a particle in the plane of the paper as shown there.

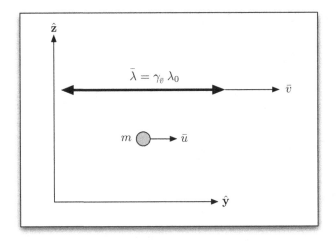

Figure 2.12 A uniform line mass of rest length λ_0 moves at constant speed $\bar{v}$ to the right in a lab. A test mass m moves to the right with speed $\bar{u}$ in the lab.

Moving line of mass

If we take the line of uniform mass from above and pull it to the right with speed $\bar{v}$ as measured in a frame $\bar{\mathcal{O}}$ (the rest frame of the laboratory), then the only change to the force on a test particle in the lab is in the Lorentz contraction of the mass per unit length of the source: $\bar{\lambda} = \gamma_{\bar{v}} \lambda_0 \left(\text{with } \gamma_{\bar{v}} \equiv \frac{1}{\sqrt{1-\bar{v}^2/c^2}}\right)$. Newtonian gravity has nothing to say about sourcing provided by the relative motion of masses, the only source for the theory is the mass density. In our lab frame $\bar{\mathcal{O}}$, then, we have the observed force on a test mass given by:

$$\bar{\mathbf{F}} = \frac{2\,m\,G\,\gamma_{\bar{v}}\,\lambda_0}{s}\,\hat{\mathbf{z}} \qquad \gamma_{\bar{v}} \equiv \frac{1}{\sqrt{1-\frac{\bar{v}^2}{c^2}}}. \tag{2.120}$$

Suppose that the test mass is itself moving to the right with speed $\bar{u}$ (as measured in $\bar{\mathcal{O}}$) – that does not change the force of Newtonian gravity felt by the mass. The situation is shown in Figure 2.12.

Analysis in the test mass rest frame

In the rest frame of the test mass m (which we will call $\mathcal{O}$), the moving line mass, which has speed $\bar{v}$ relative to $\bar{\mathcal{O}}$, has speed:

$$v = \frac{\bar{v} - \bar{u}}{1 - \frac{\bar{u}\,\bar{v}}{c^2}} \tag{2.121}$$

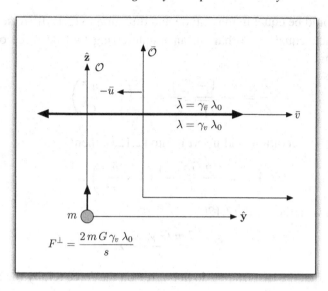

Figure 2.13   In the stationary frame of the test mass m (here called $\mathcal{O}$), the lab frame is moving to the left with speed $\bar{u}$. We can calculate the force in $\mathcal{O}$ and transform it to $\bar{\mathcal{O}}$.

relative to $\mathcal{O}$ (since in this frame, the lab is moving to the left with speed $\bar{u}$). We can then write the force on the stationary test mass (in this frame) as:

$$\mathbf{F} = \frac{2\,m\,G\,\gamma_v\,\lambda_0}{s}\,\hat{\mathbf{z}} \qquad \gamma_v \equiv \frac{1}{\sqrt{1 - \frac{v^2}{c^2}}}, \tag{2.122}$$

with v given as above.

Finally, in order to compare the force on the particle in $\mathcal{O}$ with the force on the particle in $\bar{\mathcal{O}}$, we must transform the perpendicular component of the force according to $\bar{\mathbf{F}}^{\perp} = \gamma_{\bar{u}}^{-1}\,\mathbf{F}^{\perp}$. That is, we take the force in $\mathcal{O}$ and multiply it by (one over) the boost factor associated with the relative motion of $\mathcal{O}$ and $\bar{\mathcal{O}}$: $\gamma_{\bar{u}} \equiv \frac{1}{\sqrt{1 - \frac{\bar{u}^2}{c^2}}}$. When we perform this transformation, we get the force perpendicular to the relative motion as measured in the lab frame $\bar{\mathcal{O}}$:

$$\bar{\mathbf{F}}^{\perp} = \frac{2\,m\,G\,\gamma_v\,\lambda_0}{\gamma_{\bar{u}}\,s}\,\hat{\mathbf{z}}, \tag{2.123}$$

where I leave the perpendicular reminder to distinguish this result from the direct measurement of force in $\bar{\mathcal{O}}$ represented by (2.120). Refer to Figure 2.13.

Comparison

If Newtonian gravity supported the notion that inertial frames should make identical physical predictions, even if they disagree on phenomenology, then the force in (2.123) should be equal in magnitude and direction to (2.120). To see that these are not, in fact, equal, note that (in an uninteresting tour de force of algebraic investigation):

$$\frac{\gamma_v}{\gamma_{\bar{u}}} = \frac{c^2\left(1 - \frac{\bar{u}\,\bar{v}}{c^2}\right)}{c^2\sqrt{1 - \frac{\bar{v}^2}{c^2}}} = \gamma_{\bar{v}}\left(1 - \frac{\bar{u}\,\bar{v}}{c^2}\right). \tag{2.124}$$

If we take this observation, and insert it into (2.123), then:

$$\bar{\mathbf{F}}^{\perp} = \frac{2\,m\,G\,\gamma_{\bar{v}}\,\lambda_0}{s}\left(1 - \frac{\bar{u}\,\bar{v}}{c^2}\right)\hat{\mathbf{z}}. \tag{2.125}$$

Compare this with the result (2.120):

$$\bar{\mathbf{F}} = \frac{2\,m\,G\,\gamma_{\bar{v}}\,\lambda_0}{s}\,\hat{\mathbf{z}}, \tag{2.126}$$

and it is clear that the two frames do *not* agree on the force felt by the particle.

2.6.3 Electromagnetic salvation

Let's briefly review how the dual electric and magnetic fields save the force predictions on the E&M side. The difference is in the fields associated with the $\bar{\mathcal{O}}$ frame shown in Figure 2.12 – if we now interpret λ_0 as a line charge, then we have both an electric and magnetic field in $\bar{\mathcal{O}}$, leading to both electric and magnetic forces. The fields at the particle location are:

$$\bar{\mathbf{E}} = -\frac{\gamma_{\bar{v}}\,\lambda_0}{2\,\pi\,\epsilon_0\,s}\,\hat{\mathbf{z}} \qquad \bar{\mathbf{B}} = -\frac{\mu_0\,\gamma_{\bar{v}}\,\lambda_0\,\bar{v}}{2\,\pi\,s}\,\hat{\mathbf{x}}, \tag{2.127}$$

and the force on a positive test charge is the usual $q\,\bar{\mathbf{E}} + q\,\bar{u}\,\hat{\mathbf{y}} \times \bar{\mathbf{B}}$:

$$\begin{aligned}
\bar{\mathbf{F}}^{\perp} &= \left(-\frac{q\,\gamma_{\bar{v}}\,\lambda_0}{2\,\pi\,\epsilon_0\,s} + \frac{q\,\frac{\bar{u}\,\bar{v}}{c^2}\,\gamma_{\bar{v}}\,\lambda_0}{2\,\pi\,\epsilon_0\,s}\right)\hat{\mathbf{z}} \\[2mm]
&= -\frac{q\,\gamma_{\bar{v}}\,\lambda_0}{2\,\pi\,\epsilon_0\,s}\left(1 - \frac{\bar{u}\,\bar{v}}{c^2}\right)\hat{\mathbf{z}}.
\end{aligned} \tag{2.128}$$

Transcribing the electromagnetic result (2.128) using (2.117), we see that *if* there existed a gravitational analogue of **B**, a field sourced by *moving* mass that acted on moving mass, then we would have got:

$$\bar{\mathbf{F}}^{\perp} = \frac{2\,M\,G\,\gamma_{\bar{v}}\,\lambda_0}{s}\left(1 - \frac{\bar{u}\,\bar{v}}{c^2}\right)\hat{\mathbf{z}} \tag{2.129}$$

as our force in $\bar{\mathcal{O}}$, replacing (2.120). This would then agree with the force measured in $\mathcal{O}$ (and transformed to $\bar{\mathcal{O}}$) (2.125).

2.6.4 Conclusion

The difference between this Newtonian gravitational argument and the same problem analyzed for line charges is that a moving line charge generates a "magneto-static" force in the lab frame that gives precisely the additional component found in (2.129) (i.e. in the electromagnetic case, the moving line of charge in the lab generates an electric and magnetic force on the test particle). So in E&M, the forces are identical, and the physical predictions in the two frames coincide.

The existence of a magnetic force solves the "problem", and it is tempting to put precisely an analogous "moving mass" source into Newtonian gravity. That is, we augment Newtonian gravity's electrostatic analogy with a similar magnetostatic term. This additional force is known as the "gravitomagnetic" force, and exists in almost the form implied by E&M as a linearized limit of general relativity (we shall see this explicitly in Chapter 7). There are quantitative differences beyond the obvious replacement of signs and units associated with the linearized GR limit, but the qualitative predictions that can be made by analogy with E&M provide excellent "intuition" into weak gravitational problems.

As an example, consider a spinning massive sphere with uniform charge density and constant angular velocity $\boldsymbol{\omega} = \omega\,\hat{\mathbf{z}}$. On the electromagnetic side, we expect the exterior electric and magnetic fields of a charged spinning sphere to look like Coulomb and a magnetic dipole respectively. If we took a spinning, charged test body and put it in the electromagnetic field of this configuration, we would expect the spin of the test body to precess in the presence of the magnetic dipole field (Larmor precession is an example).

In the weak-field GR limit, we expect a replacement similar to (2.117) to apply to both the electric and magnetic fields, giving us a Newtonian and gravitomagnetic contribution. If we then introduced a spinning test mass, we would expect precession just as in the electromagnetic case. In this gravitational setting, such precession is referred to as Lense–Thirring precession, and the Gravity Probe B experiment (see [21]) is currently measuring this effect in an orbit about the earth (a spinning, massive sphere).

In addition to stationary configurations like a uniformly spinning ball, our electromagnetic analogy implies the existence of gravitational waves. From the demands of special relativity, and the Newtonian gravitational starting point, we would naturally end up with gravitational fields that correspond to $\mathbf{E}$ and $\mathbf{B}$ for electro- and magnetostatics. But we know what happens in E&M to these – again appealing to (special) relativity, we get the magnetic source for $\mathbf{E}$ and the electric source for $\mathbf{B}$ – these would also exist for the gravitational analogues. Once we have the "full set of Maxwell's equations" appropriately transcribed, we know that wave-like solutions will exist, and so gravitational radiation is plausible. There are interesting parallels to be made, but be forewarned: general relativity is *not* E&M applied to gravity, even though some of the qualitative features are there. It is a new and very different theory, as we shall see.

2.7 What's next

The point of relativistic mechanics, as a motivational tool, is that it makes the four-dimensional Minkowski space-time the fundamental arena for physics. We now need to measure particles as they move through time and space in order to determine the forces acting on them. This expanded notion is required if we are to have a universal constant c, as demanded by experiment.

In addition, we know that Newtonian gravity violates the fundamental rules of special relativity, so we have also used the relativistic Lagrangian and analysis of motion to highlight some of the problems with gravity even at the level of special relativity. What we have not done is suggested that a geometric fix is in order. Given that the next two chapters are a break from explicit physical examples in order to understand the idea behind curved space-time, it would be nice to present some indication of why curved space-times are necessary. They are not, strictly speaking, necessary, this is an interpretation of the program of general relativity. We can be content working in a flat (Minkowski $\eta_{\mu\nu}$) background space-time with a field $h_{\mu\nu}$ that appears in additive combination everywhere with $\eta_{\mu\nu}$. This additive combination functions precisely as a metric, and the combination $\eta_{\mu\nu} + h_{\mu\nu}$ is not physically separable, meaning that there is no way to distinguish between the effects of $\eta_{\mu\nu}$ and $h_{\mu\nu}$, no experiment that allows us to separate the contributions to space-time provided by $\eta_{\mu\nu}$ and $h_{\mu\nu}$.

At that point, though, we may as well think of the combination as a metric defining a curved space-time geometry. The gravitational field of general relativity is precisely the metric of space-time, and the field equations relate the metric to its sources, energy. Since the metric is our target field, and (for us), the metric will define the geometry of space-time completely, we must understand tensors and curvature, and that is the goal of the next two chapters. Einstein's original motivation started from a geometric picture. Special relativity teaches us that physical processes occur identically in all laboratories moving relative to one another with constant speed. That's good, since otherwise we would need to write down our speed relative to all other objects in the universe in order to describe experimental results. But we also know that close to the earth, at least, gravity sets up forces in proportion to an object's mass, so that in free-fall, everything undergoes identical acceleration $g = 9.8$ m/s^2. Then, Einstein reasoned, the results of experiments cannot depend on that uniform identical acceleration – there is, in effect, no way for us to tell if *everything* (us, our lab, the experimental apparatus in it) is moving with identical acceleration or not. That suggests that perhaps gravity itself can be described as a feature of our space-time, rather than a discernible force. This is the result of general relativity. In order to understand how a curved space-time, generated by a mass distribution, can approximate the "force" we normally associate with gravity, we must understand curvature.

Problem 2.12

For a non-interacting dust at rest, we measure a uniform energy density ρ_0 (energy per unit volume, think of a charge density).

(a) Suppose this energy density can be thought of as a number density (number of particles per unit volume) times the relativistic energy per particle. Compute the number density as observed in a frame moving with speed v along a shared x-axis and multiply by the energy per particle as measured in that frame. This will give you the energy density as measured by the moving frame.

(b) The quantity ρ_0 should transform as the 00 component of a second-rank, symmetric tensor – in the rest frame of the dust, $T_{00} = \rho_0$ and all other components are zero. By Lorentz boosting along a shared x-axis, find the transformed component $\bar{T}_{00}$ in a moving frame, it should be identical to your result from part a.

(c) Show that the expression $\rho = T_{\mu\nu} u^\mu u^\nu$, where $T_{\mu\nu}$ is the stress tensor in the dust rest frame, and u^μ is the four-velocity of an observer (moving with speed v relative to the dust along a shared x-axis), is correct for any observer (to do this, you need only verify that you get the result from part b for an observer moving with speed v, and that you recover ρ_0 when you input the four-velocity of the stationary dust frame).

Problem 2.13

We have seen that Newton's laws (notably the second) must be updated to accommodate the predictions of special relativity. The form of Newtonian gravity must also be modified. From your work in electricity and magnetism, write the integral form of the solution to:

$$\nabla^2 \phi = 4\pi G \rho_m(\mathbf{x}). \tag{2.130}$$

What principle of special relativity is violated by the solution? How is this fixed in the context of E&M?

Problem 2.14

For the (Newtonian) gravitational field $\mathbf{G} = -\nabla\phi$, with ϕ governed by (2.130), develop the gravitational analogue of Gauss's law (which, for E&M, reads:

$\oint \mathbf{E} \cdot d\mathbf{a} = \frac{Q_{enc}}{\epsilon_0}$). Use your expression to find the gravitational field above an infinite sheet with uniform mass per unit area, σ.

Problem 2.15

We have focused on the dynamics "in the lab" for a few different forces. In each case, the proper time of the particle becomes a function of the time in the lab t (or vice versa, the function is invertible). Given the trajectory:

$$x(t) = \frac{m c^2}{F} \sqrt{1 + \frac{F^2 t^2}{m^2 c^2}} \tag{2.131}$$

for motion under the influence of a constant force F, find the ordinary differential equation relating t to τ using the invariant interval ds^2 (i.e. compare ds^2 in the lab and in the rest frame of the particle). Solve this ODE for τ as a function of t and verify that you recover (2.84). The point is, given the motion of a particle in the lab, we can find the proper time of the particle (the time in its instantaneous rest frame). Of course, there are (many) functions $x(t)$ for which this procedure will not yield "valid" (meaning, usually, real) relations between t and τ.

*Problem 2.16

Velocity addition is one of the places that special relativity plays a large role. Classically, in one dimension, given the velocity of a frame A with respect to a frame B (call it v_{AB}), and the velocity of B relative to C (v_{BC}), we can find the velocity of A relative to C via:

$$v_{AC} = v_{AB} + v_{BC}. \tag{2.132}$$

But this equation implies that the speed of light can have different measured values. That is not allowed, according to the second postulate of special relativity.

   To find the relativistic form, take the following setup: object A is moving to the right with speed v_{AB} relative to a reference frame $\bar{\mathcal{O}}$, and this reference frame is traveling to the right with speed v_{BC} relative to the "lab" frame $\mathcal{O}$ as shown in Figure 2.14.

(a) Using the Lorentz transformation in the form:

$$\begin{aligned}
\bar{x} &= \gamma (x - v t) \\
c\bar{t} &= \gamma \left(c t - \frac{x v}{c} \right),
\end{aligned} \tag{2.133}$$

together with the observation that, to describe object A, we have $\bar{x} = v_{AB}\,\bar{t}$, and the appropriate identification of v in the transformation, find v_{AC}. Verify that even if $v_{AB} = c$ *and* $v_{BC} = c$, the maximum measured speed in the lab, v_{AC}, is c.

(b) Express your result in terms of the rapidities η, ψ and α defined by:

$$\cosh \eta \equiv \frac{1}{\sqrt{1 - \frac{v_{AB}^2}{c^2}}} \qquad \cosh \psi \equiv \frac{1}{\sqrt{1 - \frac{v_{BC}^2}{c^2}}} \qquad \cosh \alpha \equiv \frac{1}{\sqrt{1 - \frac{v_{AC}^2}{c^2}}} \tag{2.134}$$

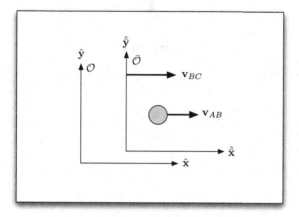

Figure 2.14 We're given the speed of object A relative to the reference frame $\bar{\mathcal{O}}$, v_{AB}. We also know the speed of $\bar{\mathcal{O}}$ relative to $\mathcal{O}$, v_{BC}, and we want to find the speed of A as measured in $\mathcal{O}$, v_{AC}.

from (2.46). So what we want is a relation between α and η, ψ. The following hyperbolic trigonometric expansions may be useful:

$$\sinh(X + Y) = \sinh(X) \cosh(Y) + \sinh(Y) \cosh(X)$$
$$\cosh(X + Y) = \sinh(X) \sinh(Y) + \cosh(X) \cosh(Y). \tag{2.135}$$

Problem 2.17

Suppose we write the Lagrangian for motion under the influence of a constant force F in the form (compare with (2.78)):

$$L = \frac{1}{2} m \, \dot{x}^{\mu} \, \eta_{\mu\nu} \, \dot{x}^{\nu} - \alpha \, \epsilon_{\mu\nu} \, \dot{x}^{\mu} \, x^{\nu}, \tag{2.136}$$

where dots refer to proper time, and we are in $1 + 1$-dimensional space-time, with t and x as coordinates.

(a) Work out the equations of motion for t and x from this Lagrangian – find the value of α that makes these identical to (2.81).

(b) Using the Legendre transformation, find the associated Hamiltonian.

3

Tensors

In special relativity, the law governing the motion of particles is $m\,\ddot{x}^\alpha = F^\alpha$ where the vectors are four-vectors, written in Cartesian coordinates (and time), and parametrization is in terms of proper time. The details of the physical principle one is studying are hidden in F^α, and potentially, its potential. That is what defines the interaction.

In general relativity, the motion of particles (under the influence of gravity alone) will be described by the geodesic equation:

$$\ddot{x}^\mu + \Gamma^\mu_{\alpha\beta}\,\dot{x}^\alpha\,\dot{x}^\beta = 0, \tag{3.1}$$

and this will also occur in a four-dimensional space-*time*. This equation of motion lacks a potential, and/or force, and can be connected in a natural way to the metric. To understand (3.1), and all that it implies, we need a more complete description of tensors and tensor calculus, which is the point of this chapter.

We defined tensor transformation induced by a coordinate transformation in Section 1.5. Let's review those results, preparatory to a more in-depth study. Here we define a general coordinate transformation $x \longrightarrow \bar{x}(x)$, where the new coordinates are $\bar{x}$, themselves invertible functions of x. Then a contravariant first-rank tensor (sometimes referred to as a vector) transforms according to:

$$\bar{f}^\alpha = \frac{d\bar{x}^\alpha}{dx^\beta}\,f^\beta. \tag{3.2}$$

Similarly, for a covariant first-rank tensor, we have:

$$\bar{f}_\alpha = \frac{dx^\beta}{d\bar{x}^\alpha}\,f_\beta. \tag{3.3}$$

The particular models we keep in mind for each of these are dx^α, the coordinate differential, for contravariant tensors, and $\partial_\mu\,\phi$, the gradient of a scalar, for covariant tensors. A straightforward application of the chain rule establishes the transformation character of each.

These letters with dangling greek letters are to be thought of as individual *components*. The usual representation we use to draw pictures is components with basis vectors. So, for example, the tensor:

$$f^\alpha \doteq \begin{pmatrix} x \\ y \end{pmatrix} \qquad (3.4)$$

is a list of two components, while:

$$\mathbf{f} = x\,\hat{\mathbf{x}} + y\,\hat{\mathbf{y}} \qquad (3.5)$$

is that list projected onto a pair of basis vectors – this distinction is important, since basis vectors can themselves be position-dependent, and our vector transformation law for f^α does *not* contain the basis vector transformation, which is separate. When we write an equation like (3.1), we are referring to the components of the tensor.

The distinction between components f^α and components in a basis $\mathbf{f}$ leads to some new ideas, along with notation to express them. We'll start with examples of linear transformations in two dimensions, where calculation and graphical representation are simple, and then look at the implications for more general transformations.

3.1 Introduction in two dimensions

The setting is a Cartesian-parametrized plane, with basis vectors $\hat{\mathbf{x}}$ and $\hat{\mathbf{y}}$ (these are position-independent), and our initial, model transformation will be rotation. First, we'll define a rotation of the axes, and the associated point labels, and then define scalars and vectors that transform in specific ways under rotation.

3.1.1 Rotation

To begin, consider a simple rotation of the usual coordinate axes through an angle θ (counterclockwise) as shown in Figure 3.1.

From Figure 3.1, we can relate the coordinates $\{\bar{x}, \bar{y}\}$ w.r.t. the new axes ($\hat{\bar{\mathbf{x}}}$ and $\hat{\bar{\mathbf{y}}}$) to coordinates $\{x, y\}$ w.r.t. the "usual" axes ($\hat{\mathbf{x}}$ and $\hat{\mathbf{y}}$) – define $\ell = \sqrt{x^2 + y^2}$, then the invariance of length allows us to write:

$$\begin{aligned} \bar{x} &= \ell \, \cos(\psi - \theta) = \ell \, \cos\psi \, \cos\theta + \ell \, \sin\psi \, \sin\theta = x \, \cos\theta + y \, \sin\theta \\ \bar{y} &= \ell \, \sin(\psi - \theta) = \ell \, \sin\psi \, \cos\theta - \ell \, \cos\psi \, \sin\theta = y \, \cos\theta - x \, \sin\theta, \end{aligned} \qquad (3.6)$$

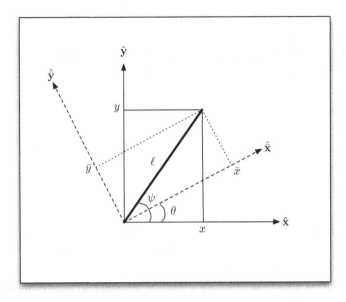

Figure 3.1 Two sets of axes, rotated through an angle θ with respect to each other. The length ℓ is the same for either set of axes.

or in matrix form:

$$\begin{pmatrix} \bar{x} \\ \bar{y} \end{pmatrix} = \underbrace{\begin{pmatrix} \cos\theta & \sin\theta \\ -\sin\theta & \cos\theta \end{pmatrix}}_{\equiv \mathbb{R} \doteq R^{\alpha}{}_{\beta}} \begin{pmatrix} x \\ y \end{pmatrix}. \tag{3.7}$$

The matrix in the middle is a "rotation" matrix, and represents the components of a mixed, second-rank tensor $R^{\alpha}{}_{\beta}$.

If we think of infinitesimal displacements centered at the origin, then we would write $d\bar{x}^{\alpha} = R^{\alpha}{}_{\beta}\, dx^{\beta}$. The tensor character of $R^{\alpha}{}_{\beta}$ (one index up, one index down) is fixed by our requirement that both dx^{α} and $d\bar{x}^{\alpha}$ are contravariant, and is reinforced by the form of the transformation. Think of the set of new coordinates as functions of the old: $\bar{x}^{\alpha}(x) = R^{\alpha}{}_{\beta}\, x^{\beta}$. Then, because the transformation is linear, we know that the derivatives of $\bar{x}^{\alpha}$ with respect to the x^{β} are constant, and must be precisely $R^{\alpha}{}_{\beta}$. Taking the derivatives explicitly gives:

$$\frac{\partial \bar{x}^{\alpha}}{\partial x^{\beta}} = \frac{\partial}{\partial x^{\beta}}\left(R^{\alpha}{}_{\gamma}\, x^{\gamma}\right) = R^{\alpha}{}_{\gamma}\, \frac{\partial x^{\gamma}}{\partial x^{\beta}} = R^{\alpha}{}_{\gamma}\, \delta^{\gamma}_{\beta} = R^{\alpha}{}_{\beta}. \tag{3.8}$$

As the "gradient" of a first-rank contravariant tensor, the index arrangement of $R^{\alpha}{}_{\beta}$ is as it must be.

Notice that, as a matrix, the inverse of $\mathbb{R}$ is:

$$\mathbb{R}^{-1} \doteq \begin{pmatrix} \cos\theta & -\sin\theta \\ \sin\theta & \cos\theta \end{pmatrix}, \tag{3.9}$$

and this is also equal to the transpose of $\mathbb{R}$, a defining property of "rotation" or "orthogonal" matrices $\mathbb{R}^{-1} = \mathbb{R}^T$ – in this case coming from the observation that one coordinate system's clockwise rotation is another's counterclockwise, and symmetry properties of the trigonometric functions.

The matrix inverse has a clear connection to the inversion of (3.8) – if we write the x and y coordinates in terms of the $\bar{x}$ and $\bar{y}$ coordinates, in terms of the inverse of $\mathbb{R}$:

$$x^\alpha = \left(R^{-1}\right)^\alpha_{\ \beta}\, \bar{x}^\beta, \tag{3.10}$$

so that:

$$\frac{\partial x^\alpha}{\partial \bar{x}^\beta} = \left(R^{-1}\right)^\alpha_{\ \beta}. \tag{3.11}$$

As a check, let's take the matrix identity $\mathbb{R}\,\mathbb{R}^{-1} = \mathbb{I}$, and write it in indexed form:

$$R^\alpha_{\ \beta} \left(R^{-1}\right)^\beta_{\ \gamma} = \frac{\partial \bar{x}^\alpha}{\partial x^\beta} \frac{\partial x^\beta}{\partial \bar{x}^\gamma} = \frac{\partial \bar{x}^\alpha}{\partial \bar{x}^\gamma} = \delta^\alpha_\gamma, \tag{3.12}$$

from the chain rule. So $\mathbb{R}^{-1}$ can be thought of directly as the inverse transformation $\frac{\partial x^\alpha}{\partial \bar{x}^\beta}$. While our use of an explicit linear transformation makes these relations easy to generate, they hold for more general transformations as well, so that a generic coordinate transformation looks like:

$$d\bar{x}^\alpha = \frac{\partial \bar{x}^\alpha}{\partial x^\beta} dx^\beta \tag{3.13}$$

as it must, and the inverse is just $dx^\beta = \frac{\partial x^\beta}{\partial \bar{x}^\alpha} d\bar{x}^\alpha$. Rotations form the most familiar example of a coordinate transformation, so we will use them as our primary example as we consider the various available tensor transformation options, replacing $R^\alpha_{\ \beta}$ with $\frac{\partial \bar{x}^\alpha}{\partial x^\beta}$ when we want to generalize.

3.1.2 Scalar

This one is easy – a scalar "does not" transform under coordinate transformations. If we have a function $\phi(x, y)$ in the first coordinate system, then we have, in the rotated (or more generally transformed) coordinate system:

$$\boxed{\bar{\phi}(\bar{x}, \bar{y}) = \phi(x(\bar{x}, \bar{y}), y(\bar{x}, \bar{y})),} \tag{3.14}$$

so that operationally, all we do is replace the x and y appearing in the definition of ϕ with the relation $x = \bar{x} \cos\theta - \bar{y} \sin\theta$ and $y = \bar{x} \sin\theta + \bar{y} \cos\theta$ (or whatever transformation is desired).

3.1.3 Vector (contravariant) and bases

With the fabulous success and ease of scalar transformations, we are led to a natural definition for an object that transforms as a *vector*. If we write a generic vector in the original coordinate system:

$$\mathbf{f} = f^x(x, y)\,\hat{\mathbf{x}} + f^y(x, y)\,\hat{\mathbf{y}} \tag{3.15}$$

(where the components are f^x and f^y and the basis vectors are $\hat{\mathbf{x}}$ and $\hat{\mathbf{y}}$), then we would like the transformation to be defined as:

$$\boxed{\bar{\mathbf{f}} = \bar{f}^x(\bar{x}, \bar{y})\,\hat{\bar{\mathbf{x}}} + \bar{f}^y(\bar{x}, \bar{y})\,\hat{\bar{\mathbf{y}}} = f^x(x, y)\,\hat{\mathbf{x}} + f^y(x, y)\,\hat{\mathbf{y}},} \tag{3.16}$$

i.e. identical to the scalar case, but with the nontrivial involvement of the basis vectors. In other words, as an object projected onto a basis, we want to simply replace all reference to x and y with their dependence on $\bar{x}$ and $\bar{y}$, and similarly replace $\{\hat{\mathbf{x}}, \hat{\mathbf{y}}\}$ with $\{\hat{\bar{\mathbf{x}}}(\hat{\bar{\mathbf{x}}}, \hat{\bar{\mathbf{y}}}), \hat{\bar{\mathbf{y}}}(\hat{\bar{\mathbf{x}}}, \hat{\bar{\mathbf{y}}})\}$ – then we will label the components sitting in front of $\hat{\bar{\mathbf{x}}}$ and $\hat{\bar{\mathbf{y}}}$ appropriately as $\bar{f}^x$ and $\bar{f}^y$. We make the distinction between the components f^α and the "components projected onto the basis" $\mathbf{f}$ – our tensor transformation laws are written for f^α, so it is the components that are of interest.

We'll again use rotation as our example to generate an explicit transformation for both the components and the basis vectors. Notice that for (3.16) to hold, both the components and basis vectors must transform, and those transformations must "undo one another", so to speak. The bold-faced $\mathbf{f}$ is what we typically draw as vector arrows attached to all points in the plane, and those little arrows are insensitive to our particular coordinate choice – only their expression in terms of coordinates will change. If the components change, the basis vectors must change in an opposite sense, so that the geometrical object $\mathbf{f}$ is identical to $\bar{\mathbf{f}}$ – this is the sentiment expressed in our target rule (3.16).

Referring to Figure 3.1, we can see how to write the basis vectors $\hat{\bar{\mathbf{x}}}$ and $\hat{\bar{\mathbf{y}}}$ in terms of $\hat{\mathbf{x}}$ and $\hat{\mathbf{y}}$:

$$\hat{\bar{\mathbf{x}}} = \cos\theta\,\hat{\mathbf{x}} + \sin\theta\,\hat{\mathbf{y}}$$
$$\hat{\bar{\mathbf{y}}} = -\sin\theta\,\hat{\mathbf{x}} + \cos\theta\,\hat{\mathbf{y}}, \tag{3.17}$$

and using this, we can write $\bar{\mathbf{f}}$ in terms of the original basis vectors. The elements in front of these will then define f^x and f^y according to our target (3.16):

$$\bar{\mathbf{f}} = \bar{f}^x \,\hat{\bar{\mathbf{x}}} + \bar{f}^y \,\hat{\bar{\mathbf{y}}} = \left(\bar{f}^x \, \cos\theta - \bar{f}^y \, \sin\theta \right) \hat{\mathbf{x}} + \left(\bar{f}^x \, \sin\theta + \bar{f}^y \, \cos\theta \right) \hat{\mathbf{y}}, \quad (3.18)$$

so that we learn (slash demand) that:

$$\begin{aligned} f^x &= \bar{f}^x \, \cos\theta - \bar{f}^y \, \sin\theta \\ f^y &= \bar{f}^x \, \sin\theta + \bar{f}^y \, \cos\theta, \end{aligned} \qquad (3.19)$$

or, inverting, we have, now in matrix form:

$$\begin{pmatrix} \bar{f}^x \\ \bar{f}^y \end{pmatrix} = \begin{pmatrix} \cos\theta & \sin\theta \\ -\sin\theta & \cos\theta \end{pmatrix} \begin{pmatrix} f^x \\ f^y \end{pmatrix}. \qquad (3.20)$$

The "component" transformation can be written as $\bar{f}^\alpha = R^\alpha{}_\beta \, f^\beta$ from the above, and this leads to the usual statement that "a (contravariant) vector transforms 'like' the coordinates themselves", compare with (3.7).[1]

Referring to the basis vector transformation (3.17), if we write $\mathbf{e}_1 = \hat{\mathbf{x}}$ and $\mathbf{e}_2 = \hat{\mathbf{y}}$, then we can write the transformation, in terms of row vectors, as:

$$\begin{pmatrix} \bar{\mathbf{e}}_1 & \bar{\mathbf{e}}_2 \end{pmatrix} = \begin{pmatrix} \mathbf{e}_1 & \mathbf{e}_2 \end{pmatrix} \underbrace{\begin{pmatrix} \cos\theta & -\sin\theta \\ \sin\theta & \cos\theta \end{pmatrix}}_{=\mathbb{R}^{-1}}. \qquad (3.21)$$

The row vector form is essentially forced on us (although it has motivation) by the demands of index-matching. We can write this matrix equation as:

$$\bar{\mathbf{e}}_\alpha = \left(R^{-1} \right)^\beta{}_\alpha \, \mathbf{e}_\beta. \qquad (3.22)$$

The rotation matrix and transformation is just one example, and the form is telling – for a more general transformation, we would replace $R^\alpha{}_\beta$ with $\frac{d\bar{x}^\alpha}{dx^\beta}$ (and $\left(R^{-1} \right)^\beta{}_\alpha$ with $\frac{\partial x^\beta}{\partial \bar{x}^\alpha}$), and so it is tempting to update the contravariant vector transformation law to include an arbitrary relation between "new" and "old" coordinates:

$$\boxed{\bar{f}^\alpha = \frac{\partial \bar{x}^\alpha}{\partial x^\beta} \, f^\beta,} \qquad (3.23)$$

precisely what we defined for a contravariant vector originally. The basis vectors, meanwhile, have general transformation (using (3.11)) given by:

$$\boxed{\bar{\mathbf{e}}_\alpha = \frac{\partial x^\beta}{\partial \bar{x}^\alpha} \, \mathbf{e}_\beta,} \qquad (3.24)$$

---

[1] In general coordinate systems, not necessarily Cartesian, the more appropriate wording would be "like the coordinate differentials", which are actual vectors: $d\bar{x}^\alpha = R^\alpha{}_\beta \, dx^\beta$.

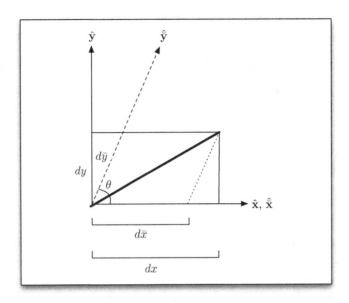

Figure 3.2   Original, orthogonal Cartesian axes, and a skewed set with coordinate differential shown.

what we recognize as a covariant transformation. The basis vectors transform so as to undo the component transformation – if we think of the indexed form of $\bar{\mathbf{f}} = \mathbf{f}$, this is clear:

$$\bar{\mathbf{f}} = \bar{f}^\alpha \, \bar{\mathbf{e}}_\alpha = \left(\frac{\partial \bar{x}^\alpha}{\partial x^\beta} \, f^\beta \right) \left(\frac{\partial x^\gamma}{\partial \bar{x}^\alpha} \, \mathbf{e}_\gamma \right) = f^\beta \, \delta^\gamma_\beta \, \mathbf{e}_\gamma = f^\beta \, \mathbf{e}_\beta = \mathbf{f} \qquad (3.25)$$

as desired.

3.1.4 Non-orthogonal axes

We will stick with linear transformations for a moment, but now let the axes be skewed – this will allow us to more easily distinguish between the contravariant transformation law in (3.23) and the covariant[2] transformation law.

Consider the two coordinate systems shown in Figure 3.2. From the figure, we can write the relation for dx^α in terms of $d\bar{x}^\alpha$:

$$\begin{pmatrix} dx \\ dy \end{pmatrix} = \begin{pmatrix} 1 & \cos\theta \\ 0 & \sin\theta \end{pmatrix} \begin{pmatrix} d\bar{x} \\ d\bar{y} \end{pmatrix} \qquad (3.26)$$

---

[2] Just as first-rank contravariant tensors go by the name "vector", covariant first-rank tensors are alternatively known as "one-forms".

or, we can write it in the more standard form (inverting the above, no longer a simple matrix transpose):

$$\begin{pmatrix} d\bar{x} \\ d\bar{y} \end{pmatrix} = \underbrace{\begin{pmatrix} 1 & -\cot\theta \\ 0 & \frac{1}{\sin\theta} \end{pmatrix}}_{\equiv \mathbb{T} \doteq T^{\alpha}_{\beta}} \begin{pmatrix} dx \\ dy \end{pmatrix}, \tag{3.27}$$

so that $d\bar{x}^{\alpha} = T^{\alpha}_{\beta} dx^{\beta}$ and the matrix $\mathbb{T}$ is the inverse of the matrix in (3.26).

Let us first check that the vector $d\mathbf{x} = dx\,\hat{\mathbf{x}} + dy\,\hat{\mathbf{y}}$ transforms appropriately. We need the basis vectors in the skewed coordinate system – from the figure, these are $\hat{\bar{\mathbf{x}}} = \hat{\mathbf{x}}$ and $\hat{\bar{\mathbf{y}}} = \cos\theta\,\hat{\mathbf{x}} + \sin\theta\,\hat{\mathbf{y}}$. Then:

$$\begin{aligned} d\bar{\mathbf{x}} &= d\bar{x}\,\hat{\bar{\mathbf{x}}} + d\bar{y}\,\hat{\bar{\mathbf{y}}} \\ &= (dx - \cot\theta\,dy)\,\hat{\mathbf{x}} + \frac{dy}{\sin\theta}(\cos\theta\,\hat{\mathbf{x}} + \sin\theta\,\hat{\mathbf{y}}) \tag{3.28} \\ &= dx\,\hat{\mathbf{x}} + dy\,\hat{\mathbf{y}}, \end{aligned}$$

as expected.

We can find the metric for this type of situation – if we take the scalar length $dx^2 + dy^2$, and write it in terms of the $d\bar{x}$ and $d\bar{y}$ infinitesimals, then:

$$dx^2 + dy^2 = d\bar{x}^2 + d\bar{y}^2 + 2\,d\bar{x}\,d\bar{y}\,\cos\theta \doteq (d\bar{x}\ \ d\bar{y}) \underbrace{\begin{pmatrix} 1 & \cos\theta \\ \cos\theta & 1 \end{pmatrix}}_{\equiv \bar{g}_{\mu\nu}} \begin{pmatrix} d\bar{x} \\ d\bar{y} \end{pmatrix},$$

$$\tag{3.29}$$

defining the metric with respect to the barred coordinates (the metric for the original, orthogonal $\hat{\mathbf{x}}$ and $\hat{\mathbf{y}}$ axes is just the Euclidean metric, represented as a 2×2 identity matrix). We have the relation $dx^{\alpha} g_{\alpha\beta} dx^{\beta} = d\bar{x}^{\alpha}\,\bar{g}_{\alpha\beta}\,d\bar{x}^{\beta}$ (the line element is a scalar), which tells us how the metric itself transforms. From the definition of the transformation, $d\bar{x}^{\alpha} = T^{\alpha}_{\beta} dx^{\beta}$, we have $\frac{\partial \bar{x}^{\alpha}}{\partial x^{\beta}} = T^{\alpha}_{\beta}$, but we can also write the inverse transformation, making use of the inverse of the matrix $\mathbb{T}$ in (3.26). Letting $\tilde{T}^{\alpha}_{\beta} \equiv \mathbb{T}^{-1}$:

$$dx^{\alpha} = \tilde{T}^{\alpha}_{\beta}\,d\bar{x}^{\beta} \qquad \frac{\partial x^{\alpha}}{\partial \bar{x}^{\beta}} = \tilde{T}^{\alpha}_{\beta}. \tag{3.30}$$

Going back to the invariant, we can write:

$$dx^{\alpha} g_{\alpha\beta} dx^{\beta} = d\bar{x}^{\alpha}\,\bar{g}_{\alpha\beta}\,d\bar{x}^{\beta} = \left(T^{\alpha}_{\gamma}\,dx^{\gamma}\right) \bar{g}_{\alpha\beta} \left(T^{\beta}_{\rho}\,dx^{\rho}\right). \tag{3.31}$$

For this expression to equal $dx^{\alpha} g_{\alpha\beta} dx^{\beta}$, we must have:

$$\bar{g}_{\alpha\beta} = \tilde{T}^{\sigma}_{\alpha}\,g_{\sigma\delta}\,\tilde{T}^{\delta}_{\beta}. \tag{3.32}$$

As a check:

$$d\bar{x}^\alpha \, \bar{g}_{\alpha\beta} \, d\bar{x}^\beta = (T^\alpha_{\ \gamma} \, dx^\gamma) \, \bar{g}_{\alpha\beta} \, (T^\beta_{\ \rho} \, dx^\rho)$$
$$= (T^\alpha_{\ \gamma} \, dx^\gamma) \left(\tilde{T}^\sigma_{\ \alpha} \, g_{\sigma\delta} \, \tilde{T}^\delta_{\ \beta}\right) (T^\beta_{\ \rho} \, dx^\rho)$$
$$= \delta^\sigma_{\ \gamma} \, \delta^\delta_{\ \rho} \, g_{\sigma\delta} \, dx^\gamma \, dx^\rho \qquad (3.33)$$
$$= dx^\sigma \, g_{\sigma\delta} \, dx^\delta.$$

The metric transformation law (3.32) is different from that of a vector – first of all, there are two indices, but more important, rather than transforming with $T^\alpha_{\ \beta}$, the covariant indices transform with the inverse of the coordinate transformation. This type of (second-rank) tensor transformation is not obvious under rotations since rotations leave the metric itself invariant ($\mathbb{R}^T \, \mathbf{g} \, \mathbb{R} = \mathbf{g}$ with $\mathbf{g}$ the matrix representation of the metric), and we don't notice the transformation of the metric.

If we move to a more general coordinate transformation, with $\tilde{T}^\alpha_{\ \beta} = \frac{\partial x^\alpha}{\partial \bar{x}^\beta}$, not necessarily the skew-axis one, then the metric transformation rule becomes:

$$\boxed{\bar{g}_{\alpha\beta} = \tilde{T}^\sigma_{\ \alpha} \, \tilde{T}^\delta_{\ \beta} \, g_{\sigma\delta} = \frac{\partial x^\sigma}{\partial \bar{x}^\alpha} \frac{\partial x^\delta}{\partial \bar{x}^\beta} \, g_{\sigma\delta}.} \qquad (3.34)$$

3.1.5 Covariant tensor transformation

The metric provides an example of a tensor transformation that is not the contravariant one. There are two factors of $\frac{\partial x}{\partial \bar{x}}$ in (3.34), appropriate to its two indices. Objects f_α that respond to coordinate transformation according to:

$$\bar{f}_\alpha = \frac{dx^\beta}{d\bar{x}^\alpha} \, f_\beta \qquad (3.35)$$

are called covariant (first-rank) tensors. The metric is a covariant second-rank tensor. The basis vectors from (3.24) represent first-rank covariant tensors.

Perpendicular and parallel projection

One can quickly show that covariant second-rank tensors, viewed as matrices, have inverses that behave as contravariant second-rank tensors, and vice versa. Given a second-rank covariant tensor (like the metric), when the matrix inverse exists, we can use the associated contravariant second-rank form to *define* the contravariant first-rank tensor $f^\alpha \equiv g^{\alpha\beta} \, f_\beta$ associated with covariant f_α. This works out functionally as described in Section 1.5. In most cases, the covariant second-rank tensor we use to define this process of raising and lowering indices is the metric, but it need not be. The advantage to using the metric is that it is almost always defined and has an inverse (almost everywhere). Then as a point of labeling, if we are given f_α, we *call* the object $g^{\alpha\beta} f_\beta$ the contravariant f^α. Similarly, given a contravariant h^α, we define the covariant form to be $h_\alpha \equiv g_{\alpha\beta} h^\beta$. The connection between up and down indices is intimately tied to the metric.

The coordinate definitions you know and love are intrinsically contravariant: $x^\mu \doteq (x, y, z)^T$, and so we ask – how has the distinction between up and down never come up in the past? We're used to writing $\mathbf{x} = x\,\hat{\mathbf{x}} + y\,\hat{\mathbf{y}}$ (in two dimensions), that doesn't appear to be either up or down.

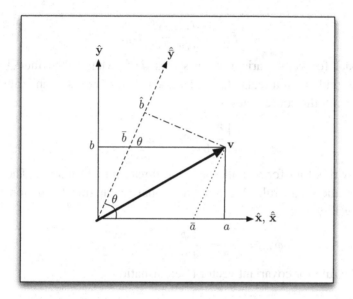

Figure 3.3 The difference between contravariant and covariant vectors in a non-orthogonal basis.

The answer can be made clear with a simple example. Suppose we again take non-orthogonal axes in two dimensions. Referring to Figure 3.3, the vector $\bar{\mathbf{v}} = \bar{a}\,\hat{\bar{\mathbf{x}}} + \bar{b}\,\hat{\bar{\mathbf{y}}}$ has components that are contravariant:

$$\bar{v}^\alpha \doteq \begin{pmatrix} \bar{a} \\ \bar{b} \end{pmatrix}. \tag{3.36}$$

We know the metric associated with the transformation to the skewed coordinate system, that's given in matrix form in (3.29). The covariant form of $\bar{v}^\alpha$ is obtained by lowering the index with the metric:

$$\bar{v}_\beta = \bar{v}^\alpha\, g_{\alpha\beta} \doteq \begin{pmatrix} 1 & \cos\theta \\ \cos\theta & 1 \end{pmatrix} \begin{pmatrix} \bar{a} \\ \bar{b} \end{pmatrix} = \begin{pmatrix} \bar{a} + \bar{b}\,\cos\theta \\ \bar{b} + \bar{a}\,\cos\theta \end{pmatrix}. \tag{3.37}$$

From the figure, the covariant component $\bar{v}_1$ corresponds to the perpendicular projection of the vector onto the $\hat{\bar{\mathbf{x}}}$-axis: $\bar{v}_1 = a$. Similarly, the covariant component $\bar{v}_2$ is labeled b in the figure and is the perpendicular projection onto the $\hat{\bar{\mathbf{y}}}$-axis. So the contravariant components of a vector are measured parallel to the axes, while the covariant components are perpendicular measurements.

If we always use orthonormal axes, the distinction never arises, since in that case, the perpendicular projection is identical to the parallel projection.

For a second-rank covariant tensor, call it $f_{\mu\nu}$, we just introduce a copy of the transformation for each index (the same is true for the contravariant transformation law):

$$\bar{f}_{\mu\nu} = \frac{\partial x^\alpha}{\partial \bar{x}^\mu} \frac{\partial x^\beta}{\partial \bar{x}^\nu} f_{\alpha\beta}. \tag{3.38}$$

Our model for contravariant vectors was dx^α, what is the model covariant vector? The gradient of a scalar. Consider a scalar $\phi(x, y)$ – we can write its partial derivatives w.r.t. the coordinates as:

$$\boxed{\frac{\partial \phi(x, y)}{\partial x^\alpha} \equiv \phi_{,\alpha}} \tag{3.39}$$

and ask how this transforms under any coordinate transformation. The answer is provided by the chain rule. We have the usual scalar transformation $\bar{\phi}(\bar{x}, \bar{y}) = \phi(x(\bar{x}, \bar{y}), y(\bar{x}, \bar{y}))$, so:

$$\bar{\phi}_{,\alpha} = \frac{\partial \bar{\phi}}{\partial \bar{x}^\alpha} = \frac{\partial \phi}{\partial x^\beta} \frac{\partial x^\beta}{\partial \bar{x}^\alpha} = \frac{\partial x^\beta}{\partial \bar{x}^\alpha} \phi_{,\beta}, \tag{3.40}$$

precisely the rule for covariant vector transformation.

Examples

It is instructive to take a simple coordinate transformation and check the covariant nature of the gradient. Consider the usual polar transformation: $x = s \cos \phi$, $y = s \sin \phi$, where we have the original coordinates $x^1 = x$, $x^2 = y$, and the new set $\bar{x}^1 = s$, $\bar{x}^2 = \phi$. Now our coordinate transformation "matrix" reads:

$$\frac{\partial \bar{x}^\alpha}{\partial x^\beta} \doteq \begin{pmatrix} \frac{x}{\sqrt{x^2+y^2}} & \frac{y}{\sqrt{x^2+y^2}} \\ -\frac{y}{x^2+y^2} & \frac{x}{x^2+y^2} \end{pmatrix} = \begin{pmatrix} \cos\phi & \sin\phi \\ -\frac{\sin\phi}{s} & \frac{\cos\phi}{s} \end{pmatrix}, \tag{3.41}$$

where we have written the matrix in both sets of coordinates, $\{x, y\}$ and $\{s, \phi\}$, for completeness.

Take the scalar $\phi(x, y) = x + y$, then the transformed scalar is just $\bar{\phi} = s (\cos \phi + \sin \phi)$. Now if we take the gradient of ϕ in each coordinate system:

$$\frac{\partial \phi}{\partial x^\alpha} \doteq \begin{pmatrix} 1 & 1 \end{pmatrix}$$
$$\frac{\partial \bar{\phi}}{\partial \bar{x}^\alpha} \doteq \begin{pmatrix} \cos\phi + \sin\phi & s(-\sin\phi + \cos\phi) \end{pmatrix}, \tag{3.42}$$

the claim is that these should be related by $\bar{\phi},_{\mu} = \frac{\partial x^{\alpha}}{\partial \bar{x}^{\mu}} \phi,_{\alpha}$. To check the relation, we need the inverse of (3.41), and this is given by:

$$\frac{\partial x^{\alpha}}{\partial \bar{x}^{\beta}} \doteq \begin{pmatrix} \cos\phi & -s\,\sin\phi \\ \sin\phi & s\,\cos\phi \end{pmatrix}. \tag{3.43}$$

The expression $\bar{\phi},_{\mu} = \frac{\partial x^{\alpha}}{\partial \bar{x}^{\mu}} \phi,_{\alpha}$ is represented by the matrix equation:

$$\begin{pmatrix} \cos\phi + \sin\phi \\ s\,(-\sin\phi + \cos\phi) \end{pmatrix}^{\mathrm{T}} = \begin{pmatrix} 1 & 1 \end{pmatrix} \begin{pmatrix} \cos\phi & -s\,\sin\phi \\ \sin\phi & s\,\cos\phi \end{pmatrix}, \tag{3.44}$$

which is a true statement. We have verified that for this particular (nonlinear) coordinate transformation, the gradient transforms as a covariant first-rank tensor.

As another example, consider the basis vectors themselves – remember that we required:

$$\mathbf{f} = f^{\alpha}\,\mathbf{e}_{\alpha} = f^{x}\,\hat{\mathbf{x}} + f^{y}\,\hat{\mathbf{y}} = \bar{f}^{\alpha}\,\bar{\mathbf{e}}_{\alpha}, \tag{3.45}$$

when defining the transformation rule for f^{α}. Here, we are using an index α to label the basis vectors: $\mathbf{e}_{1} = \hat{\mathbf{x}}$, $\mathbf{e}_{2} = \hat{\mathbf{y}}$ for a Cartesian coordinate system. The whole point of the definition is that the transformation for the components should "undo" the transformation for the basis vectors – then:

$$\mathbf{f} = \frac{\partial \bar{x}^{\alpha}}{\partial x^{\beta}}\, f^{\beta}\,\bar{\mathbf{e}}_{\alpha} = f^{\gamma}\,\mathbf{e}_{\gamma}, \tag{3.46}$$

and to make this hold, we must have:

$$\bar{\mathbf{e}}_{\alpha} = \frac{\partial x^{\gamma}}{\partial \bar{x}^{\alpha}}\,\mathbf{e}_{\gamma}, \tag{3.47}$$

as in (3.24). We can check the basis vector transformation using the skew-axis transformation shown in Figure 3.2, where we had, for the basis vectors:

$$\hat{\bar{\mathbf{x}}} = \hat{\mathbf{x}} \qquad \hat{\bar{\mathbf{y}}} = \cos\theta\,\hat{\mathbf{x}} + \sin\theta\,\hat{\mathbf{y}}, \tag{3.48}$$

or:

$$\begin{pmatrix} \hat{\bar{\mathbf{x}}} & \hat{\bar{\mathbf{y}}} \end{pmatrix} = \begin{pmatrix} \hat{\mathbf{x}} & \hat{\mathbf{y}} \end{pmatrix} \begin{pmatrix} 1 & \cos\theta \\ 0 & \sin\theta \end{pmatrix}, \tag{3.49}$$

precisely the matrix statement of $\bar{\mathbf{e}}_{\alpha} = \frac{\partial x^{\beta}}{\partial \bar{x}^{\alpha}}\,\mathbf{e}_{\beta}$. So, the basis vectors themselves transform as covariant first-rank tensors.

How does the naturally covariant expression $\phi,_{\mu}$ compare with our usual "gradient":

$$\nabla\phi = \frac{\partial \phi}{\partial x}\,\hat{\mathbf{x}} + \frac{\partial \phi}{\partial y}\,\hat{\mathbf{y}}, \tag{3.50}$$

familiar from, for example, E&M? The vector $\nabla\phi$ is being expressed in a basis, and the components here must form a contravariant first-rank tensor, like the f^γ in (3.46). But we have just belabored the idea that the coordinate derivatives of a scalar form a covariant object $\phi_{,\alpha}$. We also know that it is possible,[3] using the metric, to generate a unique contravariant object given a covariant one. So we expect that the components of $\nabla\phi$ are precisely:

$$\phi^{,\alpha} \equiv g^{\alpha\beta}\,\phi_{,\beta}. \tag{3.51}$$

This contravariant form is a distinction without a difference for a Euclidean metric. But consider our skew-axis transformation from (3.26), there we have metric:

$$\bar{g}_{\alpha\beta} \doteq \begin{pmatrix} 1 & \cos\theta \\ \cos\theta & 1 \end{pmatrix}$$

$$\bar{g}^{\alpha\beta} \doteq \begin{pmatrix} \frac{1}{\sin^2\theta} & -\frac{\cos\theta}{\sin^2\theta} \\ -\frac{\cos\theta}{\sin^2\theta} & \frac{1}{\sin^2\theta} \end{pmatrix} \tag{3.52}$$

so that:

$$\begin{aligned}
\bar{g}^{\alpha\beta}\,\bar{\phi}_{,\beta}\,\bar{\mathbf{e}}_\alpha &= \frac{1}{\sin\theta}\left(\frac{1}{\sin\theta}\frac{\partial\bar\phi}{\partial\bar x} - \frac{\cos\theta}{\sin\theta}\frac{\partial\bar\phi}{\partial\bar y}\right)\hat{\bar{\mathbf{x}}} \\
&\quad + \frac{1}{\sin\theta}\left(-\frac{\cos\theta}{\sin\theta}\frac{\partial\bar\phi}{\partial\bar x} + \frac{1}{\sin\theta}\frac{\partial\bar\phi}{\partial\bar y}\right)\hat{\bar{\mathbf{y}}},
\end{aligned} \tag{3.53}$$

and:

$$\begin{aligned}
\frac{\partial\bar\phi}{\partial\bar x} &= \frac{\partial\phi}{\partial x}\frac{\partial x}{\partial\bar x} + \frac{\partial\phi}{\partial y}\frac{\partial y}{\partial\bar x} = \frac{\partial\phi}{\partial x} \\
\frac{\partial\bar\phi}{\partial\bar y} &= \frac{\partial\phi}{\partial x}\frac{\partial x}{\partial\bar y} + \frac{\partial\phi}{\partial y}\frac{\partial y}{\partial\bar y} = \frac{\partial\phi}{\partial x}\cos\theta + \frac{\partial\phi}{\partial y}\sin\theta,
\end{aligned} \tag{3.54}$$

which we can input into (3.53):

$$\bar{g}^{\alpha\beta}\,\bar{\phi}_{,\beta}\,\bar{\mathbf{e}}_\alpha = \left(\frac{\partial\phi}{\partial x} - \cot\theta\,\frac{\partial\phi}{\partial y}\right)\hat{\bar{\mathbf{x}}} + \frac{1}{\sin\theta}\frac{\partial\phi}{\partial y}\hat{\bar{\mathbf{y}}}, \tag{3.55}$$

or, finally, replacing the unit vectors with the original set:

$$\begin{aligned}
\bar{g}^{\alpha\beta}\,\bar{\phi}_{,\beta}\,\bar{\mathbf{e}}_\alpha &= \left(\frac{\partial\phi}{\partial x} - \cot\theta\,\frac{\partial\phi}{\partial y}\right)\hat{\mathbf{x}} + \frac{1}{\sin\theta}\frac{\partial\phi}{\partial y}\hat{\mathbf{y}} \\
&= \frac{\partial\phi}{\partial x}\hat{\mathbf{x}} + \frac{\partial\phi}{\partial y}\hat{\mathbf{y}}
\end{aligned} \tag{3.56}$$

---

[3] There are deep distinctions between covariant and contravariant tensors, and we have not addressed them in full abstraction here – for our purposes, there is a one-to-one map between the two provided by the metric, and that makes them effectively equivalent.

and we see that this is the correct expression for a contravariant vector: $\bar{\nabla}\bar{\phi} = \nabla\phi$ (compare with (3.16)).

It can be important, in calculations, to keep track of the intrinsic tensor character of objects. Canonical momentum $p_\alpha = \frac{\partial L}{\partial \dot{x}^\alpha}$ is naturally covariant as *defined*, the coordinate differential is naturally contravariant, the gradient is covariant, the metric tensor is covariant. The weight you attach to the variance of an object depends on what you are given, of course. If you are provided $g^{\alpha\beta}$, then you must write $g_{\alpha\beta}$ in terms of the contravariant components before interpreting as a metric. We shall see examples in later chapters where the "natural" variance of an object must be respected.

3.2 Derivatives

We'll start our discussion of derivatives with their transformation properties. The "usual" derivative of a tensor (say) is not itself a tensor, i.e. $\frac{\partial f^\alpha}{\partial x^\beta}$ is not a mixed-rank tensor, it does not transform properly. This lack of tensorial character is easiest to show by considering derivatives in indexed form, and looking explicitly at the result of transformation, but it can also be seen from the spatial dependence of basis vectors.

So there are a couple of ways to see the issue – first think about what we mean by transformation for an object like $\frac{\partial f^\alpha}{\partial x^\beta}$. We are trying to relate $f^\alpha{}_{,\beta}$ to:

$$\bar{f}^\alpha{}_{,\beta} \equiv \frac{\partial \bar{f}^\alpha(\bar{x})}{\partial \bar{x}^\beta}. \tag{3.57}$$

If we transform back via $\bar{f}^\alpha = \frac{\partial \bar{x}^\alpha}{\partial x^\gamma} f^\gamma$ and $\bar{\partial}_\beta = \frac{\partial x^\rho}{\partial \bar{x}^\beta} \partial_\rho$ (via the chain rule, with $\partial_\mu \equiv \frac{\partial}{\partial x^\mu}$), then:

$$\bar{f}^\alpha{}_{,\beta} = \frac{\partial x^\rho}{\partial \bar{x}^\beta} \frac{\partial}{\partial x^\rho} \left(\frac{\partial \bar{x}^\alpha}{\partial x^\gamma} f^\gamma \right) = \frac{\partial x^\rho}{\partial \bar{x}^\beta} \frac{\partial \bar{x}^\alpha}{\partial x^\gamma} f^\gamma{}_{,\rho} + \frac{\partial x^\rho}{\partial \bar{x}^\beta} \frac{\partial^2 \bar{x}^\alpha}{\partial x^\rho \partial x^\gamma} f^\gamma. \tag{3.58}$$

The first term above is the appropriate one for a mixed-rank tensor transformation (as in (1.96)). But the second term spoils the tensorial property of the derivative for first-rank tensors. So we conclude, $f^\alpha{}_{,\beta}$, while it has one up index, one down index, is *not* a second-rank tensor. Our plan is to generate a derivative operator that allows us to form second-rank tensors by "taking the derivative" of first-rank tensors.

Before we do that, we can view the problematic term using the basis vectors – think of the form $\mathbf{f} = f^\alpha \mathbf{e}_\alpha$, the "gradient" of this is:

$$\frac{\partial \mathbf{f}}{\partial x^\alpha} = \frac{\partial f^\beta}{\partial x^\alpha} \mathbf{e}_\beta + f^\beta \frac{\partial \mathbf{e}_\beta}{\partial x^\alpha}, \tag{3.59}$$

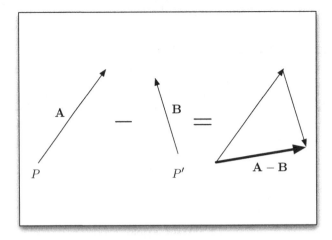

Figure 3.4   Subtracting vectors defined at different points P and P' in Cartesian coordinates (Euclidean space) – we just pick up **B** and move it.

which is the usual sort of statement. We must take the derivatives of the function *and* the basis if it depends on coordinates. Once again, the first term is basically correct, it is the derivative w.r.t. the basis vectors that is spoiling the transformation, and these terms correspond to the "bad" term in (3.58).

Finally, for a more geometrical picture, remember that derivatives are defined as limits of differences between points. In order to define this limit, we must be able to unambiguously subtract a vector at two nearby points. That's fine if we are in Cartesian coordinates – vectors (lines with little arrows) can be moved around, there is no reference to the origin, so if you want to subtract vectors defined at two different points, you just pick one up and move it over as shown in Figure 3.4.

Not so for coordinate-dependent basis vectors. If we were working in plane polar coordinates, for example, and we want to subtract $\hat{s}$ at two different points, we need to know what those points are in order to correctly evaluate the difference. In Figure 3.5, we have two different versions of $\hat{s} - \hat{s}$ – the difference in locations is important.

So if we were to use vectors that are written in terms of $\hat{s}$ and $\hat{\phi}$, we need to be a little careful when subtracting (and hence, defining the derivative).

3.2.1 What is a tensorial derivative?

The usual definition of a derivative is in terms of a limit, like:

$$f'(x) = \lim_{\delta x \to 0} \frac{f(x + \delta x) - f(x)}{\delta x}. \tag{3.60}$$

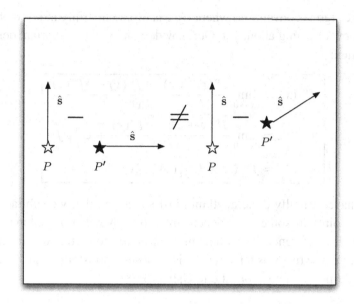

Figure 3.5 $\hat{s}(P) - \hat{s}(P')$ evaluates differently depending on the locations.

Suppose we expand a tensor $f^\alpha(x + \delta x)$ about the point P (with coordinate x), this is just the Taylor series:

$$f^\alpha(x + \delta x) = f^\alpha(x) + \delta x^\gamma f^\alpha_{,\gamma} + O(\delta x^2) \tag{3.61}$$

and we would say that the (non-tensorial) change of the vector is given by:

$$\Delta f^\alpha \equiv f^\alpha(x + \delta x) - f^\alpha(x) = \delta x^\gamma f^\alpha_{,\gamma} + O(\delta x^2). \tag{3.62}$$

In order to define a new (tensorial) derivative, we cannot use Δf^α as a starting point, so we'll introduce an additional factor. We compare $f^\alpha(x + \delta x)$ with $f^\alpha(x) + \delta f^\alpha(x)$, and determine the properties of $\delta f^\alpha(x)$ that lead to a (new) tensorial difference. For now, we will *define* the "covariant" derivative in terms of the unspecified factor δf^α:

$$f^\alpha_{;\gamma}(x) \equiv \lim_{\delta x \to 0} \frac{f^\alpha(x + \delta x) - (f^\alpha(x) + \delta f^\alpha(x))}{\delta x^\gamma}. \tag{3.63}$$

What form should this "extra bit" $\delta f^\alpha(x)$ take? Well, in the limiting case that $f^\alpha(x) = 0$, it should be zero (you *can* add the zero vector at different locations), and it should also tend to zero with $\delta x \longrightarrow 0$, i.e. if we don't move from the point P, the problem of comparing vectors at different points is side-stepped. This suggests an object of the form $\sim \delta x^\rho f^\tau$, and we still need a contravariant index to form δf^α – so we set:

$$\delta f^\alpha \equiv -C^\alpha_{\ \rho\tau} \delta x^\rho f^\tau \tag{3.64}$$

(the negative sign is convention) where $C^\alpha_{\;\rho\tau}$ is just some triply indexed object that we don't know anything about yet. Our new derivative, the "covariant derivative", will be written:

$$f^\alpha_{\;;\gamma}(x) \equiv \lim_{\delta x \to 0} \frac{f^\alpha(x + \delta x) - (f^\alpha(x) + \delta f^\alpha(x))}{\delta x^\gamma}$$

$$= \lim_{\delta x \to 0} \frac{f^\alpha(x + \delta x) - f^\alpha(x)}{\delta x^\gamma} + C^\alpha_{\;\gamma\tau} f^\tau \qquad (3.65)$$

$$= f^\alpha_{\;,\gamma}(x) + C^\alpha_{\;\gamma\tau}(x) f^\tau(x).$$

But we haven't really done anything, just suggested that we subtract a vector at different points in some $C^\alpha_{\;\beta\gamma}$-dependent way rather than by taking a straight componentwise difference. The whole motivation for doing this was to ensure that our notion of derivative was tensorial. This imposes a constraint on $C^\alpha_{\;\beta\gamma}$ (which is arbitrary right now) – let's work it out. Using (3.58):

$$\bar{f}^\alpha_{\;;\gamma} = \frac{\partial \bar{x}^\alpha}{\partial x^\beta} \frac{\partial x^\sigma}{\partial \bar{x}^\gamma} f^\beta_{\;,\sigma} + \frac{\partial^2 \bar{x}^\alpha}{\partial x^\beta \partial x^\rho} \frac{\partial x^\rho}{\partial \bar{x}^\gamma} f^\beta + \bar{C}^\alpha_{\;\gamma\beta} \frac{\partial \bar{x}^\beta}{\partial x^\sigma} f^\sigma$$

$$= \frac{\partial \bar{x}^\alpha}{\partial x^\beta} \frac{\partial x^\sigma}{\partial \bar{x}^\gamma} \left(f^\beta_{\;;\sigma} - f^\rho C^\beta_{\;\sigma\rho} \right) + \frac{\partial^2 \bar{x}^\alpha}{\partial x^\beta \partial x^\rho} \frac{\partial x^\rho}{\partial \bar{x}^\gamma} f^\beta + \bar{C}^\alpha_{\;\beta\gamma} \frac{\partial \bar{x}^\beta}{\partial x^\sigma} f^\sigma, \qquad (3.66)$$

the term in parentheses comes from noting that $f^\alpha_{\;,\gamma} = f^\alpha_{\;;\gamma} - C^\alpha_{\;\beta\gamma} f^\beta$. Collecting everything that is "wrong" in the above:

$$\bar{f}^\alpha_{\;;\gamma} = \frac{\partial \bar{x}^\alpha}{\partial x^\beta} \frac{\partial x^\sigma}{\partial \bar{x}^\gamma} f^\beta_{\;;\sigma} + \left(-\frac{\partial \bar{x}^\alpha}{\partial x^\beta} \frac{\partial x^\tau}{\partial \bar{x}^\gamma} C^\beta_{\;\tau\rho} + \frac{\partial^2 \bar{x}^\alpha}{\partial x^\rho \partial x^\tau} \frac{\partial x^\tau}{\partial \bar{x}^\gamma} + \bar{C}^\alpha_{\;\gamma\beta} \frac{\partial \bar{x}^\beta}{\partial x^\rho} \right) f^\rho, \qquad (3.67)$$

we can define the transformation rule for $C^\alpha_{\;\gamma\beta}$ so as to kill off the term in parentheses in (3.67):

$$\bar{C}^\alpha_{\;\tau\gamma} = \frac{\partial x^\lambda}{\partial \bar{x}^\tau} \frac{\partial \bar{x}^\alpha}{\partial x^\rho} \frac{\partial x^\sigma}{\partial \bar{x}^\gamma} C^\rho_{\;\sigma\lambda} - \frac{\partial^2 \bar{x}^\alpha}{\partial x^\lambda \partial x^\rho} \frac{\partial x^\rho}{\partial \bar{x}^\gamma} \frac{\partial x^\lambda}{\partial \bar{x}^\tau}, \qquad (3.68)$$

and from this we conclude that $C^\alpha_{\;\beta\gamma}$ is itself not a tensor. Anything that transforms according to (3.68) is called a "connection". What we've done in (3.65) is add two things, neither of which is a tensor, in a sum that produces a tensor. Effectively, the non-tensor parts of the two terms cancel each other (by construction, of course, that's what gave us (3.68)), we just needed a little extra freedom in our definition of derivative. For better or worse, we have it. This $f^\alpha_{\;;\beta}$ is called the covariant derivative, and plays the role, in tensor equations, of the usual $f^\alpha_{\;,\beta}$.

If we have multiple contravariant indices, we must introduce a factor of $C^{\alpha}_{\beta\gamma}$ for each when we take the derivative. For a second-rank $f^{\alpha\beta}$, the normal partial derivative will introduce two non-tensorial terms, and we need a compensating connection for each of them:

$$f^{\alpha\beta}_{;\gamma} = f^{\alpha\beta}_{,\gamma} + C^{\alpha}_{\gamma\sigma} f^{\sigma\beta} + C^{\beta}_{\gamma\sigma} f^{\alpha\sigma}. \tag{3.69}$$

The covariant derivative applied to covariant tensors takes a slightly different form, and this is to ensure that scalars work out correctly – I'll let you explore that in Problem 3.10. For a covariant first-rank tensor, the result is:

$$f_{\alpha;\beta} = f_{\alpha,\beta} - C^{\sigma}_{\alpha\beta} f_{\sigma}. \tag{3.70}$$

There is, for us, a special relation between the metric and the connection. This need not be the case – so far, we have seen that the metric is useful in, for example, forming scalars, and determining lengths. We just defined the connection as the object that restores tensorial character to the derivative. There is no reason why we cannot simply give a metric, and separately, provide a connection. However, for the spaces of interest in general relativity, the metric is all that is needed, and for these spaces, we define a "metric connection":

$$C^{\sigma}_{\alpha\beta} = \frac{1}{2} g^{\sigma\rho} \left(g_{\rho\alpha,\beta} + g_{\rho\beta,\alpha} - g_{\alpha\beta,\rho} \right). \tag{3.71}$$

This "Christoffel connection" is generally denoted $\Gamma^{\sigma}_{\alpha\beta}$.[4] The derivatives of the metric appearing in the definition should be a warning sign – or reminder, at this point, that $\Gamma^{\sigma}_{\alpha\beta}$ is not a tensor.

Problem 3.1

Generate the spherical basis vectors in terms of the Cartesian set, $e_1 = \hat{x}$, $e_2 = \hat{y}$, $e_3 = \hat{z}$, using the transformation rule $\bar{e}_{\alpha} = \frac{\partial x^{\gamma}}{\partial \bar{x}^{\alpha}} e_{\gamma}$, with:

$$x = r \sin\theta \cos\phi \qquad y = r \sin\theta \sin\phi \qquad z = r \cos\theta. \tag{3.72}$$

The basis vectors you end up with are "coordinate basis vectors", meaning that they point in the direction of increasing coordinate value, but they are not normalized. To connect this basis set to the familiar normalized set $\hat{r}$, $\hat{\theta}$ and $\hat{\phi}$, show that:

$$\hat{r} = e_r \qquad \hat{\theta} = \frac{1}{r} e_{\theta} \qquad \hat{\phi} = \frac{1}{r \sin\theta} e_{\phi} \tag{3.73}$$

where, in terms of your transformation, $e_r = \bar{e}_1$, $e_{\theta} = \bar{e}_2$ and $e_{\phi} = \bar{e}_3$.

---

[4] You should recognize the combination of metric derivatives in (3.71) from Section 1.3.

Problem 3.2

For the vector field:

$$\mathbf{E} = \frac{q}{4\pi\,\epsilon_0\,(x^2 + y^2 + z^2)^{3/2}}\,(x\,\hat{\mathbf{x}} + y\,\hat{\mathbf{y}} + z\,\hat{\mathbf{z}}) \tag{3.74}$$

in Cartesian coordinates, we would write $\mathbf{E} = E^\alpha\,\mathbf{e}_\alpha$. Using the transformation rule for E^α and $\mathbf{e}_\alpha$, transform to spherical coordinates, and write $\bar{E}^\alpha$, $\bar{\mathbf{e}}_\alpha$ separately – do you get what you expect when you form $\bar{\mathbf{E}} = \bar{E}^\alpha\,\bar{\mathbf{e}}_\alpha$?

Problem 3.3

Consider two connection fields, $C^\alpha{}_{\beta\gamma}$ and $\bar{C}^\alpha{}_{\beta\gamma}$ (not associated with a metric). Find a linear combination of these two fields that is a tensor, or prove that no such combination exists.

Problem 3.4

Properties of connections.

(a) Calculate the Christoffel connection:

$$\Gamma_{\alpha\beta\gamma} = \frac{1}{2}\left(g_{\alpha\beta,\gamma} + g_{\alpha\gamma,\beta} - g_{\beta\gamma,\alpha}\right) \tag{3.75}$$

for three-dimensional Euclidean space written in spherical coordinates.

(b) Prove that a tensor that is zero in one coordinate system is zero in all coordinate systems.

(c) Using the connection values in spherical coordinates as an example, prove that the connection is *not* a tensor.

Problem 3.5

(a) It is possible to define geometries in which structure beyond the metric is needed. Consider the second (covariant) derivative of a scalar: $\phi_{;\mu\nu}$ – calculate the difference:

$$T_{\mu\nu} = \phi_{;\mu\nu} - \phi_{;\nu\mu}, \tag{3.76}$$

without using *any* known properties of the connection (use (3.70), and note that $\phi_{;\mu\nu} \equiv \left(\phi_{;\mu}\right)_{;\nu}$). This difference is called the "torsion" of the geometry. What constraint must you place on the connection if cross-covariant-derivative equality is to hold (i.e. if the torsion vanishes)?

(b) Show that if we require that $g_{\alpha\beta;\gamma} = 0$ and further that our geometry be torsion-free, then the Christoffel connection is related to the metric via:

$$\Gamma^\rho{}_{\beta\gamma} = \frac{1}{2}g^{\alpha\rho}\left(g_{\alpha\beta,\gamma} + g_{\alpha\gamma,\beta} - g_{\beta\gamma,\alpha}\right). \tag{3.77}$$

Problem 3.6

In this problem, we will establish that the Christoffel connection transforms as a connection.

(a) Show that:

$$\frac{\partial^2 x^\rho}{\partial \bar{x}^\nu \partial \bar{x}^\mu} \frac{\partial \bar{x}^\tau}{\partial x^\rho} = -\frac{\partial^2 \bar{x}^\tau}{\partial x^\gamma \partial x^\rho} \frac{\partial x^\rho}{\partial \bar{x}^\mu} \frac{\partial x^\gamma}{\partial \bar{x}^\nu}. \tag{3.78}$$

Hint: use $\bar{\delta}^\tau_\mu = \frac{\partial \bar{x}^\tau}{\partial \bar{x}^\mu} = \frac{\partial \bar{x}^\tau}{\partial x^\rho} \frac{\partial x^\rho}{\partial \bar{x}^\mu}$ and take the $\frac{\partial}{\partial \bar{x}^\nu}$ derivative of both sides – keeping in mind that $\frac{\partial}{\partial \bar{x}^\nu} = \frac{\partial x^\gamma}{\partial \bar{x}^\nu} \frac{\partial}{\partial x^\gamma}$ via the chain rule.

(b) The metric connection (Christoffel) reads:

$$\Gamma^\alpha_{\ \mu\nu} = \frac{1}{2} g^{\alpha\beta} \left(g_{\beta\mu,\nu} + g_{\beta\nu,\mu} - g_{\mu\nu,\beta} \right). \tag{3.79}$$

Show that it transforms as a connection (i.e. via (3.68)). Refer to the transformation of Section 3.2 and, in particular, (3.58) for the starting point in developing the expression for $\bar{\Gamma}^\alpha_{\ \mu\nu}$ (this should help you evaluate $\bar{g}_{\mu\nu,\sigma}$). Also note that you will need to use the result from part a to make your expression match (3.68).

3.3 Derivative along a curve

We have our fancy new derivative, but what to do with it? In particular, how can we interpret the derivative of a vector field along a curve in terms of this new construction? We will see, using the simplest possible curved space, that the connection provides the connection (so to speak). The new element in the covariant derivative is the notion that the basis vectors themselves depend on position. This is not so surprising, we have seen it in the past (spherical or cylindrical coordinates with their position-dependent basis vectors) – what we will make explicit now is the role of $C^\alpha_{\ \beta\gamma}$ in keeping track of this change. The derivative of a vector field along a curve is, then, made up of two parts: changes to the vector field itself, and changes in the basis vectors.

3.3.1 Parallel transport

Consider a vector $f^\alpha(x)$ defined on the surface of a sphere of radius r – the coordinates there are defined to be $x^1 = \theta$, $x^2 = \phi$. Suppose that we have in mind a definite curve, C, on the sphere, parametrized by τ as shown in Figure 3.6.

From our three-dimensional point of view, the tangent to the curve is given by $\dot{x}^\alpha(\tau)$. If we have a vector field $f^\alpha(x)$ defined on the sphere, then we can define $f^\alpha(\tau)$ using the curve definition. For concreteness, let:

$$\mathbf{f} = \tilde{f}^\theta(\theta, \phi)\,\hat{\mathbf{e}}_\theta + \tilde{f}^\phi(\theta, \phi)\,\hat{\mathbf{e}}_\phi \tag{3.80}$$

in the natural orthonormal basis defined on the surface of the sphere. But the contravariant components of tensors appear as coefficients in the coordinate basis, which is not normalized. The issue here (component labeling) is only a point of

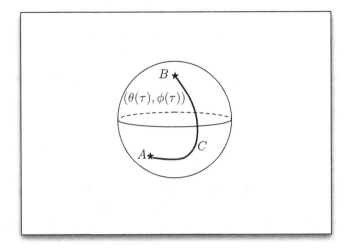

Figure 3.6 Going from point A to point B along a curve parametrized by τ defined on the surface of a sphere.

notation, but the idea of switching to the coordinate basis (and/or from it to the orthonormal basis) comes up, depending on context. To change basis, we'll use our relation (3.73):

$$\hat{\mathbf{e}}_r = \mathbf{e}_r \quad \hat{\mathbf{e}}_\theta = \frac{1}{r}\,\mathbf{e}_\theta \quad \hat{\mathbf{e}}_\phi = \frac{1}{r\,\sin\theta}\,\mathbf{e}_\phi, \tag{3.81}$$

at which point the vector in (3.80) can be written:

$$\mathbf{f} = \frac{1}{r}\,\tilde{f}^\theta(\theta, \phi)\,\mathbf{e}_\theta + \tilde{f}^\phi(\theta, \phi)\,\frac{\mathbf{e}_\phi}{r\,\sin\theta}$$
$$\equiv f^\theta\,\mathbf{e}_\theta + f^\phi\,\mathbf{e}_\phi, \tag{3.82}$$

with $f^\theta = \frac{1}{r}\,\tilde{f}^\theta$ and $f^\phi = \tilde{f}^\phi\,r^{-1}\,\sin^{-1}\theta$. The components f^θ and f^ϕ are what we have in mind when we write f^α.

And now we ask: if we go along the curve C a distance $d\tau$, how does the vector $\mathbf{f}$ change?

$$d\mathbf{f}|_C = \left[\left(f^\theta_{,\alpha}\,\mathbf{e}_\theta + f^\theta\,\frac{\partial\mathbf{e}_\theta}{\partial x^\alpha}\right) + \left(f^\phi_{,\alpha}\,\mathbf{e}_\phi + f^\phi\,\frac{\partial\mathbf{e}_\phi}{\partial x^\alpha}\right)\right]\dot{x}^\alpha\,d\tau. \tag{3.83}$$

To evaluate the above, we need the derivatives of the basis vectors – they are:

$$\frac{\partial\mathbf{e}_\theta}{\partial\theta} = -r\,\mathbf{e}_r \qquad \frac{\partial\mathbf{e}_\theta}{\partial\phi} = \cot\theta\,\mathbf{e}_\phi$$
$$\frac{\partial\mathbf{e}_\phi}{\partial\theta} = \cot\theta\,\mathbf{e}_\phi \qquad \frac{\partial\mathbf{e}_\phi}{\partial\phi} = -\sin\theta\,(\cos\theta\,\mathbf{e}_\theta + \sin\theta\,\mathbf{e}_r). \tag{3.84}$$

If we are on the surface of the sphere, there is no such thing as $\mathbf{e}_r$ – while it exists in three dimensions, there is no normal to the sphere in two. So restricting our attention, we have:

$$d\mathbf{f}|_C = \left(f^\theta_{,\alpha}\,\dot{x}^\alpha - f^\phi\,\sin\theta\,\cos\phi\,\dot{\phi}\right)\mathbf{e}_\theta\,d\tau$$
$$+\left(f^\phi_{,\alpha}\,\dot{x}^\alpha + \cot\theta\,(f^\theta\,\dot{\phi} + f^\phi\,\dot{\theta})\right)\mathbf{e}_\phi\,d\tau. \tag{3.85}$$

This can be written:

$$d\mathbf{f}|_C \doteq \begin{pmatrix} f^\theta_{,\alpha}\,\dot{x}^\alpha \\ f^\phi_{,\alpha}\,\dot{x}^\alpha \end{pmatrix}d\tau + \begin{pmatrix} -\sin\theta\,\cos\theta\,f^\phi\,\dot{\phi} \\ \cot\theta\,(f^\theta\,\dot{\phi} + f^\phi\,\dot{\theta}) \end{pmatrix}d\tau, \tag{3.86}$$

and the second term has the form $C^\alpha_{\;\beta\gamma}\,\dot{x}^\beta\,f^\gamma$ if we define:

$$C^\theta_{\;\phi\phi} = -\cos\theta\,\sin\theta \quad C^\phi_{\;\theta\phi} = C^\phi_{\;\phi\theta} = \cot\theta \quad \text{(all other components are zero).} \tag{3.87}$$

Then we can express (3.86) in terms of components, as:

$$\boxed{df^\alpha|_C = \left(f^\alpha_{\;,\gamma}\,\dot{x}^\gamma + C^\alpha_{\;\beta\gamma}\,\dot{x}^\beta\,f^\gamma\right)d\tau = \dot{x}^\gamma\,f^\alpha_{\;;\gamma}\,d\tau,} \tag{3.88}$$

so we see that we might well call $\frac{df^\alpha}{d\tau}|_C$ "the derivative along the curve parametrized by τ". This object is denoted:

$$\boxed{\left.\frac{df^\alpha}{d\tau}\right|_C = \dot{x}^\gamma\,f^\alpha_{\;;\gamma} \equiv \frac{Df^\alpha}{D\tau}.} \tag{3.89}$$

Let's take a breather – what have we done? We have shown that tensors have a notion of distance along a curve – the complication that leads to the appearance of $C^\alpha_{\;\beta\gamma}$ can either be viewed as the changing of basis vectors, or the lack of subtraction except at a point. These two views can both be used to describe the covariant derivative. In one case, we are explicitly inserting the basis vectors and making sure to take the derivative w.r.t. both the elements of f^α and the basis vectors. In the other, we are using the "fudge factor" to pick up the slack in the non-tensorial (ordinary) derivative.

Either way you like, we have a new notion of the change of a vector along a curve. Incidentally, in Cartesian coordinates, the expression equivalent to (3.89) would look like $f^\alpha_{\;,\gamma}\,\dot{x}^\gamma$, and this highlights one common procedure we will encounter in general relativity, that of "minimal replacement" (or "minimal substitution").[5] General relativity is a coordinate-independent theory, the best way to make true statements, then, is to make tensor statements. There is a lack of uniqueness that plagues the "non-tensor statements go to tensor statements" algorithm, and minimal

---

[5] We can motivate the program of minimal replacement formally, and will get to that in Chapter 6.

replacement is (sometimes) a way around this. It certainly has worked out here – we take the non-tensor $f^\alpha_{,\gamma}$ and replace it with the tensor $f^\alpha_{;\gamma}$ to get the tensor form of "$\dot{f}^\alpha$".

But wait a minute – I haven't *proved* anything, just suggested that we box up the terms that aren't familiar from (3.86) and defined a connection that has the correct form. I am entirely within my rights to do this, but I must show you that the construction of the connection has the appropriate transformation properties. You will see soon enough that the above definition (3.87) is in fact a connection, and a natural one given the surface of a sphere – it is the Christoffel connection.

Now, as with almost all studies of new derivative operators, we ask the very important question – what is a constant? In Cartesian coordinates, a vector field is constant if its components are ... constant. In the general case, we must demand constancy w.r.t. the covariant derivative, i.e. we take a vector $f^\alpha(x)$ and require:

$$\frac{Df^\alpha}{D\tau} = \dot{x}^\gamma \, f^\alpha_{;\gamma} = 0. \tag{3.90}$$

Depending on our space, we can have a "constant" vector f^α that does not have constant entries – after all:

$$\dot{x}^\gamma \, f^\alpha_{;\gamma} = \dot{x}^\gamma \, f^\alpha_{,\gamma} + C^\alpha_{\beta\gamma} \, \dot{x}^\gamma \, f^\beta = 0 \longrightarrow \frac{df^\alpha}{d\tau} = -C^\alpha_{\beta\gamma} \, \dot{x}^\gamma \, f^\beta. \tag{3.91}$$

A vector satisfying this constancy condition is said to be "parallel transported around the curve defined by $x^\alpha(\tau)$ with tangent vector $\dot{x}^\alpha(\tau)$". Remember, there is a curve lurking in the background, otherwise, we have no τ.

The flip side of this discussion is that we can *make* a vector constant (in our expanded sense, with $\frac{Df^\alpha}{D\tau} = 0$). In (3.90), we have a first-order ODE for $f^\alpha(x)$ – we could solve it given the values of $f^\alpha(P)$ at a point $P = x(0)$. This allows us to explicitly *construct* constant vectors.

Example

Working in a flat, Euclidean space with Cartesian coordinates, we can define a curve by specifying $x^\alpha(\tau)$. Then if we have a vector defined at a point P with $\tau = 0$, we can ask how, under "parallel transport", the vector changes if we go to another point along the curve, P' at $\tau = \tau'$. The answer, of course, is that the vector doesn't change at all:

$$\frac{df^\alpha}{d\tau} = f^\alpha_{,\gamma} \, \dot{x}^\gamma = 0 \longrightarrow f^\alpha(\tau) = f^\alpha(\tau = 0), \tag{3.92}$$

and the curve itself doesn't matter. We can literally (figuratively, I mean) pick up the vector and move it, placing the tail of it on the curve. In this sense, the vector is transported along the curve, always parallel to itself.

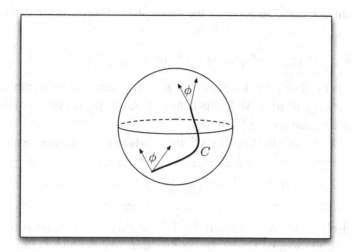

Figure 3.7 "Angle conservation" under parallel transport of two vectors.

In "flat space" (of which Euclidean space is an example), parallel transport is trivial. Still, it is good to remember this when we move to curved spaces (and space-times). We sometimes write, just to make the distinction clear:

$$\frac{Df^\alpha}{D\tau} = f^\alpha_{;\gamma}\dot{x}^\gamma = \frac{df^\alpha}{d\tau} + C^\alpha_{\beta\gamma} f^\beta \dot{x}^\gamma = 0. \tag{3.93}$$

What properties do we expect for parallel transport? From its name, and the idea that the vectors f^α with $\frac{Df^\alpha}{D\tau} = 0$ are somehow "constant", we impose the condition that two vectors that are being parallel transported around the same curve C remain at the same angle w.r.t. each other.

Same angle w.r.t. each other? We haven't even defined the angle between vectors, but it is what you expect – the length of a vector is given by $f^2 = f^\alpha f_\alpha = f^\alpha g_{\alpha\beta} f^\beta$, and the angle between two vectors, ϕ, is similarly defined to be:

$$\cos\phi = \frac{\sqrt{p^\gamma q_\gamma}}{\sqrt{p^\alpha p_\alpha}\sqrt{q^\beta q_\beta}}. \tag{3.94}$$

Refer to Figure 3.7.

So take two vectors parallel transported along a curve C, we have $\frac{Dp^\alpha}{D\tau} = \frac{Dq^\alpha}{D\tau} = 0$, then the requirement that the angle between them remain constant along the curve is summed up in the following:

$$\dot{x}^\gamma \left(g_{\alpha\beta} p^\alpha q^\beta\right)_{;\gamma} = 0. \tag{3.95}$$

The covariant derivative satisfies all the usual product rules, so we can expand the left-hand side:

$$\dot{x}^\gamma \left(g_{\alpha\beta;\gamma} \, p^\alpha \, q^\beta + g_{\alpha\beta} \, p^\alpha_{\;;\gamma} \, q^\beta + g_{\alpha\beta} \, p^\alpha \, q^\beta_{\;;\gamma} \right) = 0. \tag{3.96}$$

The two terms involving the derivatives of p^α and q^α are zero by assumption, so we must have $\dot{x}^\gamma \, g_{\alpha\beta;\gamma} \, p^\alpha \, q^\beta = 0$. If this is to be true for any curve $x^\alpha(\tau)$ and vectors p^α, q^α, then we must have $g_{\alpha\beta;\gamma} = 0$.

Well, as you showed in Problem 3.5, this leads to an interesting requirement:

$$g_{\alpha\beta;\gamma} = 0 \longrightarrow \boxed{C^\mu_{\alpha\beta} = \frac{1}{2} g^{\mu\gamma} \left(g_{\gamma\beta,\alpha} + g_{\gamma\alpha,\beta} - g_{\alpha\beta,\gamma} \right) \equiv \Gamma^\mu_{\alpha\beta}.} \tag{3.97}$$

That's precisely what we defined to be $\Gamma^\mu_{\alpha\beta}$ during our discussion of Keplerian orbits ((1.58)), the special Christoffel connection. We defined a connection to be an object that transforms in a particular way, but aside from an ad hoc construction for the sphere, we gave no examples of connections. Now we see that all we need is a metric, and then a particular combination of derivatives is itself a connection (you showed that the Christoffel connection is in fact a connection in Problem 3.6). The Christoffel connection, when computed using the metric written in spherical coordinates, has nonzero entries that are precisely (3.87).

Thinking ahead, we know that general relativity is a theory that will provide us with a metric – we see that, as far as the connection and covariant derivative are concerned, that will be enough.

3.3.2 Geodesics

The whole story becomes eerily similar to Section 1.3.2. There is a special class of curves, ones whose tangent vector is parallel transported along the curve – that is:

$$\dot{x}^\alpha_{\;;\gamma} \, \dot{x}^\gamma = 0. \tag{3.98}$$

Such curves are called geodesics. We will have more to say about this special class of curve later, but they do have the property of extremizing distance between points, that is, they are in a generalized sense, "straight lines".

Let me write out the requirement explicitly:

$$\dot{x}^\alpha_{\;;\gamma} \, \dot{x}^\gamma = \frac{d\dot{x}^\alpha}{d\tau} + \Gamma^\alpha_{\;\beta\gamma} \, \dot{x}^\beta \, \dot{x}^\gamma = 0, \tag{3.99}$$

which we sometimes write as:

$$\ddot{x}^\nu \, g_{\alpha\nu} + \Gamma_{\alpha\beta\gamma} \, \dot{x}^\beta \, \dot{x}^\gamma = 0, \tag{3.100}$$

and you should compare this equation with (1.59).

Euclidean geodesics

In three-dimensional space with Cartesian coordinates, the geodesic equation reduces to $\ddot{x}^{\mu} = 0$, or the familiar lines of force-free motion: $x^{\alpha}(t) = A^{\alpha} t + B^{\alpha}$ for constants A^{α} and B^{α}. The connection coefficients for a constant metric are zero, since there is no coordinate-dependence in the metric, and hence all coordinate derivatives vanish. We can use a nontrivial coordinate system (i.e. one with a nonvanishing connection) – take cylindrical coordinates $x^{\mu} \doteq (s, \phi, z)$ with metric and connection:

$$g_{\mu\nu} \doteq \begin{pmatrix} 1 & 0 & 0 \\ 0 & s^2 & 0 \\ 0 & 0 & 1 \end{pmatrix} \qquad \Gamma^{s}{}_{\phi\phi} = -s \qquad \Gamma^{\phi}{}_{s\phi} = \Gamma^{\phi}{}_{\phi s} = \frac{1}{s}, \tag{3.101}$$

with all other elements of $\Gamma^{\alpha}{}_{\beta\gamma} = 0$. Then the equations for geodesic motion coming from (3.99) are:

$$\ddot{s} = s\,\dot{\phi}^2$$

$$\ddot{\phi} = -\frac{2\,\dot{\phi}\,\dot{s}}{s} \tag{3.102}$$

$$\ddot{z} = 0.$$

The line solutions to this are less obvious than in the Cartesian case, $z(t) = A\,t + B$ is clear, and decoupled from the planar portion. As a quick check, $\phi = \phi_0$ (a constant angle) leads to $\ddot{s} = 0$, so straight-line "radial" motion. To find the most general solution, we can take the Cartesian case $x(t) = F\,t + B$, $y(t) = P\,t + Q$ and construct $s = \sqrt{x^2 + y^2}$ and $\phi = \tan^{-1}(y/x)$, leading to:

$$s(t) = \sqrt{(F\,t + G)^2 + (P\,t + Q)^2} \qquad \phi(t) = \tan^{-1}\left(\frac{P\,t + Q}{F\,t + G}\right), \tag{3.103}$$

which, combined with $z(t) = A\,t + B$, does indeed solve (3.102).

Geodesics on a sphere

If we work on a two-dimensional sphere, the metric is given by:

$$g_{\mu\nu} \doteq \begin{pmatrix} r^2 & 0 \\ 0 & r^2 \sin^2\theta \end{pmatrix} \tag{3.104}$$

for coordinates θ and ϕ. We already know the connection coefficients here from (3.87), and the equation of motion:

$$\ddot{x}^{\alpha} + \Gamma^{\alpha}{}_{\beta\gamma}\,\dot{x}^{\beta}\,\dot{x}^{\gamma} = 0 \tag{3.105}$$

becomes the pair:

$$\ddot{\theta} = \cos\theta\,\sin\theta\,\dot{\phi}^2$$

$$\ddot{\phi} = -2\cot\theta\,\dot{\phi}\,\dot{\theta}. \tag{3.106}$$

The general solution is difficult to see from the above – suppose we fix $\theta = \frac{1}{2}\pi$, then the first equation is satisfied, and the second becomes $\ddot{\phi} = 0$. Starting from $\phi = 0$ with zero initial velocity, the solution is $\phi = \tau$, the parameter of the curve. Then we see that the geodesics on a sphere are great circles.

Geodesics do not require three-dimensional flat space for their definition. Any metric, obtained by any means whatsoever, has a built-in set of these extremal curves. Remember that in Section 1.3.2, we constructed the same equation (cover up the potential on the right-hand side of (1.59)) by explicit generalized length extremization. Similarly, the relativistic action in Section 2.2.2 was developed by appealing to an extremal length for the Minkowski metric. The argument works for any dimension, any metric, so we see that these geodesic solutions are the natural length-extremizing curves in any context.

*Problem 3.7
Verify that the pair (3.103), with $z(t) = At + B$, satisfies (3.102). This problem is meant to keep your differentiation skills sharp.

Problem 3.8
From the geodesic equations of motion in cylindrical coordinates (3.102), we can find the geodesics associated with the (curved) surface of a cylinder of radius R.
(a) By solving the geodesic equations, find the "straight line", constrained to a cylinder (so $s = R$), connecting a point at $\phi = 0$, $z = 0$ to $\phi = 0$, $z = h$.
(b) Do the same thing for the geodesic connecting $\phi = 0$, $z = 0$ to $\phi = \frac{1}{2}\pi$, $z = 0$.

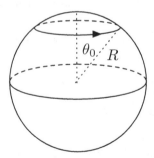

Figure 3.8 A circular path on a sphere of radius R – the path occurs at angle θ_0, and can be parametrized by ϕ.

Problem 3.9
(a) Consider the path, on the surface of a sphere ($\theta = \theta_0$, $\phi = 0 \ldots 2\pi$), i.e. we go around the sphere at angle θ_0 as shown in Figure 3.8. Construct a vector $f^\alpha(\theta, \phi)$ that is parallel transported around this path starting at $\phi = 0$ with value (α and β

are real here):

$$f^\alpha(\theta = \theta_0, \phi = 0) \doteq \begin{pmatrix} \alpha \\ \beta \end{pmatrix}. \tag{3.107}$$

What is the magnitude of f^α along the curve (assume the sphere has radius $r = 1$)?

(b) Take the two basis vectors at $(\theta = \theta_0, \phi = 0)$, i.e.

$$p^\alpha(\theta = \theta_0, \phi = 0) \doteq \begin{pmatrix} 1 \\ 0 \end{pmatrix} \quad q^\alpha(\theta = \theta_0, \phi = 0) \doteq \begin{pmatrix} 0 \\ 1 \end{pmatrix}. \tag{3.108}$$

Show explicitly that at all points along the path from part a, the parallel transported form of these initial vectors remains orthogonal. What is the angle that $p^\alpha(\phi = 0)$ makes with $p^\alpha(\phi = 2\pi)$?

Problem 3.10

We defined covariant differentiation in terms of contravariant vectors:

$$h^\alpha{}_{;\beta} = h^\alpha{}_{,\beta} + \Gamma^\alpha{}_{\beta\sigma}h^\sigma. \tag{3.109}$$

If we demand that covariant differentiation, like "normal" partial derivatives, satisfy the product rule:

$$\left(p^\alpha q^\beta\right)_{;\gamma} = p^\alpha{}_{;\gamma} q^\beta + p^\alpha q^\beta{}_{;\gamma}, \tag{3.110}$$

then show that the covariant derivative of a covariant tensor must be:

$$h_{\alpha;\beta} = h_{\alpha,\beta} - \Gamma^\sigma{}_{\alpha\beta}h_\sigma. \tag{3.111}$$

Problem 3.11

A Killing vector was defined in Section 1.7.3 by considering the infinitesimal transformation that takes "coordinates to coordinates": $\bar{x}^\alpha = x^\alpha + \epsilon \, f^\alpha(x)$ with infinitesimal generator $J = p_\alpha \, f^\alpha$, and then requiring that $[H, J] = 0$ using $H = \frac{1}{2} p_\mu \, g^{\mu\nu} \, p_\nu$ as the free-particle Hamiltonian. If we consider the "next step", we might allow linear mixing of momenta. Starting from:

$$\bar{x}^\alpha = x^\alpha + \epsilon \, p_\beta \, f^{\alpha\beta}(x), \tag{3.112}$$

with $f^{\alpha\beta}$ symmetric.

(a) Find the infinitesimal generator J that gives this transformation for $\bar{x}^\alpha$. What is the expression (in terms of $f^{\alpha\beta}$) for the transformed momenta $\bar{p}_\alpha$?

(b) Construct the Poisson bracket $[H, J]$ with your J and demand that it vanish to get the analogue of $f_{(\mu;\nu)} = 0$ (Killing's equation) for the tensor $f^{\mu\nu}(x)$. Be sure to write $[H, J] = 0$ in terms of manifestly tensorial objects, so that your final equation, like $f_{(\mu;\nu)} = 0$, is explicitly a tensor statement. The resulting equation is Killing's equation for second-rank tensors, and such tensors are called "Killing tensors".

(c) It should be obvious that $g^{\mu\nu}$ itself satisfies your tensorial Killing's equation. What is the conserved quantity J and the interpretation of the infinitesimal transformation if you set $f^{\mu\nu} = g^{\mu\nu}$?

*Problem 3.12

The Runge–Lenz vector represents a set of three additional conserved quantities for Newtonian gravity (although the set of these, together with energy and angular momentum, is not linearly independent). As a vector, it takes the form:

$$\mathbf{R} = \mathbf{p} \times \underbrace{(\mathbf{x} \times \mathbf{p})}_{\equiv \mathbf{L}} - \alpha\,\hat{\mathbf{x}} \tag{3.113}$$

where α is, for Newtonian gravity, $G\,M\,m$. This vector is made up of two components, and we are most interested in the first term, which is quadratic in the momenta. Working in Cartesian coordinates, identify the "matrix" of values $f^{\alpha\beta}$ sandwiched in between the momenta. Note that there are three such matrices here, since there are three components in $\mathbf{R}$ – write the first term as $p_\alpha\, f_i^{\alpha\beta}\, p_\beta$ for $i = 1, 2, 3$. Show that each of these is a Killing tensor: $f_{(\alpha\beta;\gamma)} = 0$ (for symmetric $f_{\alpha\beta}$).

Problem 3.13

Given a covariant second-rank tensor $h_{\mu\nu}$ (with matrix inverse $h^{\mu\nu}$), we can form its determinant by interpreting the tensor elements as matrix elements. One useful definition for $h \equiv \det(h_{\mu\nu})$ is:

$$h\,\delta_\alpha^\gamma = h_{\alpha\beta}\,C^{\beta\gamma}, \tag{3.114}$$

where $C^{\beta\gamma}$ is the cofactor matrix (and is independent of the element $h_{\alpha\beta}$).

(a) Using this definition, find the derivatives of the determinant w.r.t. the covariant components $h_{\mu\nu}$, i.e. what is $\frac{\partial h}{\partial h_{\mu\nu}}$. (Hint: check your result with a two-by-two matrix.)

(b) What about $\frac{\partial h}{\partial h^{\mu\nu}}$? Write this derivative in terms of $h = \det(h_{\mu\nu})$ and $h_{\mu\nu}$.

Problem 3.14

(a) The Kronecker delta is a tensor with numerical value:

$$\delta_\beta^\alpha = \begin{cases} 1 & \alpha = \beta \\ 0 & \alpha \neq \beta \end{cases}, \tag{3.115}$$

in Cartesian coordinates (say). Show that this numerical relation holds in *any* coordinate system.

(b) For a second-rank tensor $h_{\mu\nu}$ and the contravariant form defined by its matrix inverse, write the derivative $\frac{\partial h_{\mu\nu}}{\partial h_{\alpha\beta}}$ in terms of Kronecker deltas. Using this result, write $\frac{\partial h^{\mu\nu}}{\partial h_{\alpha\beta}}$ in terms of contravariant $h^{\rho\gamma}$?

4

Curved space

We have, so far, studied classical mechanics from a variety of different points of view. In particular, the notion of length extremization, and its mathematical formulation as the extremum of a particular action, were extensible to define the dynamics of special relativity. As we went along, we developed notation sufficient to discuss the generalized motion of particles in an arbitrary metric setting. Most of our work on dynamics has been done in an explicitly "flat" space, or in the case of special relativity, space-time. But, we have pushed the geometry as far as we can without any fundamentally new ideas. Now it is time to turn our attention to the characterization of geometry that will allow us to define "flat" versus "curved" space-times. In order to do this, we need less familiar machinery, and will develop that machinery as we push on toward the ultimate goal: to find equations governing the generation of curvature from a mass distribution that mimic, in some limit, the Newtonian force of gravity.

Remember that it is gravity that is the target here. We have seen that the classical gravitational force violates special relativity, and that a modified theory that shares some common traits with electricity and magnetism would fix this flaw. For reasons that should become apparent as we forge ahead, such a modified theory cannot capture all of the observed gravitational phenomena, and, as it turns out, a much larger shift in thinking is necessary. We will begin by defining structures that ensure special relativity holds locally, and look at how the larger geometric structure could be used as a replacement for Newtonian gravity, while predicting new, observable phenomena.

For all that follows, we specialize to a Riemannian manifold[1] – this is a special class of manifolds (points with labels) which have the nice, physical property that

---

[1] Technically, a Riemannian manifold must have a positive-definite metric at each point, and ours has Minkowski at each point. Such generalizations are referred to as pseudo-Riemannian manifolds, although I will typically drop this (important) feature in what follows.

at any point, we can introduce a coordinate system in which space-time looks like Minkowski space-time – an important feature given our earth-based laboratories (where space-time is described via the Minkowski metric). In such space-times, the metric defines a connection (for us, they will anyway), the connection defines the Riemannian curvature, and the Riemannian curvature is, by Einstein's equations, related to sources. So for Einstein's theory, the metric is enough.[2] After we have defined Riemannian curvature, and looked at its physical significance, we will follow Einstein and employ equivalence to constrain the allowed geometries by their mass distributions, arriving, in the end, at Einstein's equation.

4.1 Extremal lengths

Let's review the geodesic target one last time – we have done this in the context of both classical and (special) relativistic mechanics, but there, the metric was known. Here we manifestly employ the symbols without any reference to dimension or Pythagoras. Going back to our notion of length, one of the defining features of Riemannian geometry is its line element – we know that the metric defines everything in our subset of these geometries, so the "distance" between two points always has the same quadratic structure:

$$ds^2 = dx^\mu \, g_{\mu\nu} \, dx^\nu. \tag{4.1}$$

This tells us, again, how to measure lengths given a particular position (via the coordinate-dependence of $g_{\mu\nu}$) in space-time. How can we get an invariant measure of length along a curve? For a curve, parametrized by $x^\mu(\tau)$, we can write the "small displacement" dx^μ as $\dot{x}^\mu \, d\tau$. Then the line element (squared) along the curve is given by:

$$ds^2 = \dot{x}^\mu \, g_{\mu\nu} \, \dot{x}^\nu \, d\tau^2, \tag{4.2}$$

just as in classical mechanics (where $g_{\mu\nu}$ is the Euclidean metric) and special relativistic mechanics (with $g_{\mu\nu}$ the Minkowski metric).

As always, for a generic $g_{\mu\nu}$ the total "length" of the curve is obtained by integrating. We can take our action to be:

$$S = \int_a^b \sqrt{-\dot{x}^\mu \, g_{\mu\nu} \, \dot{x}^\nu} \, d\tau \tag{4.3}$$

---

[2] There are other geometric indicators available in general, but space-times with these more complicated structures (torsion is an example) do not produce theories of gravity that are "better" (predict more, more accurate, or easier to use) than GR itself.

(as with the Minkowski case, we assume that the metric has a time-like direction, hence the minus sign underneath the square root) or, equivalently, in arc length parametrization:[3]

$$\tilde{S} = \int_a^b \dot{x}^\mu g_{\mu\nu} \dot{x}^\nu \, d\tau. \tag{4.4}$$

The point is, suppose we find the equations of motion by varying the action w.r.t. x^μ, and setting the result to zero:

$$\delta\tilde{S} = \int d\tau \left(2\ddot{x}_\gamma + 2 g_{\gamma\alpha,\beta} \dot{x}^\alpha \dot{x}^\beta - g_{\alpha\beta,\gamma} \dot{x}^\alpha \dot{x}^\beta \right) \delta x^\gamma = 0. \tag{4.5}$$

This result, with familiar massaging as in Section 1.3.2 (think of the left-hand side of (1.59)) can be written to incorporate the connection explicitly:

$$\boxed{\ddot{x}_\gamma + \Gamma_{\gamma\alpha\beta} \dot{x}^\alpha \dot{x}^\beta = 0.} \tag{4.6}$$

We have what we would call the "equation of motion for a free particle", in Euclidean space, a line. In our generic space (or as generic as we can get given the restriction of our study to metric spaces), the interpretation is the same.

Here again, we can see how important it is that we differentiate between coordinate choices and geometry – the trajectory of a free particle cannot depend on the coordinate system, but *must* depend on the geometry in which it defines an extremal curve. Such curves are called geodesics.

4.2 Cross derivative (in)equality

The definition of the Riemann tensor arises naturally when we ask how the parallel transport of vectors depends on the path we take. Note that in a Euclidean space (or Minkowski space-time, for that matter), parallel transport of vectors is independent of path – we pick up a vector and move it parallel to itself. This is well-defined in these flat spaces, but for an arbitrary metric, parallel transport is defined by an ODE: $f^\alpha_{;\beta} \dot{x}^\beta = 0$ for a contravariant vector field f^α and a particular curve with tangent $\dot{x}^\beta$.

We will see that interpretation of the Riemann tensor in Section 4.3 – for its definition, we turn to cross-derivative inequality for the covariant derivative

---

[3] Arc length parametrization is just the usual proper time parametrization from special relativity in units where $c = 1$ – amounts to defining a unit tangent vector to the curve: $\sqrt{-\dot{x}^\mu g_{\mu\nu} \dot{x}^\nu} = 1$.

(amounts to the same thing as path-dependence, here). But as long as we're asking about cross-derivative equality, why not go back to scalars themselves?

4.2.1 Scalar fields

Think of the usual partial derivative of a scalar field ϕ in flat space – we know that $\phi_{,\alpha\beta} = \phi_{,\beta\alpha}$, it doesn't matter if we take the x-derivative or y-derivative first or second, for example. Is this true of the covariant derivative? We are interested because in our spaces, partial derivatives do not, in general, lead to tensor behavior.

The first derivative of a scalar is a covariant vector – let $f_\alpha \equiv \phi_{,\alpha}$. Fine, but the second derivative is now a covariant derivative acting on f_α:

$$f_{\alpha;\beta} = \phi_{,\alpha\beta} - \Gamma^\sigma_{\alpha\beta}\,\phi_{,\sigma}. \tag{4.7}$$

Then the difference is:

$$\phi_{;\alpha\beta} - \phi_{;\beta\alpha} = f_{\alpha;\beta} - f_{\beta;\alpha} = -\left(\Gamma^\sigma_{\alpha\beta} - \Gamma^\sigma_{\beta\alpha}\right)\phi_{,\sigma}. \tag{4.8}$$

Now we know from the connection of the connection to the metric (no pun intended) that $\Gamma^\sigma_{\alpha\beta} = \Gamma^\sigma_{\beta\alpha}$, but this is a negotiable point for more general manifolds. If the connection is not symmetric, its antisymmetric part can be associated with the "torsion" of the manifold. For our metrically connected spaces, the torsion is zero. We conclude that, for us:

$$\boxed{\phi_{;\alpha\beta} = \phi_{;\beta\alpha}.} \tag{4.9}$$

Scalars have cross-covariant-derivative equality.

4.2.2 Vector fields

Given a vector field $f^\alpha(x)$, we know how to form $f^\alpha(x)_{;\beta}$ but what about the second derivatives? If $f^\alpha_{,\beta\gamma} = f^\alpha_{,\gamma\beta}$, is it the case that $f^\alpha_{;\beta\gamma} = f^\alpha_{;\gamma\beta}$? It is by no means obvious that it should be, given all the non-tensorial $\Gamma^\alpha_{\beta\gamma}$ lying around, and indeed it isn't true in general.

What we want to know is: "by how much is cross-derivative equality violated?" Following the pattern for the scalar case, we will work slowly and explicitly. Let $h^\alpha_{\ \beta} \equiv f^\alpha_{;\beta}$, then:

$$h^\alpha_{\ \beta;\gamma} = h^\alpha_{\ \beta,\gamma} + \Gamma^\alpha_{\gamma\sigma}\,h^\sigma_{\ \beta} - \Gamma^\sigma_{\beta\gamma}\,h^\alpha_{\ \sigma}, \tag{4.10}$$

so that:

$$f^\alpha_{;\beta\gamma} = \left(f^\alpha_{,\beta} + \Gamma^\alpha_{\beta\sigma} f^\sigma\right)_{,\gamma} + \Gamma^\alpha_{\gamma\sigma}\left(f^\sigma_{,\beta} + \Gamma^\sigma_{\beta\rho} f^\rho\right) - \Gamma^\sigma_{\beta\gamma}\left(f^\alpha_{,\sigma} + \Gamma^\alpha_{\sigma\rho} f^\rho\right)$$

$$f^\alpha_{;\gamma\beta} = \left(f^\alpha_{,\gamma} + \Gamma^\alpha_{\gamma\sigma} f^\sigma\right)_{,\beta} + \Gamma^\alpha_{\beta\sigma}\left(f^\sigma_{,\gamma} + \Gamma^\sigma_{\gamma\rho} f^\rho\right) - \Gamma^\sigma_{\gamma\beta}\left(f^\alpha_{,\sigma} + \Gamma^\alpha_{\sigma\rho} f^\rho\right).$$

(4.11)

Now noting that $\Gamma^\alpha_{\gamma\beta} = \Gamma^\alpha_{\beta\gamma}$ and cross-derivative equality for ∂_μ, we can simplify the difference. It is:

$$\boxed{\begin{aligned} f^\alpha_{;\beta\gamma} - f^\alpha_{;\gamma\beta} &= \left(\Gamma^\alpha_{\gamma\sigma}\Gamma^\sigma_{\beta\rho} - \Gamma^\alpha_{\beta\sigma}\Gamma^\sigma_{\gamma\rho} + \Gamma^\alpha_{\beta\rho,\gamma} - \Gamma^\alpha_{\gamma\rho,\beta}\right) f^\rho \\ &\equiv R^\alpha_{\rho\gamma\beta} f^\rho, \end{aligned}}$$

(4.12)

with $R^\alpha_{\rho\gamma\beta}$ the "Riemann tensor".

There is a long tradition of definition here (for good accounts from the mathematical end, see [25, 28, 35]). That the above mess in parentheses is relevant enough to warrant its own letter and dangling indices should not be particularly clear to us at this point. I am going to try to show you how one might approach it in another way, and by so doing, give meaning to this object. One thing that should be somewhat surprising, and indicate the deep significance of the Riemann tensor, is that we started with a discussion of a vector f^α, took some derivatives, and found that the result depended only linearly on f^α itself – i.e. the Riemann tensor is interesting in that it is independent of f^α, any vector is proportional to the same deviation from cross-derivative equality.

Another impressive aspect of this tensor is its complicated relationship to the metric – if we input the Christoffel connection in terms of the metric and its derivatives, we have terms $\sim g^2$ as well as second derivatives of the metric.

Well, that's fine, we are defining the Riemann tensor (four indices) to be the extent to which covariant derivatives don't commute, not exactly breathtaking. Let me just write it explicitly without the baggage:

$$\boxed{R^\alpha_{\rho\gamma\beta} \equiv \Gamma^\alpha_{\gamma\sigma}\Gamma^\sigma_{\beta\rho} - \Gamma^\alpha_{\beta\sigma}\Gamma^\sigma_{\gamma\rho} + \Gamma^\alpha_{\beta\rho,\gamma} - \Gamma^\alpha_{\gamma\rho,\beta}.}$$

(4.13)

We are going to start by looking at the path-dependent interpretation of the Riemann tensor. The basic question we want to answer is: given a vector at some starting location, how does the vector change when parallel transported along two different paths to the same ending location? That this should be related to cross-derivative equality (or lack thereof) is not surprising – cross-derivative inequality implies that moving first in x and then in y is different from moving first in the y direction, then in the x. So we expect path-*dependence* to be a hallmark of curvature.

The important result we are heading toward lies in a comparison of the path-dependence of geodesics in our Riemannian space-time and the deviation of paths of test particles undergoing acceleration due to the Newtonian force of gravity.

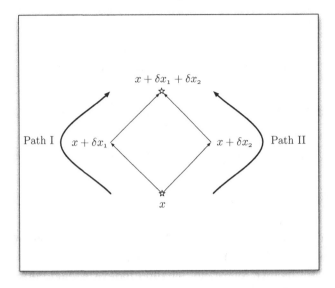

Figure 4.1   The two paths used to propagate a vector from x to $x + \delta x_1 + \delta x_2$.

This comparison will provide the connection between allowed metrics and matter distributions, which is the whole point. While the motivation for such a program may or may not be obvious, it should be clear that the comparison must involve the Riemann tensor or objects derived from it. The full implications, even formulation, of general relativity cannot proceed without the Riemann tensor in place, and so it will be beneficial to work with this new, fourth-rank mixed tensor, and understand its geometrical significance.

4.3 Interpretation

With our parallel transport law, we know how to move vectors around from point to point, but what path should we take? In flat space with Cartesian coordinates, the path doesn't matter, moving the vector is simply a matter of adding constants to its components, and we never even talk about the path along which we move, we just move the thing. In a curved space, though, we can imagine scenarios where the path *does* matter, and so we must be able to quantify the effects of taking different paths. After all, if we make a carbon copy of ourselves standing around with our vector f^α at point P, and then we go around, proudly showing off our vector at neighboring points, it would be a little embarrassing to return to P and find that our vector was rotated without knowing why.

We will construct the difference between a vector moved along two different paths. Referring to Figure 4.1, we take path I from $x \longrightarrow x + \delta x_1 \longrightarrow (x + \delta x_1) + \delta x_2$ and path II from $x \longrightarrow x + \delta x_2 \longrightarrow (x + \delta x_2) + \delta x_1$. Then we can just subtract the two vectors at the endpoint (subtraction of two vectors at the same point is allowed) to find the difference.

At $x + \delta x_1$ along path I, we have, by Taylor expansion:

$$f^\alpha(x + \delta x_1) = f^\alpha(x) + \delta x_1^\gamma \, f^\alpha_{,\gamma}(x) + O(\delta x_1^2), \tag{4.14}$$

and we have been careful to parallel transport our vector so that along the path connecting the two points, we have $\delta x^\gamma f^\alpha_{;\gamma} = 0$. Because of this, partial derivatives in the above can be replaced via $f^\alpha_{,\gamma} \, \delta x_1^\gamma = -\Gamma^\alpha_{\gamma\beta} \, f^\beta \, \delta x_1^\gamma$. Putting this in, our parallel transported vector takes on the value, at $x + \delta x_1$:

$$f^\alpha(x + \delta x_1) = f^\alpha(x) - \Gamma^\alpha_{\beta\gamma}(x) \, f^\beta(x) \, \delta x_1^\gamma \tag{4.15}$$

(the substitution that we made here highlights another aspect of parallel transport – it moves the vector in such a way that the coordinate derivatives cancel the basis derivatives).

Now starting from $x + \delta x_1$, we use Taylor expansion and parallel transport again to get to $x + \delta x_1 + \delta x_2$:

$$f^\alpha((x + \delta x_1) + \delta x_2) = f^\alpha(x + \delta x_1) + f^\alpha_{,\gamma}(x + \delta x_1) \, \delta x_2^\gamma$$
$$= f^\alpha(x + \delta x_1) - \Gamma^\alpha_{\beta\gamma}(x + \delta x_1) \, f^\beta(x + \delta x_1) \, \delta x_2^\gamma. \tag{4.16}$$

We have to expand the Christoffel symbol about the point x (using guess what) and we can also input the result from (4.15) to write the vector at $x + \delta x_1 + \delta x_2$ along path I entirely in terms of elements evaluated at x:

$$f^\alpha((x+\delta x_1)+\delta x_2) = \left(f^\alpha(x) - \Gamma^\alpha_{\beta\gamma}(x) \, f^\beta(x) \, \delta x_1^\gamma\right)$$
$$- \left(\Gamma^\alpha_{\beta\gamma}(x) + \delta x_1^\sigma \, \Gamma^\alpha_{\beta\gamma,\sigma}(x)\right)\left(f^\beta(x) - \Gamma^\beta_{\gamma\sigma}(x) \, f^\sigma(x)\right) \delta x_2^\gamma. \tag{4.17}$$

Finally, we can expand to order δx^2 (everything evaluated at x):

$$f^\alpha((x + \delta x_1) + \delta x_2) = f^\alpha - \Gamma^\alpha_{\beta\gamma} \, f^\beta \left(\delta x_1^\gamma + \delta x_2^\gamma\right)$$
$$- \delta x_1^\gamma \, \delta x_2^\rho \, f^\sigma \left(\Gamma^\alpha_{\beta\gamma} \, \Gamma^\beta_{\sigma\rho} - \Gamma^\alpha_{\sigma\gamma,\rho}\right) \tag{4.18}$$

(plus terms of order δx^3). Taking $\delta x_1 \leftrightarrow \delta x_2$, we can write the equivalent expression for path II:

$$f^\alpha((x + \delta x_2) + \delta x_1) = f^\alpha - \Gamma^\alpha_{\beta\gamma} \, f^\beta \left(\delta x_1^\gamma + \delta x_2^\gamma\right)$$
$$- \delta x_2^\gamma \, \delta x_1^\rho \, f^\sigma \left(\Gamma^\alpha_{\beta\gamma} \, \Gamma^\beta_{\sigma\rho} - \Gamma^\alpha_{\sigma\gamma,\rho}\right). \tag{4.19}$$

The difference in final value (written in terms of the point x) between (4.18) and (4.19) is:

$$
\begin{aligned}
f_{II}^{\alpha} - f_{I}^{\alpha} &= -\delta x_1^{\rho}\, \delta x_2^{\gamma}\, f^{\sigma} \left(\Gamma^{\alpha}{}_{\beta\gamma}\, \Gamma^{\beta}{}_{\sigma\rho} - \Gamma^{\alpha}{}_{\beta\rho}\, \Gamma^{\beta}{}_{\sigma\gamma} - \Gamma^{\alpha}{}_{\sigma\gamma,\rho} + \Gamma^{\alpha}{}_{\sigma\rho,\gamma} \right) \\
&= \delta x_1^{\rho}\, \delta x_2^{\gamma}\, f^{\sigma}\, R^{\alpha}{}_{\sigma\rho\gamma}.
\end{aligned}
\tag{4.20}
$$

We see that the difference between the vector f^{α} at the point $x + \delta x_1 + \delta x_2$ as calculated along the two different paths is proportional to the Riemann tensor as defined in (4.13).

4.3.1 Flat spaces

Part of the utility of the Riemann tensor is its ability to distinguish between flat and curved spaces. A space is defined to be "flat" if there exists a coordinate system in which the metric has only ± 1 along its diagonal,[4] zero elsewhere (this is true for Euclidean space, and Minkowski space-time, for example). A space that is not flat is "curved". If we are given a coordinate-dependent metric, according to the definition of "flat", we must exhibit a transformation into coordinates in which the metric takes the simplified form:

$$
g_{\mu\nu} \doteq \begin{pmatrix}
\pm 1 & 0 & 0 & 0 \\
0 & \pm 1 & 0 & 0 \\
0 & 0 & \pm 1 & 0 \\
0 & 0 & 0 & \ddots
\end{pmatrix},
\tag{4.21}
$$

or show that no such transformation exists. That procedure sounds difficult, and it is, in general. The alternative we will now prove: a space is flat if and only if its Riemann tensor vanishes. The problem of determining whether or not a space is flat is reduced to a straightforward computational task.

The theorem is: "a space-time is flat if and only if its Riemann tensor vanishes". One direction is trivial (flat space has zero Riemann tensor – obvious since its connection vanishes everywhere), now we go the other direction.[5]

Suppose we have $R_{\alpha\beta\gamma\delta} = 0$ – then parallel transport of vectors is path-independent, and we can uniquely define a vector field $f_{\alpha}(x)$ at every point simply by specifying its value at a single point. The PDE we must solve is:

$$
f_{\alpha,\gamma} - \Gamma^{\sigma}{}_{\alpha\gamma}\, f_{\sigma} = 0.
\tag{4.22}
$$

---

[4] When a flat metric is brought to this form, with ± 1 along the diagonal, we define the *signature* of the metric to be the number of positive entries minus the number of negative ones. So, for example, the Minowski metric has signature $+2$ with our convention.

[5] This argument comes from the Dirac lecture notes [10], and his treatment is concise and simple – so much so that I cannot resist repeating it here. An expanded version can be found in [9].

Take f_α to be the gradient of a scalar ϕ, then:

$$\phi_{,\alpha\gamma} = \Gamma^\sigma_{\alpha\gamma}\, \phi_{,\sigma} \tag{4.23}$$

and we know this PDE is also path-independent (meaning that we can solve it without reference to a particular path) since the Christoffel connection is symmetric. Now choose four (in four dimensions) independent scalar solutions to the above, ϕ^ρ, and use these to define a new set of coordinates $\bar{x}^\rho = \phi^\rho$, so these new coordinates themselves are solutions to:

$$\bar{x}^\rho_{,\mu\nu} = \Gamma^\sigma_{\mu\nu}\, \bar{x}^\rho_{,\sigma} \qquad \bar{x}^\rho_{,\sigma} \equiv \frac{\partial \bar{x}^\rho}{\partial x^\sigma}. \tag{4.24}$$

The metric transforms in the usual way, and we can write the original metric in terms of the new one via:

$$g_{\alpha\beta} = \frac{\partial \bar{x}^\mu}{\partial x^\alpha}\, \frac{\partial \bar{x}^\nu}{\partial x^\beta}\, \bar{g}_{\mu\nu}. \tag{4.25}$$

Our goal is to show that the new metric has vanishing partial derivatives, hence must consist of constant entries and is therefore flat[6] – if we take the x^γ derivative of both sides of the metric transformation, we have:

$$g_{\alpha\beta,\gamma} = \bar{x}^\mu_{,\alpha}\, \bar{x}^\nu_{,\beta}\, \frac{\partial \bar{g}_{\mu\nu}}{\partial x^\gamma} + \left(\bar{x}^\mu_{,\alpha\gamma}\, \bar{x}^\nu_{,\beta} + \bar{x}^\mu_{,\alpha}\, \bar{x}^\nu_{,\beta\gamma} \right) \bar{g}_{\mu\nu}, \tag{4.26}$$

and using the definition of the coordinates (4.24) gives:

$$\begin{aligned}
g_{\alpha\beta,\gamma} &= \bar{x}^\mu_{,\alpha}\, \bar{x}^\nu_{,\beta}\, \frac{\partial \bar{g}_{\mu\nu}}{\partial x^\gamma} + \left[\Gamma^\sigma_{\alpha\gamma}\, \bar{x}^\mu_{,\sigma}\, \bar{x}^\nu_{,\beta} + \Gamma^\sigma_{\beta\gamma}\, \bar{x}^\mu_{,\alpha}\, \bar{x}^\nu_{,\sigma} \right] \bar{g}_{\mu\nu} \\
&= \bar{x}^\mu_{,\alpha}\, \bar{x}^\nu_{,\beta}\, \frac{\partial \bar{g}_{\mu\nu}}{\partial x^\gamma} + \Gamma^\sigma_{\alpha\gamma}\, g_{\sigma\beta} + \Gamma^\sigma_{\beta\gamma}\, g_{\alpha\sigma} \\
&= \bar{x}^\mu_{,\alpha}\, \bar{x}^\nu_{,\beta}\, \frac{\partial \bar{g}_{\mu\nu}}{\partial x^\gamma} + g_{\alpha\beta,\gamma},
\end{aligned} \tag{4.27}$$

where the second line follows from (4.25), and the final line comes from the relationship of the connection to derivatives of the metric. We have, finally:

$$\bar{x}^\mu_{,\alpha}\, \bar{x}^\nu_{,\beta}\, \frac{\partial \bar{g}_{\mu\nu}}{\partial x^\gamma} = 0, \tag{4.28}$$

and assuming the transformation is invertible, this tells us that $\bar{g}_{\mu\nu,\gamma} \equiv \frac{\partial \bar{g}_{\mu\nu}}{\partial \bar{x}^\gamma} = 0$, so the metric is constant. This program is carried out for polar coordinates below.

---

[6] The additional jump from constant entries to flat form will be made in Problem 4.1.

Example

Suppose we were given metric and coordinates in two (spatial) dimensions:

$$g_{\mu\nu} \doteq \begin{pmatrix} 1 & 0 \\ 0 & s^2 \end{pmatrix} \quad x^\mu \doteq \begin{pmatrix} s \\ \phi \end{pmatrix}, \tag{4.29}$$

then the nonzero connection coefficients are:

$$\Gamma^s_{\phi\phi} = -s \quad \Gamma^\phi_{s\phi} = \Gamma^\phi_{\phi s} = \frac{1}{s}. \tag{4.30}$$

We want to solve the PDE (4.23) (replacing the scalar ϕ with ψ for obvious reasons) – written out in matrix form, this provides three independent equations:

$$\begin{pmatrix} \frac{\partial^2 \psi}{\partial s^2} & \frac{\partial^2 \psi}{\partial \phi \partial s} \\ \frac{\partial^2 \psi}{\partial \phi \partial s} & \frac{\partial^2 \psi}{\partial \phi^2} \end{pmatrix} = \begin{pmatrix} 0 & \frac{1}{s} \frac{\partial \psi}{\partial \phi} \\ \frac{1}{s} \frac{\partial \psi}{\partial \phi} & -s \frac{\partial \psi}{\partial s} \end{pmatrix}. \tag{4.31}$$

From $\frac{\partial^2 \psi}{\partial s^2} = 0$, we learn that $\psi = a(\phi) + s\, b(\phi)$, and from the lower right-hand equation, we have:

$$s\,(b'' + b) = -a'' \tag{4.32}$$

which, since both a and b depend only on ϕ, tells us that $a(\phi) = a_0\, \phi + a_1$ for constants a_0 and a_1, and $b(\phi) = b_0\, \cos\phi + b_1\, \sin\phi$. The final scalar solution reads, once a_0 is set to zero to satisfy the off-diagonal equation:

$$\psi = s\,(b_0\, \cos\phi + b_1\, \sin\phi). \tag{4.33}$$

Next, we take two of these solutions to define the new coordinates – how about $\bar{x}^1 = s\, \cos\phi$ and $\bar{x}^2 = s\, \sin\phi$? This is precisely the usual two-dimensional Cartesian coordinates for flat space. If we calculate the transformation of the metric:

$$\bar{g}_{\mu\nu} = \frac{\partial x^\alpha}{\partial \bar{x}^\mu} \frac{\partial x^\beta}{\partial \bar{x}^\nu}\, g_{\alpha\beta}, \tag{4.34}$$

we find

$$\bar{g}_{\mu\nu} = \begin{pmatrix} 1 & 0 \\ 0 & 1 \end{pmatrix}. \tag{4.35}$$

The moral: if someone hands you a metric in whatever coordinates they want, and you want to know if the underlying space is flat, you could try to exhibit a coordinate transformation that puts the metric in constant diagonal form, or you could just calculate the Riemann tensor for the space and verify that it is or is not zero.

Problem 4.1

We have been using the implication:

$$g_{\mu\nu,\beta} = 0 \longrightarrow g_{\mu\nu} \doteq \begin{pmatrix} \pm1 & 0 & 0 & \cdots \\ 0 & \pm1 & 0 & \cdots \\ 0 & 0 & \pm1 & \cdots \\ & & & \\ 0 & 0 & \cdots & \ddots \end{pmatrix}. \qquad (4.36)$$

But, technically, all $g_{\mu\nu,\beta} = 0$ really implies is that $g_{\mu\nu}$ is coordinate-independent (and we assume symmetric). Finish the argument: prove that any symmetric, constant, invertible real metric can be brought to the form on the right in (4.36) – diagonal with ones along the diagonal (assume purely spatial coordinates, so don't worry about possible temporal directions).

4.4 Curves and surfaces

The Riemann tensor is an important tool for characterizing spaces, but we can say more about the physical interpretation of it by considering its contractions. Specifically, we will look at the definition of the Ricci curvature – formally this is the "only" natural scalar we can form from the full Riemann tensor, so it is of some interest as an object, but more explicitly, the curvature scalar (or "Ricci scalar") does measure the "curvature" (in an everyday sense) of a space.

In the space-time of general relativity, we lose the obvious pictorial representation we can attach to "curves" because the space that is curved isn't purely spatial, and in addition, the interpretation we will discuss in this section relies on an embedding in Euclidean space. We don't generally think of a curved space-time as embedded in some larger flat space (although it *is*, see [28]), so the curvature ideas discussed here are only meant to get us thinking, not to be taken in a literal sense later on.

4.4.1 Curves

Consider an arbitrary, smooth curve, $x^\alpha(\lambda)$, in three dimensions parametrized by λ. We trace out the curve as λ goes from λ_0 (which we'll generally take to be zero) to some final value, λ_f, as shown in Figure 4.2.

Our first goal is to determine the length of the curve – this is a simple geometric quantity obtained by summing up the line element $d\tau$:

$$d\tau = \sqrt{\left(\frac{dx^\alpha}{d\lambda} d\lambda\right) g_{\alpha\beta} \left(\frac{dx^\beta}{d\lambda} d\lambda\right)} \longrightarrow \tau(\lambda_f) = \int_0^{\lambda_f} \sqrt{\frac{dx^\alpha}{d\lambda} \frac{dx_\alpha}{d\lambda}} \, d\lambda. \qquad (4.37)$$

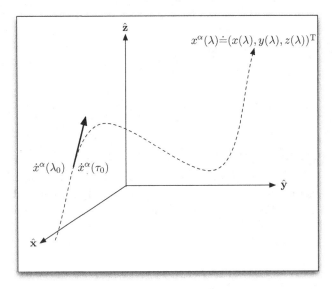

Figure 4.2   A curve, parametrized by λ.

We can parametrize the curve by τ rather than λ[7] – from the chain rule, we have:

$$\frac{dx^\alpha}{d\tau} = \frac{dx^\alpha}{d\lambda}\frac{d\lambda}{d\tau} = \frac{\dot{x}^\alpha}{\sqrt{\dot{x}^\beta\,\dot{x}_\beta}}, \tag{4.38}$$

where $\dot{x}^\alpha \equiv \frac{dx^\alpha(\lambda)}{d\lambda}$. When a curve is parametrized in terms of τ defined by (4.37), the "arc length", we say that the curve is in "arc-length parametrization".

The λ-derivative of a curve at λ_0 is tangent to the curve at λ_0 by definition of the derivative, and simply changing the parametrization doesn't change this property – so the τ-derivative of the curve is tangent to the curve at the point τ_0. But the τ-derivative has the nice property that:

$$\frac{dx^\alpha}{d\tau}\frac{dx_\alpha}{d\tau} = \frac{\dot{x}^\alpha\,\dot{x}_\alpha}{\left(\sqrt{\dot{x}^\beta\,\dot{x}_\beta}\right)^2} = 1, \tag{4.39}$$

so that $\frac{dx^\alpha}{d\tau} \equiv t^\alpha$ is a *unit* tangent vector. That is an interesting shift – it is simple to generate a unit tangent vector in λ parametrization, just take $\frac{\dot{x}^\alpha}{\sqrt{\dot{x}^\beta\,\dot{x}_\beta}}$, where dots refer to λ-derivatives. We are replacing explicit normalization with an astute choice of parameter.

The derivative of $t^\alpha(\tau)$ w.r.t. τ, call it $k^\alpha \equiv \frac{dt^\alpha}{d\tau}$, is perpendicular to t^α itself:

$$k^\alpha\,t_\alpha = \frac{dt^\alpha}{d\tau}\,t_\alpha = \frac{1}{2}\frac{d}{d\tau}(t^\alpha\,t_\alpha) = \frac{1}{2}\frac{d}{d\tau}(1) = 0, \tag{4.40}$$

so $k^\alpha(\tau)$ is everywhere perpendicular to the curve $x^\alpha(\tau)$.

---

[7] This should look familiar – precisely the move we make in special relativity from an arbitrary parametrization to proper time.

What is this vector in terms of the derivatives of x^α w.r.t. λ, the original parameter?

$$
\boxed{
\begin{aligned}
k^\alpha &= \frac{dt^\alpha}{d\tau} = \frac{d\lambda}{d\tau}\frac{dt^\alpha}{d\lambda} = \frac{1}{\sqrt{\dot{x}^\beta \dot{x}_\beta}}\frac{d}{d\lambda}\left(\frac{\dot{x}^\alpha}{\sqrt{\dot{x}^\gamma \dot{x}_\gamma}}\right)\\
&= \frac{\ddot{x}^\alpha}{\dot{x}^\beta \dot{x}_\beta} - \frac{\dot{x}^\alpha\left(\ddot{x}^\gamma \dot{x}_\gamma\right)}{(\dot{x}^\beta \dot{x}_\beta)^2}.
\end{aligned}
}
\tag{4.41}
$$

This normal vector, even though it is w.r.t. τ, does not have unit magnitude. Indeed its magnitude is special, it is called the "curvature" of the curve. Let's see how this definition relates to our usual ideas about curviness with a simple example (for more on the definition of curvature in this one-dimensional setting, see [25, 35]).

Example

Take the curve to be a circle of radius r parametrized by $\lambda = \theta$:

$$
x^\alpha(\theta) = \begin{pmatrix} r\cos\theta \\ r\sin\theta \\ 0 \end{pmatrix} \longrightarrow \dot{x}^\alpha(\theta) = \begin{pmatrix} -r\sin\theta \\ r\cos\theta \\ 0 \end{pmatrix}.
\tag{4.42}
$$

We don't know what the arc-length parametrization of this curve is just by looking (or do we?), but we can easily generate the unit tangent vector $t^\alpha = \frac{\dot{x}^\alpha}{r}$, and the normal vector, from (4.41), can be calculated:

$$
\ddot{x}^\alpha(\theta) = \begin{pmatrix} -r\cos\theta \\ -r\sin\theta \\ 0 \end{pmatrix}
\tag{4.43}
$$

$$
\ddot{x}^\gamma \dot{x}_\gamma = 0,
$$

so that we have only the first term:

$$
k^\alpha = \frac{\ddot{x}^\alpha}{\dot{x}^\beta \dot{x}_\beta} = \begin{pmatrix} -\frac{\cos\theta}{r} \\ -\frac{\sin\theta}{r} \\ 0 \end{pmatrix}
\tag{4.44}
$$

and this has $\kappa^2 \equiv k^\alpha k_\alpha = \frac{1}{r^2}$ so the "curvature" is just $\kappa = r^{-1}$. That makes sense: if we blow the circle up, so that it has large radius, the curve looks locally pretty flat. The inverse of the curvature is called the "radius of curvature", $\kappa^{-1} = r$, and indicates the distance to the center of a circle. Locally, the radius of curvature for a more generic curve can be thought of as the distance to the center of the circle tangent to the arc segment, as shown in Figure 4.3.

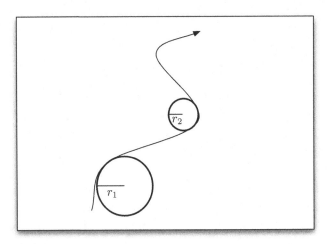

Figure 4.3   A curve with two example circles, tangent at different points. The radius of curvature amounts to the radius of the circles tangent at each point along the curve.

Finally, it is pretty easy to make the connection between the arc length and the parametrization for a circle. If we take the curve parametrized by θ, then the distance we have traveled along the curve, for a particular value of θ, is $s = r\,\theta$, so we can parametrize in terms of s directly:

$$x^\alpha(s) = \begin{pmatrix} r\,\cos(s/r) \\ r\,\sin(s/r) \\ 0 \end{pmatrix} \longrightarrow \dot{x}^\alpha(s) = \begin{pmatrix} -\sin(s/r) \\ \cos(s/r) \\ 0 \end{pmatrix}, \qquad (4.45)$$

and we see that arc-length parametrization has automatically given us a unit tangent vector: $\dot{x}^\alpha \dot{x}_\alpha = 1$, which is to be expected – this is the spatial version of our proper time requirement: $dx^\alpha \eta_{\alpha\beta}\, dx^\beta = -c^2\, d\tau^2$ from special relativity.

4.4.2 Higher dimension

That's fine for one-dimensional curves, but in more than one dimension, our notion of λ and τ gets confusing – there could be more than one parameter describing a surface. In addition, our ability to draw things like tangent vectors can get complicated or impossible.

The fundamental scalar invariant for a Riemannian manifold is called the "Ricci scalar", and is formed from the Riemann tensor by contracting indices. This scalar can be defined when $g_{\mu\nu}$ is given, since the metric is enough to determine the Christoffel connections, and hence the Riemann tensor itself. From the Riemann tensor, we define the Ricci tensor:

$$R_{\mu\nu} \equiv R^{\alpha}{}_{\mu\alpha\nu}, \tag{4.46}$$

a manifestly symmetric, second-rank tensor (write it out). Then, taking the trace of the Ricci tensor gives the Ricci scalar:

$$R \equiv R_{\mu\nu} g^{\mu\nu} = R^{\mu}{}_{\mu} = R^{\alpha\mu}{}_{\alpha\mu}. \tag{4.47}$$

These may seem like arbitrary contractions, but think about trying to contract any other pair of indices, the most you can get from the Riemann tensor is a sign change for R.

The goal now is to show that the Ricci scalar plays the same role as κ in generic spaces. It certainly has the right basic form – κ was constructed out of second derivatives of a curve w.r.t. τ (what we might think of as a one-dimensional "coordinate"), and the Ricci scalar, generated out of the Riemann tensor, will involve second derivatives of the metric defining the space of interest. The similarities are nice, but the Ricci scalar defines an "intrinsic" curvature rather than the "extrinsic" curvature associated with κ. The difference is that κ requires a higher-dimensional (flat) space to set the curvature – we needed two dimensions to define the curvature (which is a vector length), while the intrinsic curvature makes no reference to an external space. The intrinsic curvature of a curve (which is necessarily one-dimensional) is zero, meaning that the curve is flat. This makes sense if you think about being trapped on the curve, with no knowledge of the exterior space – you can go forward (or back, amounting to a sign), there is only one direction, so the curve is equivalent to a line (think of the implications for the path-dependence of parallel transport).

The Ricci scalar curvature is clearly the one of most interest to us in our study of metric spaces, since we will not have any obvious embedding – we get the metric, which tells us the dimension, we do not get a higher-dimensional flat space from which to view. Keep this in mind as we go, while we will make contact with the κ curvature, the Ricci scalar is a fundamentally different object.

Computing the Ricci scalar

We begin by thinking about the simplest two-dimensional surface, a sphere. In three dimensions, spherical coordinates still represent flat space, we are not trapped on the surface of the sphere, so our distance measurements proceed according to the usual Pythagorean rule, albeit written in funny coordinates. For a true two-dimensional surface, we cannot measure radially, and our notion of distance relies on the two-dimensional underlying metric.

In this simplified setting, we will go through the process of calculating the Ricci scalar. It is an involved computation, but care and practice render such calculations tractable. We'll go relatively slowly in this case, starting from the metric for a sphere, computing the connection coefficients (which we have done previously), and then generating the Riemann tensor, and Ricci tensor and scalar via contraction.

We can put ourselves on the surface of a sphere of radius r by eliminating dr from the line element written in spherical coordinates:

$$ds^2 = r^2 \, d\theta^2 + r^2 \sin^2\theta \, d\phi^2 \longrightarrow g_{\mu\nu} \doteq \begin{pmatrix} r^2 & 0 \\ 0 & r^2 \sin^2\theta \end{pmatrix}. \tag{4.48}$$

This metric has nonzero connection coefficients given by:

$$\Gamma^\theta_{\phi\phi} = -\cos\theta \, \sin\theta$$
$$\Gamma^\phi_{\theta\phi} = \Gamma^\phi_{\phi\theta} = \cot\theta. \tag{4.49}$$

The Riemann tensor is:

$$R^\alpha_{\ \sigma\gamma\rho} = \Gamma^\alpha_{\beta\gamma} \Gamma^\beta_{\sigma\rho} - \Gamma^\alpha_{\beta\rho} \Gamma^\beta_{\sigma\gamma} - \Gamma^\alpha_{\sigma\gamma,\rho} + \Gamma^\alpha_{\sigma\rho,\gamma}, \tag{4.50}$$

take $\alpha = \theta$:

$$R^\theta_{\ \sigma\gamma\rho} = \Gamma^\theta_{\beta\gamma} \Gamma^\beta_{\sigma\rho} - \Gamma^\theta_{\beta\rho} \Gamma^\beta_{\sigma\gamma} - \Gamma^\theta_{\sigma\gamma,\rho} + \Gamma^\theta_{\sigma\rho,\gamma}. \tag{4.51}$$

Looking at the first term, only $\beta = \gamma = \phi$ contributes, so we can expand the sums over β in the first two terms:

$$R^\theta_{\ \sigma\gamma\rho} = \Gamma^\theta_{\phi\gamma} \Gamma^\phi_{\sigma\rho} - \Gamma^\theta_{\phi\rho} \Gamma^\phi_{\sigma\gamma} - \Gamma^\theta_{\sigma\gamma,\rho} + \Gamma^\theta_{\sigma\rho,\gamma}. \tag{4.52}$$

By the symmetries of the Riemann tensor (it is antisymmetric in $\gamma \leftrightarrow \rho$ interchange), we cannot have $\rho = \gamma$, so let's set $\rho = \theta$, $\gamma = \phi$, the only choice (keeping in mind that $\rho = \phi$, $\gamma = \theta$ is just the negative of this) is:

$$R^\theta_{\ \sigma\phi\theta} = \Gamma^\theta_{\phi\phi} \Gamma^\phi_{\sigma\theta} - \Gamma^\theta_{\sigma\phi,\theta}, \tag{4.53}$$

from which we get potentially two terms:

$$R^\theta_{\ \theta\phi\theta} = \Gamma^\theta_{\phi\phi} \Gamma^\phi_{\theta\theta} - \Gamma^\theta_{\theta\phi,\theta} = 0$$
$$R^\theta_{\ \phi\phi\theta} = \Gamma^\theta_{\phi\phi} \Gamma^\phi_{\phi\theta} - \Gamma^\theta_{\phi\phi,\theta}$$
$$= (-\cos\theta \, \sin\theta)(\cot\theta) + \sin^2\theta - \cos^2\theta = -\sin^2\theta \tag{4.54}$$
$$= -R^\theta_{\ \phi\theta\phi}.$$

For $\alpha = \phi$ in (4.50), we have:

$$R^\phi_{\ \sigma\gamma\rho} = \Gamma^\phi_{\beta\gamma} \Gamma^\beta_{\sigma\rho} - \Gamma^\phi_{\beta\rho} \Gamma^\beta_{\sigma\gamma} - \Gamma^\phi_{\sigma\gamma,\rho} + \Gamma^\phi_{\sigma\rho,\gamma} \tag{4.55}$$

and as before, our only option is $\rho = \theta$, $\gamma = \phi$:

$$R^{\phi}{}_{\sigma\phi\theta} = \Gamma^{\phi}{}_{\beta\phi}\,\Gamma^{\beta}{}_{\sigma\theta} - \Gamma^{\phi}{}_{\beta\theta}\,\Gamma^{\beta}{}_{\sigma\phi} - \Gamma^{\phi}{}_{\sigma\phi,\theta} + \Gamma^{\theta}{}_{\sigma\theta,\phi}. \tag{4.56}$$

The first term is zero for both $\beta = \{\theta, \phi\}$, and the fourth term is zero since there is no ϕ-dependence in the metric. We have (the second term above only contributes for $\beta = \phi$):

$$R^{\phi}{}_{\sigma\phi\theta} = -\Gamma^{\phi}{}_{\phi\theta}\,\Gamma^{\phi}{}_{\sigma\phi} - \Gamma^{\phi}{}_{\sigma\phi,\theta}, \tag{4.57}$$

which can only be nonzero for $\sigma = \theta$, then the only nonzero component left is:

$$R^{\phi}{}_{\theta\phi\theta} = -\Gamma^{\phi}{}_{\phi\theta}\,\Gamma^{\phi}{}_{\theta\phi} - \Gamma^{\phi}{}_{\theta\phi,\theta} = -\cot^2\theta + \frac{1}{\sin^2\theta} = 1. \tag{4.58}$$

Collecting our results, we have two nonzero components of the Riemann tensor:

$$
\boxed{
\begin{aligned}
R^{\theta}{}_{\phi\phi\theta} &= -R^{\theta}{}_{\phi\theta\phi} = -\sin^2\theta \\
R^{\phi}{}_{\theta\phi\theta} &= -R^{\phi}{}_{\theta\theta\phi} = 1.
\end{aligned}
} \tag{4.59}
$$

The Ricci tensor is obtained from the Riemann tensor via:

$$R_{\sigma\rho} = R^{\alpha}{}_{\sigma\alpha\rho}, \tag{4.60}$$

so that in matrix form, we have:

$$
\begin{aligned}
R_{\sigma\rho} &\doteq
\begin{pmatrix}
R^{\theta}{}_{\theta\theta\theta} + R^{\phi}{}_{\theta\phi\theta} & R^{\theta}{}_{\theta\theta\phi} + R^{\phi}{}_{\theta\phi\phi} \\
R^{\theta}{}_{\phi\theta\theta} + R^{\phi}{}_{\phi\phi\theta} & R^{\theta}{}_{\phi\theta\phi} + R^{\phi}{}_{\phi\phi\phi}
\end{pmatrix} \\
&= \begin{pmatrix} 1 & 0 \\ 0 & \sin^2\theta \end{pmatrix}
\end{aligned} \tag{4.61}
$$

and then finally, the Ricci scalar is:

$$
\begin{aligned}
R &= g^{\mu\nu}\,R_{\mu\nu} = g^{\theta\theta}\,R_{\theta\theta} + g^{\phi\phi}\,R_{\phi\phi} \\
&= \frac{1}{r^2} + \frac{1}{r^2\,\sin^2\theta}\,\sin^2\theta \\
&= \frac{2}{r^2}.
\end{aligned} \tag{4.62}
$$

That was a pretty explicit calculation, and we went through it carefully this time so you could see how such arguments go – with all the indices and entries, the use of symmetries and a quick scan of the relevant indices that will play a role in the final answer is important.

Notice that $R = \frac{2}{r^2}$ is a quantity we could measure from the surface of the sphere itself – we could measure the Riemann tensor by taking a vector and going around different paths, and then it would be a simple matter to discover the curvature of our two-dimensional space. Alternatively, we could make distance measurements in different directions in order to construct the metric, and then use that to find the curvature. Either way, we make no reference to any higher-dimensional space in which we are embedded. The Ricci scalar, R, is similar to the radius of curvature, $\kappa = r^{-1}$, we established for a circle – R is small when the radius (of the sphere, now) is large, and vice versa.

Problem 4.2
For the parabolic curve $y = x^2$ in two dimensions, find the curvature κ, plot this in the vicinity of zero.

Problem 4.3
Using the standard (p, e) parametrization of radius for an ellipse:

$$r(\phi) = \frac{p}{1 + e \, \cos \phi}, \tag{4.63}$$

find the curvature of the ellipse as a function of ϕ – plot for $\phi = 0 \longrightarrow 2\pi$ with $p = 1, e = \frac{1}{2}$. Note: feel free to use `Mathematica` for this one, the derivatives can get . . . involved.

Problem 4.4
Write the equation for a straight line in two dimensions: $y = A\,x$ in the form
$x^\alpha(s) \doteq \begin{pmatrix} x(s) \\ y(s) \end{pmatrix}$, where s is the arc-length parameter, so that $\dot{x}^\alpha \, \dot{x}_\alpha = 1$.

Problem 4.5
Find the Riemann tensor for the surface of an infinitely long cylinder (hint: do not calculate the Riemann tensor directly).

4.5 Taking stock

Let's summarize and discuss the properties of the Riemann tensor (and its role as a geometric indicator) that we have been defining and working on. In particular, we shall review the covariant derivative, its role in defining geodesics, and also in determining the curvature of space.

Our one and two-dimensional curves and surfaces have been useful in understanding concepts like curvature, but now we kick the scaffold over and are on our own. We will not be referring to the space(-times) in GR as embedded in some higher-dimensional space, so things like Ricci curvature are what we get.

4.5.1 Properties of the Riemann tensor

We started by defining the Riemann tensor in terms of a lack of commutativity in second (covariant) derivatives. Let's review, starting from the definition of covariant derivative:

$$f^{\alpha}{}_{;\beta} = f^{\alpha}{}_{,\beta} + \Gamma^{\alpha}{}_{\beta\gamma} f^{\gamma}$$
$$f_{\alpha;\beta} = f_{\alpha,\beta} - \Gamma^{\gamma}{}_{\alpha\beta} f_{\gamma},$$

(4.64)

where the upper and lower forms are defined by (1) our desire to have a tensorial derivative (i.e. one that transforms appropriately as a tensor):

$$\bar{f}^{\alpha}{}_{;\beta} = \frac{\partial \bar{x}^{\alpha}}{\partial x^{\gamma}} \frac{\partial x^{\sigma}}{\partial \bar{x}^{\beta}} f^{\gamma}{}_{;\sigma}$$

(4.65)

and (2) the requirement that $(f^{\alpha} f_{\alpha})_{;\gamma} = (f^{\alpha} f_{\alpha})_{,\gamma}$.

These two ideas lead to derivatives that redefine our notion of displacement (that's what derivatives *do* after all), and in particular, we generated a new type of "constant" vector – one that is parallel transported along a curve (a set of points parametrized by τ: $x^{\alpha}(\tau)$, with tangent $\dot{x}^{\gamma}$). The hallmark of parallel transport is the condition:

$$\frac{D}{D\tau} f^{\alpha} \equiv f^{\alpha}{}_{;\beta} \dot{x}^{\beta} = 0,$$

(4.66)

a set of ordinary differential equations that tells us what the value of $f^{\alpha}(x(\tau))$ is given $f^{\alpha}(x(0))$ – in flat space, the corresponding vector is just a constant, but here in curved space, we must consider how the vector transforms from point-to-point.

There is a special class of curve – geodesics, defined in terms of parallel transport as "curves whose tangent vector is parallel transported around themselves". Formally, this amounts to replacing f^{α} with $\dot{x}^{\alpha}$ in (4.66):

$$\frac{D}{D\tau} \dot{x}^{\alpha} = \dot{x}^{\alpha}{}_{;\beta} \dot{x}^{\beta} = \dot{x}^{\beta} \left(\dot{x}^{\alpha}{}_{,\beta} + \Gamma^{\alpha}{}_{\beta\gamma} \dot{x}^{\gamma} \right) = \ddot{x}^{\alpha} + \Gamma^{\alpha}{}_{\beta\gamma} \dot{x}^{\beta} \dot{x}^{\gamma} = 0$$

(4.67)

and geometrically, the same equation comes from a variational principle $\delta \int ds = 0$ indicating that solutions are the successors to our flat-space notion of "straight lines".

After that, we developed the requirement that the angle between vectors transported in the above manner should be constant along the curve, and used this to uniquely determine $\Gamma^{\alpha}{}_{\beta\gamma}$ in terms of our metric $g_{\mu\nu}$:

$$g_{\mu\nu;\gamma} = 0 \longrightarrow \Gamma^{\alpha}{}_{\beta\gamma} = \frac{1}{2} g^{\alpha\rho} \left(g_{\rho\beta,\gamma} + g_{\rho\gamma,\beta} - g_{\beta\gamma,\rho} \right).$$

(4.68)

With a definite relationship between the (derivatives of the) metric and the connection, we were able to ask the question: how is $f^{\alpha}{}_{;\beta\gamma}$ related to $f^{\alpha}{}_{;\gamma\beta}$? and answer it in terms of more derivatives of the metric. This led us to the definition of the Riemann tensor in (4.13):

$$R^{\alpha}_{\ \beta\gamma\delta} \equiv \Gamma^{\alpha}_{\ \sigma\gamma}\, \Gamma^{\sigma}_{\ \beta\delta} - \Gamma^{\alpha}_{\ \sigma\delta}\, \Gamma^{\sigma}_{\ \beta\gamma} + \Gamma^{\alpha}_{\ \beta\delta,\gamma} - \Gamma^{\alpha}_{\ \beta\gamma,\delta}. \tag{4.69}$$

We also saw how this definition is connected to the difference induced in a vector when transported along two different curves.

Looking at (4.69), it is clear that, for example, there is a symmetry w.r.t. the index interchange $\gamma \leftrightarrow \delta$ – the situation becomes more interesting when we lower the α:

$$R_{\alpha\beta\gamma\delta} \equiv \Gamma_{\alpha\sigma\gamma}\, \Gamma^{\sigma}_{\ \beta\delta} - \Gamma_{\alpha\sigma\delta}\, \Gamma^{\sigma}_{\ \beta\gamma} + g_{\alpha\tau}\left(\Gamma^{\tau}_{\ \beta\delta,\gamma} - \Gamma^{\tau}_{\ \beta\gamma,\delta}\right), \tag{4.70}$$

and we ask: what are the symmetries here? Clearly, $\gamma \leftrightarrow \delta$ is still antisymmetric, but what more can we say?

4.5.2 Normal coordinates

Some properties of the Riemann tensor are best uncovered in a particular coordinate system. Here, we will develop the notion of "normal coordinates". These are defined by the behavior of the metric and Christoffel connection, rather than constructed explicitly. We will use this local coordinate system to make observations that we can then turn into tensor statements that must hold everywhere.

The procedure of taking an explicit equality in a specific coordinate system and transforming it to generic coordinates is not unfamiliar – going back to E&M, the electric field of a dipole $\mathbf{p} = p_0\,\hat{\mathbf{z}}$ sitting at the origin is:

$$\mathbf{E} = \frac{p_0}{4\,\pi\,\epsilon_0\,r^3}(2\,\cos\theta\,\hat{\mathbf{r}} + \sin\theta\,\hat{\boldsymbol{\theta}}). \tag{4.71}$$

From this expression, we can generate the coordinate-free form:

$$\mathbf{E} = \frac{1}{4\,\pi\,\epsilon_0\,r^3}(3\,(\mathbf{p}\cdot\hat{\mathbf{r}})\,\hat{\mathbf{r}} - \mathbf{p}), \tag{4.72}$$

and we have turned a result in a specific coordinate system into a coordinate-independent statement.

The same is true for a lot of expressions that come up in general relativity: we find some easy set of coordinates, prove whatever it is we want in terms of those, and then make the non-tensor statement (usually) into a tensor statement at which point it's true in any coordinate system. The "normal coordinates" that we are about to construct represent a particularly nice coordinate system for studying the Riemann tensor.

Consider a point P_0 in space (or space-time). Suppose we know the geodesics of the space $x^{\alpha}(\tau)$, and we travel along one of them from P_0 (a point on the geodesic with parameter $\tau = 0$) to a nearby point P_τ. The geodesic satisfies the defining equation (4.67). The Taylor expansion near P_0 is given by:

$$x^\alpha(\tau) = x^\alpha(0) + \tau\,\dot{x}^\alpha(0) - \frac{1}{2}\tau^2\,\Gamma^\alpha{}_{\beta\gamma}(P_0)\,\dot{x}^\beta(0)\,\dot{x}^\gamma(0) + O(\tau^3), \qquad (4.73)$$

and this expansion satisfies the geodesic equation to first order in τ. Consider new coordinates at P_0 defined by $\bar{x}^\alpha(\tau) = \tau\,\dot{x}^\alpha(0)$ (notice that this definition implicitly relies on P_τ since it uses the tangent vector to the geodesic connecting P_0 and P_τ), we can use (4.73) to verify that this is an invertible transformation at P_0. All geodesics emanating from P_0 obey the geodesic equation for the new coordinates, since it is a tensor equation:

$$\frac{d\dot{\bar{x}}^\alpha}{d\tau} + \bar{\Gamma}^\alpha{}_{\beta\gamma}\,\dot{\bar{x}}^\beta\,\dot{\bar{x}}^\gamma = 0. \qquad (4.74)$$

Using the transformation, $\frac{d\bar{x}^\alpha}{d\tau} = \dot{x}^\alpha(0)$, a constant, so the first term above is zero, and what remains is:

$$\bar{\Gamma}^\alpha{}_{\beta\gamma}\,\dot{x}^\beta(0)\,\dot{x}^\gamma(0) = 0. \qquad (4.75)$$

This can only hold, for a generic geodesic (with arbitrary $\dot{x}^\beta(0)$) if $\bar{\Gamma}^\alpha{}_{\beta\gamma} = 0$ in the vicinity of P_0. You'll do the explicit transformation of the Christoffel symbols in Problem 4.9. As for the metric, we have, in any coordinate system, $g_{\mu\nu;\gamma} = 0$, and if we write this out at P_0:

$$\bar{g}_{\mu\nu;\gamma}(P_0) = \bar{g}_{\mu\nu,\gamma} - \bar{\Gamma}^\sigma{}_{\mu\gamma}\,\bar{g}_{\sigma\nu} - \bar{\Gamma}^\sigma{}_{\nu\gamma}\,\bar{g}_{\mu\sigma} = \bar{g}_{\mu\nu,\gamma} = 0, \qquad (4.76)$$

so that $\bar{g}_{\mu\nu}$ is constant at the point P_0. I have omitted the error terms, but keep in mind that the above statements hold only in the vicinity of P_0, i.e. there are corrections that depend on $O(\bar{x}^\alpha\,\bar{x}_\alpha)$.

Actually calculating the normal coordinates for a given system is not necessarily easy to do. It requires, for example, that you have some nice form for the geodesic trajectories. But just knowing that such a coordinate system exists can get us pretty far.

The prescription is: set the connection to zero at a point, set the metric to the identity matrix (with ± 1 along the diagonal) to put the equation in normal coordinates. More than that, one cannot do, the second derivatives of the metric are not in general zero (unless the space is flat). For example, consider the Riemann tensor, defined in (4.13). In these normal coordinates, the two terms quadratic in Γ

are zero, but the derivatives are not, so we have:

$$R^\alpha{}_{\rho\gamma\beta}|_{NC} = \Gamma^\alpha{}_{\beta\rho,\gamma} - \Gamma^\alpha{}_{\gamma\rho,\beta}$$

$$R_{\alpha\rho\gamma\beta}|_{NC} = \Gamma_{\alpha\beta\rho,\gamma} - \Gamma_{\alpha\gamma\rho,\beta} \tag{4.77}$$

$$= \frac{1}{2}\left(g_{\alpha\beta,\rho\gamma} - g_{\beta\rho,\alpha\gamma} - g_{\alpha\gamma,\rho\beta} + g_{\gamma\rho,\alpha\beta}\right).$$

If we can make tensor statements, then our choice of normal coordinates is a moot point – that's the real idea here. Well, looking at the above, I see that in these coordinates, the Riemann tensor is antisymmetric under $\alpha \leftrightarrow \rho$ and $\gamma \leftrightarrow \beta$ and symmetric under $(\alpha, \rho) \leftrightarrow (\gamma, \beta)$. The tensor statement is, for example:

$$R_{\alpha\rho\gamma\beta}|_{NC} + R_{\rho\alpha\gamma\beta}|_{NC} = 0, \tag{4.78}$$

and since this is a tensor equation (addition of two Riemann tensors), and it's zero in one coordinate system, it must be zero in all – we can drop the NC identifier.

The symmetries just mentioned are true for the generic expression (4.13), but here we can see them more clearly. There are more – consider cyclic permutations, again referring to (4.77), we see that:

$$\boxed{R_{\alpha\rho\gamma\beta} + R_{\alpha\gamma\beta\rho} + R_{\alpha\beta\rho\gamma} = 0} \tag{4.79}$$

(this is not as obvious as the rest, but can be worked out relatively quickly). Again, since it is a tensor statement, this holds in all coordinate systems.

We have exhausted the symmetries of the Riemann tensor, and we're finally ready to do the famous counting argument – how many independent elements are there in $R_{\mu\nu\alpha\beta}$? As a four-indexed object in D dimensions, we have *a priori* D^4 independent components. The symmetries will reduce this number, and we want to start with the pair interchange symmetry:

$$R_{\alpha\beta\gamma\delta} = R_{\gamma\delta\alpha\beta}, \tag{4.80}$$

and antisymmetry under interchange within pairs:

$$R_{\beta\alpha\gamma\delta} = -R_{\alpha\beta\gamma\delta} \qquad R_{\alpha\beta\delta\gamma} = -R_{\alpha\beta\gamma\delta}. \tag{4.81}$$

To count the freedom of the pair interchange, think of the two sets of indices grouped together, so let $A = \{\alpha\beta\}$ and $B = \{\gamma\delta\}$, then $R_{AB} = R_{BA}$ is a statement of the symmetry, and this is in two-dimensional form, i.e. its content is equivalent to a symmetric matrix of dimension N, so there are $\frac{1}{2}N(N+1)$ independent components. To find N, we note that $R_{\{\alpha\beta\}\,B} = -R_{\{\beta\alpha\}\,B}$ has the same content as a $D \times D$-dimensional antisymmetric matrix, i.e. there are $\frac{1}{2}D(D-1)$ choices for A, and similarly for B. Then $N = \frac{1}{2}D(D-1)$. The total number of components so far is:

$$R_{\alpha\beta\gamma\delta} \text{ components} = \frac{1}{2} \underbrace{\left(\frac{1}{2}D(D-1)\right)}_{N} \underbrace{\left(\frac{1}{2}D(D-1)+1\right)}_{N+1} - \text{number in (4.79)}.$$

(4.82)

To count the number of constraints imposed by (4.79), notice that if one sets any two components equal, we get zero identically by the symmetries already in place (for example, take $\rho = \alpha$: one term goes away by antisymmetry, the other two cancel), so only for $\{\alpha\rho\gamma\beta\}$ distinct do we get a constraint. Well, there are "D choose 4" ways to arrange the indices, so the final counting is:

$$\text{components} = \frac{1}{2}\left(\frac{1}{2}D(D-1)\right)\left(\frac{1}{2}D(D-1)+1\right) - \binom{D}{4}$$

$$= \frac{1}{8}D(D-1)(D(D-1)+2) - \frac{1}{24}D(D-1)(D-2)(D-3)$$

$$= \boxed{\frac{1}{12}D^2(D^2-1).}$$

(4.83)

In three-dimensional space, then, there are only six independent components in the Riemann tensor, whereas in four-dimensional space-time, there are 20.

Finally, I mention the derivative relationship for the Riemann tensor, this is also easiest to show in the normal coordinates we have been considering. If we take a derivative in (4.77), then:

$$R_{\alpha\rho\gamma\beta,\delta} = \frac{1}{2}\left(g_{\alpha\beta,\rho\gamma\delta} - g_{\beta\rho,\alpha\gamma\delta} - g_{\alpha\gamma,\rho\beta\delta} + g_{\gamma\rho,\alpha\beta\delta}\right)$$

(4.84)

and we can cyclically permute the last three indices:

$$R_{\alpha\rho\gamma\beta,\delta} + R_{\alpha\rho\delta\gamma,\beta} + R_{\alpha\rho\beta\delta,\gamma} = 0.$$

(4.85)

Now this is not a tensor statement since partial derivatives are not tensorial, but in normal coordinates, a normal derivative is equal to a covariant one (since the connection vanishes), so replacing the commas in the above with semicolons, we have the "Bianchi identity":

$$\boxed{R_{\alpha\rho\gamma\beta;\delta} + R_{\alpha\rho\delta\gamma;\beta} + R_{\alpha\rho\beta\delta;\gamma} = 0.}$$

(4.86)

This relation is important in deriving Einstein's equation.

4.5.3 Summary

In setting up these elements of tensor analysis, we have at certain points specialized to a class of spaces. The arena of general relativity is a space with zero torsion (Christoffel symbols are symmetric) and a "metric connection" (connection related to derivatives of the metric) which automatically implies that the metric has zero covariant derivative. One thing that can be shown for these spaces is that at a point P, moving to normal coordinates, the Christoffel connection can itself be made zero, and the metric manifestly flat. So at any point P, we can set up coordinates such that:

$$g_{\alpha\beta;\gamma} = 0 \quad \text{with} \quad \Gamma^{\alpha}_{\ \beta\gamma} = 0 \longrightarrow g_{\alpha\beta,\gamma} = 0, \tag{4.87}$$

from which diagonal form with ± 1 along the diagonal is achievable. The existence of normal coordinates is important for doing physics, since we believe, for the most part, that our local environment is flat Minkowksi space-time. This belief comes from wide experience and *must* be built in, in some manner, to any physical theory. It is a defining property of Riemannian manifolds that in the vicinity of any point, a locally flat metric can be used (and if that metric is not positive definite, we call these pseudo-Riemannian manifolds) – so these are the natural spaces for the theory.

Problem 4.6

(a) For a torus, parametrized by the two angles θ and ϕ as indicated in Figure 4.4 (with R the radius to the center of the tube, a the radius of the tube), find the metric, compute the connection coefficients, the Riemann $R^{\alpha}_{\ \beta\gamma\delta}$ and Ricci tensors and finally, the Ricci scalar.

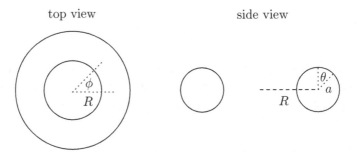

top view   side view

Figure 4.4   Top and side views of our torus – the radius R is the distance from the center to the middle of the "tube", and a is the radius of the tube. The two angles used to define points on the torus are shown.

(b) Our formula for the number of elements of the Riemann tensor indicates that in two dimensions, there should be only one independent element – why does it appear we ended up with two for the torus? What is the one independent element?

Problem 4.7

(a) Show that for any two-dimensional space with positive definite metric (i.e. $g > 0$), there is a coordinate choice that allows the metric to take the form:

$$g^{\mu\nu} \doteq f(X, Y) \begin{pmatrix} 1 & 0 \\ 0 & 1 \end{pmatrix}, \tag{4.88}$$

with coordinates $\{X, Y\}$. You may assume that the solution to the generalized Laplace's equation $(\sqrt{g}\,g^{\mu\nu}\phi_{,\mu})_{,\nu} = 0$ exists (note that $g \equiv \det g_{\mu\nu}$) – also assume that any metric is symmetric. Any metric that can be written as $g^{\mu\nu} = F(x)\,\delta^{\mu\nu}$ is called "conformally flat" – "conformal flatness" can be established by a tensor test analogous to the Riemann tensor test for flatness. The relevant fourth-rank tensor is known as the "Weyl tensor", and can be constructed from the trace-free part of the Riemann tensor (see, for example, [9]). Evidently, all two-dimensional spaces are conformally flat.

(b) Construct an example of a two-dimensional space that has zero Ricci scalar, but nonzero Riemann tensor or show that no two-dimensional space has this property.

Problem 4.8

In this problem, we will construct a set of coordinates in which the sphere with line element:

$$ds^2 = r^2\,d\theta^2 + r^2 \sin^2\theta\,d\phi^2 \tag{4.89}$$

has conformally flat metric.

(a) The "problem" (for conformal flatness) is the $\sin^2\theta$ term in front of $d\phi^2$ – we want to define a new coordinate $\bar\theta(\theta)$ so that there is a common factor of $\sin^2\theta$ in front of both $d\bar\theta^2$ and $d\phi^2$. By transforming the above ds^2, find the ODE that will determine $\bar\theta(\theta)$.

(b) Integrate your ODE to find $\bar\theta$ (set the constant of integration to zero, corresponding to $\theta = \frac{1}{2}\pi$ when $\bar\theta = 0$), and write the two-dimensional space in terms of this new coordinate.

Problem 4.9

Another way to generate normal coordinates comes from considering the simplest transformation from a set of coordinates x^α at a particular point P – define:

$$\bar x^\alpha = x^\alpha + F^\alpha_{\beta\gamma}\,x^\beta x^\gamma \tag{4.90}$$

at P, for $F^\alpha_{\beta\gamma}$ a collection of numbers used to get quadratic dependence on x^α.

Find the Christoffel connection associated with this transformation using (3.68). Set $F^\alpha_{\beta\gamma}$ appropriately so that the connection vanishes in this new coordinate system.

Note that you should assume that $x^\alpha|_P = 0$, so that P defines the origin of the original coordinate system, and hence, from the form of (4.90), the new coordinate system as well, i.e. $\bar{x}^\alpha|_P = 0$.

Problem 4.10

As an example of actually constructing normal coordinates, consider a sphere of radius r – we have a point with $\theta = \frac{1}{4}\pi$ and $\phi = 0$ that has nonzero connection values. In order to use your result from the previous problem, we'll start in a coordinate system in which this point is at $\Theta = 0$ (so that $\Theta = \theta + \pi/4$) – define the metric from the line element, describing lengths measured on the surface of a sphere of radius r:

$$ds^2 = r^2 \, d\Theta^2 + r^2 \, \sin^2\left(\Theta + \frac{1}{4}\pi\right) d\phi^2. \tag{4.91}$$

(a) Compute the connection coefficients at the point P defined by $\Theta = 0$, $\phi = 0$ (don't recompute the connection, use (4.49) with appropriate θ).

(b) Using these values, write a new set of coordinates, $\bar{\Theta}$ and $\bar{\phi}$, that has connection vanishing at P (use (4.90) with appropriate $F^\alpha_{\beta\gamma}$, so that the point P is at $\bar{\Theta} = 0$ and $\bar{\phi} = 0$ in the new coordinates).

4.6 Equivalence principles

We have the machinery of tensor analysis, at least enough to discuss the physics – we are in a four-dimensional space-time with a metric. From the metric we can calculate connections and curvature, but how to relate these to the physical picture of masses orbiting central bodies? That is: how does mass *generate* curvature? Perhaps a step back from this is the question, *why* does mass generate curvature? Or even: *does* mass generate curvature?

This final question is the most reasonable starting point – my goal here is to make this question precise and begin to answer it. Historically, the important observation (for the physics) is the weak equivalence principle (see [8, 11, 39, 40]).

4.6.1 Newtonian gravity

Newtonian gravity consists of two parts: a theory of motion under the influence of a potential (Newton's second law):

$$\boxed{m\,\ddot{x}^\alpha = -m\,\phi_{,}{}^\alpha} \tag{4.92}$$

and a theory of the potential that relates it to a mass distribution:

$$\boxed{\nabla^2 \phi = 4\pi\,\rho} \tag{4.93}$$

(we set $G = 1$, in units where $c = 1$), very much equivalent to the potential and charge density relation in electrostatics. The first equation is a statement about the effect of the field on a particle of mass m, and the second is a statement about the generation of the field by a source with mass density ρ. Our studies thus far are well suited to a general relativistic form of (4.92).

Given a metric $g_{\mu\nu}$, we can compute the geodesics. The idea is to take the motion induced by a gravitational field, normally thought of as a forced trajectory, and reinterpret it as a geodesic on a Riemannian manifold with some metric. The metric itself will become the "field" of the theory, and Einstein's equation, relating the metric to sources, replaces (4.93). Riemannian space-times are precisely the ones of physical interest: they are completely defined by a metric, with a metric connection and (hence) Riemann tensor related to derivatives of the metric. The metric is the *only* input, and it is the field of interest. We have specialized to four dimensions with one "temporal" coordinate, but that doesn't change much.

Before we begin with equivalence principles, it is important to reiterate the underlying problem with each of the above components (4.92) and (4.93) of the Newtonian theory – the equations of motion and the forces that source them (in (4.92)) lead to different predictions of force in different frames (related by Lorentz transformation), while the relation of ρ to ϕ through the Laplacian (as in (4.93)) violates finite propagation speed (time is not involved at all). One can write a (special) relativistic theory for a potential, but there is a third problem, with the "field" equation of Newtonian gravity: it can only couple to explicitly massive sources. That is to say, we have no way of translating energy into ρ here. We must do this, since mass and energy are equivalent, every form of energy has got to couple to gravity.

Each of these will be addressed as we develop the complete general relativistic form of gravitation – we expect to have to modify the equations of motion for a test particle, and the field equations themselves. In addition, the field equations must allow coupling to all forms of energy.

4.6.2 Equivalence

Weak equivalence principle

Newton's second law gives us a relation between mass, acceleration, and some external forcing. In E&M, for example, we have:

$$m_i \, \mathbf{a} = q \, \mathbf{E} \tag{4.94}$$

for an electric field $\mathbf{E}$. It is clear, from units alone, that the (inertial) mass of the body m_i and the charge of the body q are independent entities.

When we first encounter gravity near the earth, we write:

$$m_i \, \mathbf{a} = m_p \, \mathbf{g}, \tag{4.95}$$

where $\mathbf{g} = -g\,\hat{\mathbf{z}}$ (and $g = 9.8 \, \text{m/s}^2$) pointing down. Here, the "passive" mass of the body m_p is playing the role of the "charge" q in (4.94). The experimental observation, surprising in our early study of physics, is that to great experimental accuracy, $m_i \approx m_p$. That is, all objects, regardless of mass, undergo identical acceleration near the surface of the earth.

This experimental observation forms the basis of the weak equivalence principle, where as a matter of natural law, we set $m_i = m_p$. It is then theoretically impossible to distinguish between a local gravitational field (like the one near the surface of the earth, with $a = g$) and an accelerated frame. If you are at rest and you drop a ball, it falls toward the earth with constant acceleration, and you say it falls because of gravity (a force). If the ball is at rest and you move with constant acceleration relative to the ball, then you observe the same motion, but this time, it is your constant acceleration, not gravity, that is the cause. For the gravitational field near the earth, all masses fall with the same acceleration, and therefore, that acceleration is undetectable (if you fall with the same constant acceleration as the ball, then it does not move relative to you).

How is it that we ever see anything falling, then? If everything: people, laboratories, and the balls in them, fall with the same acceleration, how do we notice balls falling at all? The earth is the anomaly here – usually, when we drop a ball, we are standing at rest – the earth is preventing us from undergoing constant acceleration. If we were falling with the balls, we would observe nothing since we all fall at the same rate. So we can ask the question – does a ball fall because of the force of gravity, or "just" due to a uniform acceleration that we do not experience? This question represents the fundamental shift allowed by the weak equivalence principle – it is possible that what we are calling the force of gravity is really unforced motion in a uniformly accelerating frame. Refer to Figure 4.5.

That shift, from force of nature to feature of the arena, is the beginning of the geometrization of gravity, and allows us to ask any number of interesting questions (if a whole frame moves with constant acceleration, what about the light in it?). How seriously can we take the idea that gravity can be replaced by frame acceleration? After all, we know that gravity near the surface of the earth is only an approximation to the actual (Newtonian) force of gravity associated with a sphere with the mass of the earth.

In order to distinguish between uniform acceleration and Newtonian gravity, we must look at very large distances. Two balls falling toward a common center

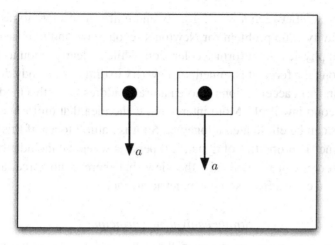

Figure 4.5 Two views: the ball on the left falls due to the force of gravity, the ball on the right falls because it is in a box undergoing constant acceleration.

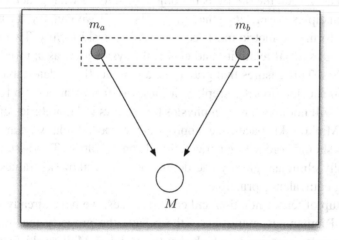

Figure 4.6 For an extremely large laboratory, and over extreme distances, two balls in a box approach each other as shown here, falling toward a common center defined by the spherical Newtonian potential – evidently, to distinguish between constant acceleration and gravity one needs to compare *two* falling objects.

implied by the full Newtonian force:

$$\mathbf{F} = -\frac{G\,M\,m}{r^2}\,\hat{\mathbf{r}},$$ (4.96)

would approach each other as in Figure 4.6. So if we see two balls approaching one another, that is the hallmark of gravitational interaction – that would not happen for two balls falling together under uniform acceleration.

The distinction between *force* of gravity and uniform acceleration is important, a force of gravity is no problem for Newton's second law, and falls neatly into the framework of physics. A uniform acceleration, while it clearly amounts to the same thing, is no longer a force, it is something new (try to draw a free body diagram with it). Indeed, uniform acceleration is more naturally addressed with a modification of Newton's second law itself. More interesting is the idea that uniform acceleration of all objects can be eliminated altogether. So it is natural to ask if this shift (from force to geometric properties of dynamics) persists when we include the complete Newtonian force, can we still take the view that there is no force, and that the perceived force comes from some geometric artifact?

Strong equivalence principle

The plan is to try to understand even the full force of Newtonian gravity as a geometric constraint on motion. Einstein's ultimate idea was to view gravitational motion as geodesic motion on a curved background – a background whose only input is a metric. Then the metric is the object of study – what generates it? What equations/principles govern its generation? About all we can say for now is that whatever its form, the metric must reduce to Minkowski locally. That way, all of our ideas about special relativity and classical physics arise as approximations in "little" patches. This ensures that gravity is, again locally, undetectable as it must be if uniform acceleration is to replace it far away from sources. So for the most part, gravity will not play a role in physics that occurs in laboratories on the earth, we just use Minkowski space-time, appropriate to special relativity, and all of our observations should hold without gravitational consideration. This assumption, that we can locally eliminate gravity and define inertial Minkowski frames, is known as the strong equivalence principle.

In our setup of tensor notation and curved spaces, we have already specialized to (pseudo-)Riemannian manifolds – these have the property that we can find, locally, coordinate systems in which the metric takes Minkowski form (flat), so that these manifolds appropriately support the strong equivalence principle. In addition, Riemannian manifolds are completely described by their metric – given the metric, we can form the Christoffel connection, and from that the Riemann tensor, Ricci tensor, and Ricci scalar. So the metric is the only input needed.[8]

Now all we need to do is figure out what sources to use, and how to relate the metric to them. We will begin this process by following Einstein's original line of thinking. In later chapters, we will see that the ultimate relation between the

---

[8] We could imagine more exotic configurations, in which the metric and torsion, for example, were required to specify the geometry of the manifold.

metric and the appropriate sources comes directly from some relatively simple field theoretic considerations.

4.7 The field equations

I have suggested that we need to look to deviations of two balls being dropped (for example) to distinguish between a frame of constant acceleration and any sort of gravity. We'll analyze this deviation for the case of Newtonian gravity and demand that the motion be interpretable as a deviation between geodesics in a curved space-time. This process will lead to Einstein's equation. In the end, you can't *derive* Einstein's equation any more than you can derive Newton's laws or Maxwell's equations – they can be motivated, but they must be verified by physical observations.

4.7.1 Equations of motion

Let's turn these equivalence principles into some physics. One of the implications of the strong equivalence principle is that we should write down all our physics as tensor equations. That way, no matter where we are, we get the proper transformations out of the physical laws. Seems sensible – let's do the easiest case. A free particle in a flat space-time has acceleration:

$$\frac{dv^\alpha}{d\tau} = 0. \tag{4.97}$$

Because it involves non-covariant derivatives, this is not a viable equation, but the correction (as we have belabored) is easy:

$$\frac{Dv^\alpha}{D\tau} = 0, \tag{4.98}$$

and this just says that free motion occurs along geodesics of the curved space-time – that is also true for the non-covariant form, but the geodesics of flat space are straight lines. Other spaces have different geodesics – great circles on a sphere are "straight lines" in that curved spatial setting.

What should we do about the metric field? How can we combine the equivalence principles to make a concrete statement about the metric? Another question which would be natural to ask (if we hadn't narrowed our view of geometry already) is: is the metric field the only thing that matters? The answer in our setting is "yes", and this again is by construction. Given the metric, we can construct its derivatives to get $\Gamma^\alpha{}_{\beta\gamma}$ and take *its* derivatives to get $R^\alpha{}_{\beta\gamma\delta}$, then work our way through the rest of the list of important tensors. We are looking, then, for a field equation to set constraints on $g_{\mu\nu}$. The equation of geodesic motion:

$$\ddot{x}^\alpha + \Gamma^\alpha{}_{\beta\gamma}\, \dot{x}^\beta \dot{x}^\gamma = 0 \tag{4.99}$$

involves the metric, and presumably, the deviation between two geodesics would as well.

We claim that locally, gravity isn't observable, everyone travels along their geodesics and no one is the wiser. But this is only true in small patches. Suppose we are on an elevator freely falling toward the center of the earth, we drop two balls along with ourselves. We would see the balls (which look, over a small range, like they are falling parallel to one another) begin to approach each other (think of a radially directed field pointing away from each ball to some external center) as in Figure 4.6. Similarly, in a curved space, the way in which the balls approach each other would tell us something about the curvature of the space.

Let's be more explicit. A geodesic in flat space is a straight line – if we have two straight lines that cross at a point, and you and a friend start there and walk along your geodesics, you notice the distance between the two of you increasing (linearly). For two people trapped on the surface of a sphere, you start together at the south pole, say, and walk along your geodesics – the distance between you increases at first, but then decreases until you meet again at the north pole. In a sense, we can measure curvature in this manner, by violating the "small local patch" assumption of the equivalence principle. So we consider the classical and relativistic form of the "geodesic deviation equations".

4.7.2 Newtonian deviation

On the classical side, we have a gravitational field defined by a potential ϕ and connected to a matter distribution ρ by $\nabla^2\phi = 4\pi\rho$. We take two test particles in this field defined by their coordinate vectors $\mathbf{x}_1(t)$ and $\mathbf{x}_2(t)$, then the equations of motion for the two particles are:

$$\ddot{\mathbf{x}}_1 = -\nabla\phi(\mathbf{x}_1)$$
$$\ddot{\mathbf{x}}_2 = -\nabla\phi(\mathbf{x}_2). \tag{4.100}$$

Consider the situation at time t_0, then the particles are at $\mathbf{x}_1(t_0)$ and $\mathbf{x}_2(t_0)$ and the accelerations read:

$$\ddot{\mathbf{x}}_1(t_0) = -\nabla\phi(\mathbf{x}_1(t_0))$$
$$\ddot{\mathbf{x}}_2(t_0) = -\nabla\phi(\mathbf{x}_2(t_0)) = -\nabla\phi(\mathbf{x}_1(t_0) + \boldsymbol{\eta}(t_0)), \tag{4.101}$$

where we have defined $\boldsymbol{\eta}(t_0) \equiv \mathbf{x}_2(t_0) - \mathbf{x}_1(t_0)$, the separation vector as shown in Figure 4.7. Assuming the curves are relatively close together, we take $\boldsymbol{\eta}$ to be small, and expand the potential about the first point $\mathbf{x}_1(t_0)$:

$$\ddot{\mathbf{x}}_2(t_0) = -\nabla\phi(\mathbf{x}_1(t_0)) + \boldsymbol{\eta} \cdot \nabla(\nabla\phi(\mathbf{x}_1(t_0))) = \ddot{\mathbf{x}}_1(t_0) - \boldsymbol{\eta} \cdot \nabla(\nabla\phi(\mathbf{x}_1(t_0))), \tag{4.102}$$

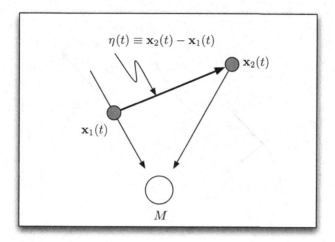

Figure 4.7   The separation vector η between two different gravitationally forced trajectories has time evolution given by (4.103).

from which we conclude that classically:

$$\ddot{\eta} = -\eta \cdot \nabla\left(\nabla\phi\right).$$

(4.103)

4.7.3 Geodesic deviation in a general space-time

For the same problem treated on a curved Riemannian manifold, we need to do a little more setup (basically, it's just a bit harder to do the Taylor expansion). Consider two geodesics "close together", as shown in Figure 4.8 – we have two natural vectors, the tangent to the curve at constant σ (σ is a curve-selecting parameter): $\dot{x}^\alpha(\tau,\sigma) \equiv \frac{\partial x^\alpha(\tau,\sigma)}{\partial\tau}$, and the orthogonal σ derivative: $x'^\alpha(\tau,\sigma) \equiv \frac{\partial x^\alpha(\tau,\sigma)}{\partial\sigma}$. We should be careful – these two directions form a two-dimensional surface on our space (read space-time), that they are orthogonal is something we need to establish. Well, suppose they were orthogonal at some point τ_0, then by our metric connection $g_{\alpha\beta;\gamma} = 0$, we know that they will be orthogonal at all points along (either) curve. I claim (without proof) that these two vector fields can be made orthogonal at a point and hence are orthogonal everywhere.

With this coordinate system in place (two-dimensional, a τ-direction along the geodesic and a σ-direction moving us to some other geodesic), we can calculate the displacement vector η^α at a constant τ:

$$\eta^\alpha = \frac{\partial x^\alpha}{\partial\sigma}\,d\sigma.$$

(4.104)

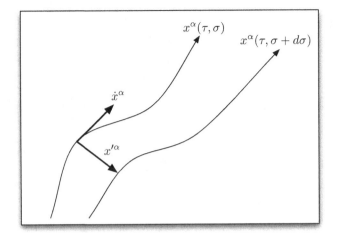

Figure 4.8   Two geodesic curves: $x^\alpha(\tau, \sigma)$ and $x^\alpha(\tau, \sigma + d\sigma)$.

We are interested in how this vector changes as we move along the curve $x^\alpha(\tau, \sigma)$, that's a question answered by the covariant derivative:

$$
\frac{D\eta^\alpha}{D\tau} = \frac{D}{D\tau}\left(\frac{\partial x^\alpha}{\partial \sigma}\right) d\sigma = \left(\frac{\partial^2 x^\alpha}{\partial \sigma \partial \tau} + \Gamma^\alpha_{\beta\gamma} \frac{\partial x^\beta}{\partial \sigma} \frac{\partial x^\gamma}{\partial \tau}\right) d\sigma
$$
$$
= \frac{D}{D\sigma}\left(\frac{\partial x^\alpha}{\partial \tau}\right) d\sigma.
$$

(4.105)

This is what we would call the "velocity" of the deviation, to get the acceleration (in order to compare with the Newtonian form) we need to take one more τ-derivative:

$$
\frac{D}{D\tau}\left(\frac{D\eta^\alpha}{D\tau}\right) = \left(\frac{D}{D\tau}\left(\frac{D\dot{x}^\alpha}{D\sigma}\right) - \frac{D}{D\sigma}\left(\frac{D\dot{x}^\alpha}{D\tau}\right) + \frac{D}{D\sigma}\left(\frac{D\dot{x}^\alpha}{D\tau}\right)\right) d\sigma, \quad (4.106)
$$

where we have added zero by adding and subtracting the same term. The first two terms in (4.106) look a lot like a commutator of covariant derivatives. Indeed, for a vector $f^\alpha(\tau, \sigma)$ defined on our surface via parallel transport, we have:

$$
\frac{D}{D\tau}\left(\frac{Df^\alpha}{D\sigma}\right) = f^\alpha_{;\beta\gamma} x'^\beta \dot{x}^\gamma + f^\alpha_{;\gamma} \frac{Dx'^\gamma}{D\tau}
$$
$$
\frac{D}{D\sigma}\left(\frac{Df^\alpha}{D\tau}\right) = f^\alpha_{;\beta\gamma} \dot{x}^\beta x'^\gamma + f^\alpha_{;\gamma} \frac{D\dot{x}^\gamma}{D\sigma}
$$

(4.107)

but from (4.105), the last term on the right in each of the above are equal, so when we subtract, we get:

$$\frac{D}{D\tau}\left(\frac{Df^\alpha}{D\sigma}\right) - \frac{D}{D\sigma}\left(\frac{Df^\alpha}{D\tau}\right) = \left(f^\alpha_{;\beta\gamma} - f^\alpha_{;\gamma\beta}\right)\dot{x}^\beta x'^\gamma$$

$$= -R^\alpha_{\ \rho\gamma\beta}x'^\gamma \dot{x}^\beta f^\rho,$$

(4.108)

where the last line follows from the definition of the Riemann tensor. Now, setting $f^\alpha = \dot{x}^\alpha$ we can finish off (4.106):

$$\boxed{\frac{D^2\eta^\alpha}{D\tau^2} = -R^\alpha_{\ \rho\gamma\beta}\dot{x}^\beta \dot{x}^\rho \eta^\gamma,}$$

(4.109)

and the last term in (4.106) is zero since $\dot{x}^\alpha$ is parallel transported along x^α because these are geodesics.

Remember that the goal here is to associate Newtonian trajectory deviation with geodesic deviation, so we want to compare (4.109) to (4.103), but what we've got now are two different expressions for the deviation vector η in two different spaces (three-dimensional flat space, and some unspecified four-dimensional space-time). If we squint at (4.103), then we could write $\ddot{\eta}_\beta = -\eta_\gamma\,\partial^\gamma\partial_\beta\phi = -\eta^\alpha\,\phi_{,\alpha\beta}$, and we might convince ourselves that a reasonable association is $\phi_{,\alpha\beta} \approx R_{\alpha\rho\beta\gamma}\dot{x}^\gamma\dot{x}^\rho$.

That's all well and good, but what should we do with the "source" equation – the Newtonian potential ϕ that appears in (4.103) is sourced by a matter distribution ρ via Poisson's equation $\nabla^2\phi = 4\pi\rho$. The matter density, ρ, can be written in terms of a stress–energy tensor $T_{\alpha\beta}$ via $\rho \approx T_{\alpha\beta}u^\alpha u^\beta$ for an "observer" (the particle reacting to ρ) traveling with four-velocity u^α.[9] This leads to the final association:

$$\nabla^2\phi = \phi^{\ \alpha}_{,\alpha} \approx R^\alpha_{\ \rho\alpha\gamma}\dot{x}^\gamma\dot{x}^\rho \approx 4\pi T_{\alpha\beta}\dot{x}^\alpha\dot{x}^\beta \Rightarrow \boxed{R_{\rho\gamma} \approx 4\pi T_{\rho\gamma}.}$$

(4.110)

I leave the $\approx$ symbol in because, and I want to stress this, the above equation is incorrect. You can't deny that it is exactly what we want, though – precisely a relationship between the metric, its derivatives, and a matter source. Indeed, the jump from the above to Einstein's final equation (and apparently he wrote down the above as well) is a short one. Regardless of your outlook, in terms of how we got here, let its accuracy be our guide, we will be convinced when we see the correct physics come out.

---

[9] In special relativity, we learn how to make an observation of density – and we will see this later on, but the ρ component of the most general distribution $T^{\mu\nu}$ is observed by contracting the stress tensor with the observer's four-velocity (the local temporal basis vector in the observer's rest frame). This means that an energy density measurement, ρ, must involve the four-velocities of particles measuring it.

Aside: stress-energy tensors

We know from Newtonian gravity that a (stationary) matter distribution $\rho(x)$ (energy density, more generally) generates a Newtonian gravitational field. In E&M, a charge distribution generates the electric field and current generates the magnetic – in four-vector language we have $J^\alpha(x)$ (zero component is $\rho(x)$, spatial component is $\mathbf{J}(x)$). Remember, from Section 2.6, that in order for (even) Newtonian gravity to be correct with respect to special relativity, we had to include a moving mass source, analogous to current. So it is reasonable that the full gravitational theory, GR, should have both mass(/energy) density and moving mass/energy as sources.

Finally, again going back to E&M, the fields $\mathbf{E}$ and $\mathbf{B}$ *also* carry energy and momentum, and must be considered as a possible source for gravity. In the end, then, we are looking for an object that includes both ρ, and $\rho\,\mathbf{v}$ (taken together to form J^α), and in addition, is a second-rank symmetric tensor (this is necessary since the left-hand side of Einstein's equation is a symmetric, second-rank tensor). The natural choice is the stress–energy tensor $T^{\alpha\beta}$ – we know how to construct this for E&M (see, for example, [24]), and for fluids, and will learn how to construct stress–energy tensors for other physical theories later on.

4.8 Einstein's equation

The geodesic deviation correspondence that we have reads:[10]

$$R_{\mu\nu} \approx 4\pi\, T_{\mu\nu}. \tag{4.111}$$

For electrodynamics, the full four-dimensional stress–energy tensor gives us a set of conservation laws encapsulated in the divergence condition $T^{\mu\nu}{}_{;\mu} = 0$. These four equations, which behave like $J^\mu{}_{;\mu} = 0$ for charge conservation, effectively give us the relations between energy density and the Poynting vector, and the Poynting vector and Maxwell stress tensor (the spatial components of $T^{\mu\nu}$) in the language of E&M. If we take the covariant divergence of both sides of (4.111), then we should get zero (by the definition of energy–momentum tensors, this property is not unique to the Maxwell case). What can we say about the derivatives of the Ricci tensor?

Remember we had the Bianchi identity for the Riemann tensor:

$$R_{\alpha\rho\gamma\beta;\delta} + R_{\alpha\rho\delta\gamma;\beta} + R_{\alpha\rho\beta\delta;\gamma} = 0, \tag{4.112}$$

but we want the Ricci tensor form of this statement, so hit the above with $g^{\alpha\gamma}$:

$$R_{\rho\beta;\delta} - R_{\rho\delta;\beta} + R^\gamma{}_{\rho\beta\delta;\gamma} = 0 \tag{4.113}$$

---

[10] We developed this with some squinting that comes from, for example, [39] – for a detailed treatment with appropriate basis vector projections, see [9].

and multiplying this by $g^{\rho\delta}$, for example, gives:

$$
\begin{aligned}
0 &= R^{\rho}{}_{\beta;\rho} - R^{\rho}{}_{\rho;\beta} + R^{\gamma\rho}{}_{\beta\rho;\gamma} \\
&= R^{\rho}{}_{\beta;\rho} - R_{;\beta} + R^{\gamma}{}_{\beta;\gamma} \\
&= 2 R^{\rho}{}_{\beta;\rho} - R_{;\beta}
\end{aligned}
\tag{4.114}
$$

so we learn that $R^{\mu}{}_{\nu;\mu} = \frac{1}{2} R_{;\nu}$. This is not, in general, zero, and that violates our expanded notion of conservation for a stress–energy tensor – if we take the divergence of both sides of (4.111), we will not achieve equality. The situation is similar to quasistatic electricity and magnetism (where we lack the $\frac{\partial \mathbf{E}}{\partial t}$ term as a source for $\mathbf{B}$), charge conservation is violated unless we include the full Maxwell source terms. The "fix" is also similar – in E&M we just add back in the residual term.

We can write the Bianchi identity, boiled down to apply to the Ricci tensor divergence, as:

$$
R^{\mu\nu}{}_{;\mu} = \frac{1}{2} (R\, g^{\mu\nu})_{;\mu} \longrightarrow \left(R^{\mu\nu} - \frac{1}{2} g^{\mu\nu} R \right)_{;\mu} = 0.
\tag{4.115}
$$

So the cute trick: suppose we replace the lone Ricci tensor on the left of (4.111) with the combination above (and to get the correct limiting behavior, we also multiply the right-hand side by 2):

$$
\boxed{R_{\mu\nu} - \frac{1}{2} g_{\mu\nu} R = 8\pi\, T_{\mu\nu}.}
\tag{4.116}
$$

This is Einstein's equation – the tensor on the left is called the Einstein tensor:

$$
\boxed{G_{\mu\nu} \equiv R_{\mu\nu} - \frac{1}{2} g_{\mu\nu} R,}
\tag{4.117}
$$

and (4.116) tells us, given a distribution of "source" (as typified by the energy–momentum tensor), how to construct $g_{\mu\nu}$ – notice that although it is simple to write down the equation, the left-hand side involves quadratic derivatives in the metric and is highly nonlinear (remember that the Riemann tensor itself has terms like $\Gamma\Gamma$, which are quadratic in first derivatives of the metric). In general, solving Einstein's equation exactly as written is almost impossible. So as we go along, we will do two important things to make life tractable. The first thing we can do is simplify the form of the metric based on physical arguments (symmetries, for example). Then there are vacuum solutions, where we move away from the matter itself (or other fields) and consider the source-free solutions.

We will return to the problem of solving Einstein's equation, in vacuum and elsewhere, after a discussion of the field theoretic formulation of GR. If you are more interested in solutions, by all means, skip to Chapter 7. We are going to approach Einstein's equation from a very different direction, and one that I feel illuminates some of the interpretive issues of the theory. But, in order to discuss GR as a model theory for second-rank symmetric tensors, we need some machinery from classical field theory, and it is this set of ideas that we will look at next. The interesting result is that, just as Maxwellian electrodynamics follows almost uniquely from the most general first-rank, linear, (special) relativistic tensor field theory one can form, GR – with all of its complexity and interpretation – is again "almost" uniquely the simplest consistent second-rank tensor theory. This lends an element of manifest destiny to the equations themselves, and gives us a way to understand the solutions without an *a priori* geometric identification.

Problem 4.11

Two identical masses (mass m) are falling in toward a central body of mass M along the same radial line. The mass that is closest to the central body is at r_1, and the second mass is at $r_2 = r_1 + \eta$. Find the acceleration of each body at these points – construct the difference $\ddot{r}_2 - \ddot{r}_1 = \ddot{\eta}$, and expand for η small. Verify that you get the same expression from the equation of Newtonian deviation (4.103) applied to this configuration.

Problem 4.12

We can consider an object with "negative mass" – the equality of inertial and passive mass in Newtonian gravity leads to a prediction for the direction of acceleration of a pair of masses with opposite signs that is different from the analogous electrical configuration. For two masses m_ℓ and m_r at locations x_ℓ and x_r on the x-axis, with $x_\ell < x_r$ (the subscripts refer to the left and right masses), find the force on m_ℓ due to m_r and the force on m_r due to m_ℓ. Show that if $m_\ell < 0$ and $m_r > 0$, the forces point in opposite directions, but using Newton's second law, the accelerations point in the *same* direction. Evidently, objects with opposite mass sign will chase each other (see [29] and [3], for the general relativistic form of this negative mass motion).

Problem 4.13

Starting from the field equation for Newtonian gravity:

$$\nabla^2 \phi = 4 \pi \, G \, \rho_m, \tag{4.118}$$

with ρ_m the mass density, rewrite in terms of ρ_e, energy density.

Suppose you now made the argument that $T_{\mu\nu} u^\mu u^\nu \sim \rho_e$, as we did in Section 4.7, but without using $G = c = 1$ – what factor of c must you introduce so that $T_{\mu\nu} u^\mu u^\nu$ has units of energy density?

By following the identification, $R_{\mu\nu} u^\mu u^\nu \sim \nabla^2 \phi$ (from Section 4.7), write (4.118) in terms of $R_{\mu\nu}$ and $T_{\mu\nu}$ with all the units in place (i.e. insert the correct factors of G

and *c*). Note that the only thing you can't get out of this is the factor of 2 that takes the 4π above to 8π in Einstein's equation.

Problem 4.14
We can measure mass in meters – find the conversion factor α below:

$$M_{\text{in meters}} = \alpha \, M_{\text{in kilograms}} . \tag{4.119}$$

Using this factor, find the mass of the sun in meters. What is α when $G = c = 1$ (this should establish that when we set $G = c = 1$, we are measuring mass in meters).

Problem 4.15
Download the `Mathematica` package "EinsteinVariation.m" and associated workbook. Using the definitions and functions you find there:
(a) Verify that Minkowski space-time expressed in spherical coordinates is flat.
(b) Find the curvature scalar (R) for the torus from Problem 4.6.

Problem 4.16
Prolate spheroidal coordinates are related to spherical coordinates – there are two angles (called θ and ϕ) and a hyperbolic trigonometric coordinate ξ related to the spherical r. In order to maintain the units of length, the parameter a is used to define the coordinate system in terms of the Cartesian set:

$$x = a \, \sinh(\xi) \, \sin(\theta) \, \cos(\phi)$$
$$y = a \, \sinh(\xi) \, \sin(\theta) \, \sin(\phi) \tag{4.120}$$
$$z = a \, \cosh(\xi) \, \cos(\theta).$$

(a) From the line element:

$$ds^2 = dx^2 + dy^2 + dz^2, \tag{4.121}$$

with dx, dy, and dz written in terms of $d\xi, d\theta$, and $d\phi$, extract the metric $g_{\mu\nu}$ for this coordinate system (use the ordering $d\bar{x}^\mu \doteq (d\xi, d\theta, d\phi)^T$).
(b) Find the nonzero Christoffel connection values.
Use the `Mathematica` package for this problem.

Problem 4.17
You showed, previously, that there is a set of coordinates on the sphere which demonstrates conformal flatness explicitly – the line element in both spherical coordinates and this new system is:

$$ds^2 = r^2 \left(d\theta^2 + \sin^2 \theta \, d\phi^2\right) = r^2 \operatorname{sech}(\bar{\theta})^2 \left(d\bar{\theta}^2 + d\phi^2\right). \tag{4.122}$$

We know, by construction, that these two sets of coordinates, $\{\theta, \phi\}$ and $\{\bar{\theta}, \phi\}$, describe the same space. Use the `Mathematica` package to establish that R, the

Ricci scalar, is the same in each set of coordinates (since R is a scalar, it must be coordinate-independent).

Problem 4.18

Einstein's equation:

$$R_{\mu\nu} - \frac{1}{2} g_{\mu\nu} R = 8\pi \, T_{\mu\nu} \tag{4.123}$$

can be written in the equivalent form, in $D = 4$:

$$R_{\mu\nu} = 8\pi \left(T_{\mu\nu} - \frac{1}{2} g_{\mu\nu} T \right) \tag{4.124}$$

for $T \equiv T^\alpha_\alpha = g_{\alpha\beta} T^{\alpha\beta}$, the trace of the stress tensor. Show that this is true, and find the analogue of this equation for arbitrary dimension.

*Problem 4.19

(a) The Bianchi identity (4.86) for the Riemann tensor was a cyclic relation involving the derivatives of $R_{\alpha\beta\gamma\delta}$. The electromagnetic field strength tensor is defined as $F_{\mu\nu} = \partial_\mu A_\nu - \partial_\nu A_\mu$ for four-potential A^μ. Find the analogue of the Bianchi identity for $F_{\mu\nu}$ (assume we're in flat Minkowski space-time) – i.e. a cyclic relation involving the derivatives of $F_{\mu\nu}$.

(b) Using the Bianchi identity for the Riemann tensor, what is the relation between the gradient of the Ricci scalar, $R_{,\gamma}$, and the "divergence" of the Ricci tensor, $R^\mu_{\nu;\mu}$?

5

Scalar field theory

The ultimate goal of the next two chapters is to connect Einstein's equation to relativistic field theory. The original motivation for this shift away from geometry, and toward the machinery of field theory, was (arguably) the difficulty of unifying the four forces of nature when one of these is not a force. There was a language for relativistic fields, and a way to think about quantization in the context of that language. If gravity was to be quantized in a manner similar to E&M, it needed a formulation more like E&M's.

We will begin by developing the Lagrangian description for simple scalar fields, this is a "continuum"-ization of Newton's second law, and leads immediately to the wave equation. Once we have the wave equation, appropriate to, for example, longitudinal density perturbations in a material, we can generalize to the wave equation of empty space, which, like materials, has a natural speed. This gives us the massless Klein–Gordon scalar field theory. We will explore some of the Lagrangian and Hamiltonian ideas applied to fields, and make the connection between these and natural continuum forms of familiar (point) classical mechanics. In the next chapter, we will extend to vector fields, discuss electricity and magnetism, and move on to develop the simplest second-rank symmetric tensor field theory. This is, almost uniquely, Einstein's general relativity, and the field of interest is the metric of a (pseudo-)Riemannian space-time.

5.1 Lagrangians for fields

Consider a spring connecting two masses in one dimension (see Figure 5.1). The location of the left mass we'll call x_{-1} and the location of the right, x_1. We want to find the time-dependence of the motion of the two masses: $x_1(t)$ and $x_{-1}(t)$.

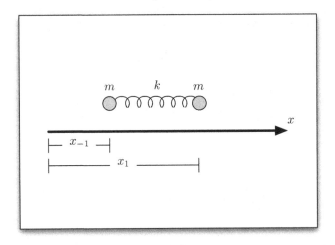

Figure 5.1   Two identical masses connected by a spring.

We can form the Lagrangian, the kinetic energy is just:

$$T = \frac{1}{2} m \, \dot{x}_1^2 + \frac{1}{2} m \, \dot{x}_{-1}^2, \tag{5.1}$$

as always for a two-mass system. The potential energy depends on the extension of the spring, $x_1 - x_{-1}$, and the equilibrium length of the spring, call it a:

$$U = \frac{1}{2} k \left((x_1 - x_{-1}) - a \right)^2, \tag{5.2}$$

so that:

$$L = T - U = \frac{1}{2} m \left(\dot{x}_1^2 + \dot{x}_{-1}^2 \right) - \frac{1}{2} k \left((x_1 - x_{-1}) - a \right)^2. \tag{5.3}$$

The equations of motion can be obtained in the usual way:

$$0 = \left(\frac{d}{dt} \frac{\partial L}{\partial \dot{x}_1} - \frac{\partial L}{\partial x_1} \right) = m \, \ddot{x}_1 + k \left((x_1 - x_{-1}) - a \right)$$

$$0 = \left(\frac{d}{dt} \frac{\partial L}{\partial \dot{x}_{-1}} - \frac{\partial L}{\partial x_{-1}} \right) = m \, \ddot{x}_{-1} - k \left((x_1 - x_{-1}) - a \right). \tag{5.4}$$

If we introduce more springs, the same basic procedure holds – take three masses, now we have three coordinates, labeled x_{-1}, x_0, and x_1 (see Figure 5.2). The Lagrangian undergoes the obvious modification:

$$L = \frac{1}{2} m \left(\dot{x}_{-1}^2 + \dot{x}_0^2 + \dot{x}_1^2 \right) - \frac{1}{2} k \left(\left((x_1 - x_0) - a \right)^2 + \left((x_0 - x_{-1}) - a \right)^2 \right), \tag{5.5}$$

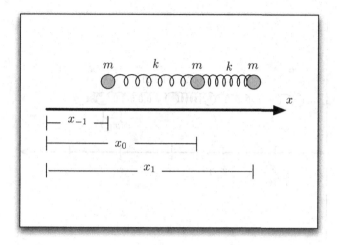

Figure 5.2   Three masses connected by two identical springs.

and the equations of motion follow. In particular, consider the equation for $x_0(t)$:

$$m\,\ddot{x}_0 - k\,(x_{-1} - 2\,x_0 + x_1) = 0. \tag{5.6}$$

Notice that for this "internal" mass, there is no reference to the equilibrium position a. As we add more springs and masses, more of the equations of motion will depend only on the relative displacements on the left and right. In the end, only the boundary points (the left and right-most masses, with no compensating spring) will show any a-dependence.

What we want to do, then, is switch to new coordinates that make the Lagrangian manifestly independent of a. We can accomplish this by defining a set of local relative positions. Take a uniform grid $\{\bar{x}_j\}_{j=-N}^{N}$ with $\bar{x}_j \equiv j\,a$ for a the grid spacing, and introduce the variables $\{\phi(x_j, t)\}_{j=-N}^{N}$ to describe the displacement from $\bar{x}_j$ as shown in Figure 5.3, so that:

$$\phi(x_j, t) \equiv x_j(t) - \bar{x}_j \tag{5.7}$$

in terms of our previous position coordinates $\{x_j(t)\}_{j=-N}^{N}$.

Transforming the Lagrangian is simple – for each of the terms in the potential, we have (ignoring endpoints):

$$((x_j(t) - x_{j-1}(t)) - a)^2 = (\phi(x_j, t) - \phi(x_{j-1}, t))^2, \tag{5.8}$$

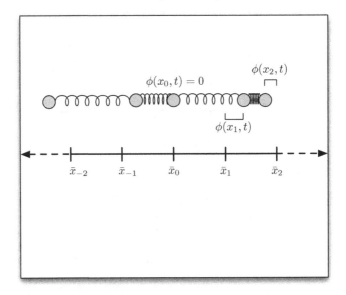

Figure 5.3   A grid with uniform spacing a and a few of the displacement variables $\phi(x_j, t)$ shown.

and the kinetic terms are replaced via $\dot{x}_j(t) = \dot{\phi}(x_j, t)$. So, the Lagrangian in these new coordinates is:

$$L = \frac{1}{2} m \sum_{j=-N}^{N} \dot{\phi}(x_j, t)^2 - \frac{1}{2} k \sum_{j=-N+1}^{N} \left(\phi(x_j, t) - \phi(x_{j-1}, t)\right)^2. \qquad (5.9)$$

5.1.1 The continuum limit for equations of motion

The equations of motion become, ignoring the boundary points (i.e. assume we are looking at a finite segment of an infinite chain):

$$m \ddot{\phi}(x_j, t) - k \left(\phi(x_{j+1}, t) - 2\phi(x_j, t) + \phi(x_{j-1}, t)\right) = 0, \qquad (5.10)$$

capturing the sentiment of (5.6).

When we move to a continuous system, we will take $\phi(x_j, t) \longrightarrow \phi(x, t)$, with $\phi(x, t)$ the displacement of the patch of mass that is in equilibrium at x, relative to x. The equations of motion (5.10), written with a continuous x index ($x_j \longrightarrow x$, $x_{j\pm1} \longrightarrow x \pm a$), are:

$$m \ddot{\phi}(x, t) - k \left(\phi(x + a, t) - 2\phi(x, t) + \phi(x - a, t)\right) = 0. \qquad (5.11)$$

We are en route to sending $a \longrightarrow 0$, and in preparation for that move, we can make the observation that, as a function, $\phi(x \pm a, t)$ can be written in terms of $\phi(x, t)$

via Taylor expansion (for small a):

$$\phi(x+a,t) - 2\phi(x_j,t) + \phi(x-a,t)$$

$$\approx \left(\phi(x,t) - a\,\frac{\partial\phi(x,t)}{\partial x} + \frac{1}{2}a^2\,\frac{\partial^2\phi(x,t)}{\partial x^2} \right)$$

$$- 2\phi(x,t) + \left(\phi(x,t) + a\,\frac{\partial\phi(x,t)}{\partial x} + \frac{1}{2}a^2\,\frac{\partial^2\phi(x,t)}{\partial x^2} \right) \qquad (5.12)$$

$$= a^2\,\frac{\partial^2\phi(x,t)}{\partial x^2} + O(a^4),$$

so that we may write the equation of motion approximately as:

$$m\,\ddot{\phi}(x,t) - k\,a^2\,\phi''(x,t) = 0. \qquad (5.13)$$

This is an approximate result for a, the grid spacing, small, and written with a continuous index x. We are now prepared to take the final step, passing to the limit $a \longrightarrow 0$, so that we are describing a continuum of springs and balls. The mathematical move is relatively simple here, but the moral point is considerable: we have promoted $\phi(x_j,t)$, a continuous function of t labeled by x_j, to $\phi(x,t)$, a continuous function of x, i.e. position. We now have a scalar *field*, a continuous function that assigns, to each point x, at each time t, a displacement value. In order to take the limit of (5.13), we need to write m and k in terms of variables that are fixed as $a \longrightarrow 0$ (we have omitted an implicit $O(a^4)$ in (5.13) to avoid clutter, so that the limit process will additionally eliminate detritus from the expansion (5.12)).

Suppose we consider a finite segment of an infinite chain with fixed total length $L = a\,(2\,N)$, modeling a flexible "rod". We let $\mu = \frac{m}{a}$ be the constant mass density (here just mass per unit length) of the chain. Looking at the equation of motion (5.13), we see that dividing through by a will give $\frac{m}{a}$ for the first term, that's precisely the μ we want:

$$\frac{m}{a}\,\ddot{\phi}(x,t) - k\,a\,\phi''(x,t) = 0. \qquad (5.14)$$

The second term in (5.14) has the factor $k\,a$ – evidently, if we want to take $a \longrightarrow 0$ to yield a reasonable limit, we must shrink down the mass of each individual ball so as to keep $\frac{m}{a} \equiv \mu$ constant, and at the same time, increase the spring constant of the springs connecting the balls so that $k\,a$ remains constant.

It is pretty reasonable to assume a finite length and finite mass as we shrink $a \longrightarrow 0$, but what physical description should we give to the stiffness of the rod? We are building a continuous object out of springs, so shouldn't the rod itself have

some associated spring constant?[1] We can assign to the rod a net spring constant
K that we will keep fixed.

Springs in series

Think of a set of Q identical springs in series. If we extend the set a distance Δx, then
the energy stored in the configuration is:

$$U = \frac{1}{2} K \Delta x^2. \tag{5.15}$$

We want to know, given that the serial combination has constant K, what the spring
constant of the constituent springs is. Assuming the combination is homogenous, each
spring stores one Qth of the total energy, and the displacement of each spring is $\frac{\Delta x}{Q}$.
We can find the individual spring constant k by setting the energy in a single spring
equal to $\frac{1}{Q} U$:

$$\frac{1}{Q} \frac{1}{2} K \Delta x^2 = \frac{1}{2} k \left(\frac{\Delta x}{Q} \right)^2 \longrightarrow k = K Q. \tag{5.16}$$

For our current setup, with $2N + 1$ balls and $2N$ springs, we have $Q = 2N$, and
$K = \frac{k}{2N}$. Now we can see that as $N \longrightarrow \infty$, we must send $k \longrightarrow \infty$ to keep K
constant. This is just what we get from the usual spring-addition formula for springs
in series, $k_{tot} = (1/k_1 + 1/k_2)^{-1}$, for our system of $2N$ identical springs.

We can replace the local constants m, k, and a in (5.14) with the continuum
constants $\mu = \frac{m}{a}$, $K = \frac{k}{2N}$, and $L = 2N a$:

$$\mu \ddot{\phi}(x, t) - \frac{L}{2N} (2N) K \phi''(x, t) = 0. \tag{5.17}$$

Now for the limit, take $N \longrightarrow \infty$ (corresponding to $a \longrightarrow 0$) – all that happens is
we lose the $O(a^4)$ terms from (5.12), which we have dropped anyway:

$$\ddot{\phi}(x, t) = \frac{K L}{\mu} \phi''(x, t). \tag{5.18}$$

Notice that the constant $K L$ looks like a force (tension, for example) and $K L/\mu$
has units of velocity squared. What we have is a wave equation representing
longitudinal propagation of displacement, a slinky if you like. If we define $v^2 \equiv \frac{K L}{\mu}$,
then the solutions to the above are the usual:

$$\phi(x, t) = f_\ell(x + v t) + f_r(x - v t), \tag{5.19}$$

---

[1] This physical property is called "Young's modulus" in the theory of elasticity.

so that given some initial displacement function, $\phi(x, 0) = f_\ell(x) + f_r(x)$, we know how the displacement propagates as a function of time. We started with balls and springs, and ended with a function $\phi(x, t)$ that tells us, for each time t, by how much a bit of slinky is displaced from its "equilibrium", the local compression and stretching of the slinky as a function of time. More concretely, $\phi(x, t)$ tells us where the mass that *should* be at x (in equilibrium) *is* relative to x, at time t.

We have taken a discrete chain, and turned its equations of motion into a continuous wave equation. This is a standard move when developing the equations of motion for a field, and can be found in [16] and [26] (for example). We now back up, from the familiar wave equation, to the Lagrangian that generated it in the first place.

5.1.2 The continuum limit for the Lagrangian

Given that we have an equation of motion for a simple scalar field, it is natural to return to the Lagrangian, and try to find a variational principle that generates the field equation. It is reasonable to ask why we are interested in a Lagrangian – after all, we have the field equation (5.18), and its most general solution, why bother with the Lagrangian from whence it came? The situation is analogous to mechanics – in the normal progression of mechanics, one starts with free-body diagrams and Newton's laws, and then moves to a Lagrangian description (then Hamiltonian, etc.) – the motivation is coordinate freedom. On the field side, the same situation holds – Lagrangians and variational principles provide a certain generality, in addition to a more compact description.

From a modern point of view, field theories are *generated* by their Lagrangians. Most model building, extensions of different theories, occurs at the level of the theory's action, where interpretation is somewhat simpler, and the effect of various terms well-known in the field equations. Finally, perhaps the best motivation of all, we recover a field-theoretic application of Noether's theorem.

So, we return to (5.9):

$$L = \frac{1}{2} m \sum_{j=-N}^{N} \dot{\phi}(x_j, t)^2 - \frac{1}{2} k \sum_{j=-N+1}^{N} \left(\phi(x_j, t) - \phi(x_{j-1}, t) \right)^2 . \qquad (5.20)$$

In order to take the limit, we will introduce our constant, physical values, μ, K and L in the Lagrangian (don't be confused by the use of L on the right to refer to the length of the rod, and L on the left to refer to the Lagrangian):

$$L = \frac{1}{2} \mu \sum_{j=-N}^{N} \dot{\phi}(x_j, t)^2 a - \frac{1}{2} \frac{KL}{a} \sum_{j=-N+1}^{N} \left(\phi(x_j, t) - \phi(x_{j-1}, t) \right)^2 . \qquad (5.21)$$

The sums above will approximate integrals in the limit $a \longrightarrow 0$. The first term, as a sum, approximates the definite integral on the right below:

$$\sum_{j=-N}^{N} \dot{\phi}(x_j, t)^2 a = \int_{-\frac{1}{2}L}^{\frac{1}{2}L} \dot{\phi}(x, t)^2 \, dx + O(a^2), \qquad (5.22)$$

where we have again moved from $x_j \longrightarrow x$, representing our continuum view of the function $\phi(x, t)$. In the limit, the sum will converge to the integral (under appropriate smoothness restrictions). The second sum in (5.21) is approximately:

$$\sum_{j=-N+1}^{N} \left(\phi(x_j, t) - \phi(x_{j-1}, t) \right)^2 \approx \int_{-\frac{1}{2}L}^{\frac{1}{2}L} a \left(\phi'(x, t) \right)^2 \, dx, \qquad (5.23)$$

since the difference $\phi(x_j, t) - \phi(x_{j-1}, t) \approx \phi'(x, t) a$, and the $a \longrightarrow dx$ in the sum-goes-to-integral process, leaving us with an additional factor of a in the integrand (just right to kill the factor of a^{-1} sitting in front of the relevant term in (5.21)). Since we have written all constants m, k and a in terms of constants that will remain fixed under the limit in (5.21), the Lagrangian becomes, as $a \longrightarrow 0$:

$$\begin{aligned} L &= \frac{1}{2} \mu \int_{-L/2}^{L/2} \dot{\phi}(x, t)^2 \, dx - \frac{\mu v^2}{2} \int_{-L/2}^{L/2} \phi'(x, t)^2 \, dx \\ &= \int_{-L/2}^{L/2} \frac{\mu}{2} \left(\dot{\phi}(x, t)^2 - v^2 \phi'(x, t)^2 \right) \, dx, \end{aligned} \qquad (5.24)$$

using $K L = \mu v^2$ from the definition of characteristic speed.

This is the Lagrangian for the continuum form of the ball and spring problem. Now our mass is smeared out continuously over the length of the rod (we have chosen a constant μ to describe this, but we could, in theory, allow μ to have position-dependence), and our spring constant is the Young's modulus of the material.

The Lagrangian we use in classical mechanics has equations of motion that come naturally from the extremization of an action. The utility of the Lagrangian is that we can obtain equations of motion directly from it, and these equations are adapted to coordinate systems that we choose. In our current work, we have both the equations of motion for a scalar field, and a Lagrangian for the scalar field – can we again appeal to some action extremization in order to connect the two more directly?

5.2 Multidimensional action principle

We begin from the Lagrangian:

$$L = \frac{\mu}{2} \int_{-L/2}^{L/2} \left(\dot{\phi}(x, t)^2 - v^2 \phi'(x, t)^2 \right) \, dx, \qquad (5.25)$$

and will form an action $S = \int L \, dt$ that will naturally be an integral over time and space together. The integrand of that action is more naturally a "Lagrangian" in field theory (because it lacks integrals), and the variation of the action will produce precisely the field equation we saw earlier, but entirely within the continuum context.

Consider the action for the Lagrangian (5.25):

$$S[\phi(x, t)] = \int_{t_0}^{t_f} \int_{-L/2}^{L/2} \frac{\mu}{2} \left(\dot{\phi}(x, t)^2 - v^2 \, \phi'(x, t)^2 \right) dx dt. \qquad (5.26)$$

Time and space are both integrated, and S is, as always, a functional – it takes a function of (x, t) to the real numbers. Our usual particle action does the same thing:

$$S_o[x(t)] = \int_{t_0}^{t_f} \left(\frac{1}{2} m \, \dot{x}(t)^2 - U(x) \right) dt \qquad (5.27)$$

but the integration is only over t, and $x(t)$ is a function only of t. Just as we call the integrand of $S_o[x(t)]$ the Lagrangian L, it is customary to call the integrand of (5.26) $\mathcal{L}$, the "Lagrange density":

$$\mathcal{L} = \frac{\mu}{2} \left(\dot{\phi}(x, t)^2 - v^2 \, \phi'(x, t)^2 \right). \qquad (5.28)$$

This is a density in the sense that we integrate it over space, in addition to time, so that $\mathcal{L}$ has the units of Lagrangian-per-unit-length.

5.2.1 One-dimensional variation

Remember the variational principle from Chapter 1 – we vary a trajectory (in that case), keeping the endpoints fixed. That is, we add to $x(t)$ an arbitrary function $\eta(t)$ that vanishes at t_0 and t_f, so that $\tilde{x}(t) = x(t) + \eta(t)$ has the same endpoints as $x(t)$, as shown in Figure 5.4.

Our variational requirement is that S is, to first order, unchanged by an arbitrary $\eta(t)$. This is the usual "stationary" or "extremal" action statement – let's review the idea. If we expand S_o about $\eta(t) = 0$:

$$S_o[x(t) + \eta(t)] = \int_{t_0}^{t_f} \left(\frac{1}{2} m \, (\dot{x} + \dot{\eta})^2 - U(x(t) + \eta(t)) \right) dt$$

$$\approx \int_{t_0}^{t_f} \left(\frac{1}{2} m \, (\dot{x}^2 + 2 \, \dot{x} \, \dot{\eta}) - \left(U(x(t)) + \frac{dU}{dx} \eta(t) \right) \right) dt, \qquad (5.29)$$

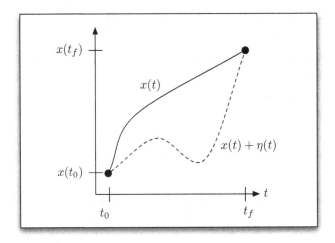

Figure 5.4   A trajectory $x(t)$ with a perturbed trajectory $x(t) + \eta(t)$ sharing the same endpoints.

where we keep only those terms that are first order in $\eta(t)$ and $\dot{\eta}(t)$, then we can collect a portion that looks like $S_\circ[x(t)]$ and terms that represent deviations:

$$S_\circ[x(t) + \eta(t)] \approx S_\circ[x(t)] + \underbrace{\int_{t_0}^{t_f} \left(m\, \dot{x}\, \dot{\eta} - \frac{dU}{dx}\, \eta \right) dt}_{\equiv \delta S} . \tag{5.30}$$

The extremization condition is that $\delta S = 0$ for any $\eta(t)$, but in order to see what this implies about $x(t)$, we must first write the perturbation entirely in terms of $\eta(t)$, i.e. we must get rid of $\dot{\eta}(t)$ in favor of $\eta(t)$. An integration by parts on the first term of δS does the job:

$$\int_{t_0}^{t_f} \dot{x}\, \dot{\eta}\, dt = \underbrace{\dot{x}(t)\, \eta(t) \Big|_{t=t_0}^{t_f}}_{=0} - \int_{t_0}^{t_f} \ddot{x}\, \eta(t)\, dt$$

$$= - \int_{t_0}^{t_f} \ddot{x}\, \eta(t)\, dt, \tag{5.31}$$

since the boundary term vanishes at both endpoints by assumption ($\eta(t_0) = \eta(t_f) = 0$).

So, transforming the $\dot{\eta}$ term in δS:

$$\delta S = \int_{t_0}^{t_f} \left(-m\, \ddot{x} - \frac{\partial U}{\partial x} \right) \eta(t)\, dt, \tag{5.32}$$

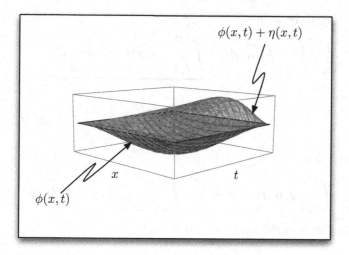

Figure 5.5   Variation in two dimensions – we have a surface $\phi(x, t)$ and the perturbed surface $\phi(x, t) + \eta(x, t)$ sharing the same boundaries in both x and t.

and for this to be zero for any $\eta(t)$, the integrand itself must be zero, and we recover:

$$m\,\ddot{x} = -\frac{\partial U}{\partial x}. \tag{5.33}$$

5.2.2 Two-dimensional variation

Variation in two dimensions, x and t, is no different, although there is extra functional dependence to keep track of. Referring to Figure 5.5, we have some function $\phi(x, t)$ with associated action $S[\phi(x, t)]$ and we want this action to be unchanged (to first order) under the introduction of an arbitrary additional function $\eta(x, t)$ with $\phi(x, t)$ and $\phi(x, t) + \eta(x, t)$ sharing the same boundary conditions, i.e. $\eta(x, t)$ vanishes on the boundaries.

The action, to first order in $\eta(x, t)$ and its derivatives, is:

$$S[\phi(x, t) + \eta(x, t)] \approx \int_{t_0}^{t_f} \int_{-L/2}^{L/2} \frac{\mu}{2} \left(\dot{\phi}^2 + 2\dot{\phi}\,\dot{\eta} - v^2 \left(\phi'^2 + 2\phi'\,\eta' \right) \right) dx\,dt$$

$$= S[\phi(x, t)] + \underbrace{\int_{t_0}^{t_f} \int_{-L/2}^{L/2} \mu \left(\dot{\phi}\,\dot{\eta} - v^2 \phi'\,\eta' \right) dx\,dt}_{\equiv \delta S},$$

$$\tag{5.34}$$

where, as above, the second term here is δS, and we want this to vanish for arbitrary $\eta(x, t)$. Again, we use integration by parts, now on two terms, to rewrite $\dot{\eta}$ and η'

in terms of η itself. Note that:

$$\int_{t_0}^{t_f} \int_{-L/2}^{L/2} \dot{\phi}\,\dot{\eta}\,dxdt = \underbrace{\int_{-L/2}^{L/2} \dot{\phi}(x,t)\,\eta(x,t)\Big|_{t=t_0}^{t_f} dx}_{=0}$$

$$- \int_{t_0}^{t_f} \int_{-L/2}^{L/2} \ddot{\phi}(x,t)\,\eta(x,t)\,dxdt, \qquad (5.35)$$

and, similarly, that:

$$\int_{t_0}^{t_f} \int_{-L/2}^{L/2} \phi'\,\eta'\,dxdt = - \int_{t_0}^{t_f} \int_{-L/2}^{L/2} \phi''(x,t)\,\eta(x,t)\,dxdt. \qquad (5.36)$$

We can write the perturbation to the action as:

$$\delta S = \int_{t_0}^{t_f} \int_{-L/2}^{L/2} \mu \left(-\ddot{\phi} + v^2\,\phi''\right)\,\eta(x,t)\,dxdt \qquad (5.37)$$

and, as before, for this to hold for all $\eta(x,t)$, the integrand must vanish, giving:

$$\boxed{\ddot{\phi}(x,t) = v^2\,\phi''(x,t),} \qquad (5.38)$$

the wave equation.

Where is the potential here? It is interesting that what started as particles connected by springs appears to have become a "free" scalar field in the sense that there is no external "force" in the field equations, nor a term like $U(\phi)$ in the action. In other words, there is no "source" for the field here. For now, our job is to build the field-theoretic analogue to the Euler–Lagrange equations for a more general action. In addition, we will express the action in a way that makes extension to higher dimension and vacuum natural.

5.3 Vacuum fields

We have a particular example of a free scalar field, and this serves as a model for more general situations. An action like (5.26) depends on the derivatives of the field $\phi(x,t)$. Notice that it depends on the derivatives in an essentially antisymmetric way. This suggests that if we introduce the coordinates $x^\mu \doteq (v\,t, x)^T$ for $\mu = 0, 1$, and a metric:

$$g_{\mu\nu} \doteq \begin{pmatrix} -1 & 0 \\ 0 & 1 \end{pmatrix}, \qquad (5.39)$$

then the action can be written as:

$$S = - \int \int \frac{\mu\,v}{2}\,\phi_{,\mu}\,g^{\mu\nu}\,\phi_{,\nu}\,d(v\,t)dx, \qquad (5.40)$$

with $\phi_{,\mu} \equiv \frac{\partial \phi}{\partial x^\mu} \equiv \partial_\mu \phi$ as always. In this form, the Lagrange density is:

$$\mathcal{L} = -\frac{\mu v}{2} \phi_{,\mu} g^{\mu v} \phi_{,v},$$
(5.41)

and one typically writes $d\tau \equiv dx^0 dx^1 = d(vt)dx$, so the action takes the form:

$$S = \int \mathcal{L} d\tau.$$
(5.42)

From (5.42), we can make the jump from fields that exist within a physical medium with a characteristic speed v, to fields in vacuum. The difference in notation comes from replacing the coordinates with $x^\mu \doteq (c\,t, x)^T$, so that the fundamental speed is now the speed associated with the vacuum. The metric in (5.39) already looks like the Minkowski metric. Our fields are now, typically, associated with forces of some sort (the scalar and vector potentials of E&M, for example, satisfy the wave equation in vacuum, and directly inform the force on moving, charged test bodies), and can exhibit interesting dynamics of their own. So while we will now be thinking about fields that travel with speed c in empty space-time, don't let the physical shift from "real" fields in a medium to fields in vacuum pass you by – they are identical in notation, but not in spirit. We shall not address the deep questions: why does the vacuum have a definite speed attached to it?, what agency sets the scale of that speed? beyond advertising that they exist.

Returning to the model – a scalar field ϕ with Lagrange density given by (5.41) (with $v \longrightarrow c$), on a Minkowski "background" (meaning we assume the metric is Minkowski). Because we know the metric here, we can generalize the density to any dimension we like – in particular, we will work in $D = 4$ space-time. Then there is a temporal coordinate and three spatial coordinates, our field equation is the wave equation.

But from the starting point (5.41), we can also consider more complicated Lagrange densities. Our current example involves only the derivatives of ϕ, so we would write $\mathcal{L}(\phi_{,\mu})$, but what happens to the field equations that come from a Lagrange density of the form $\mathcal{L}(\phi, \phi_{,\mu})$?

For the moment, let's ignore the overall constants in the action, those are important, but the free-field equations do not depend on them. If we had, in any $D + 1$-dimensional space, the action:

$$S[\phi(x^\mu)] = \int \mathcal{L}(\phi, \phi_{,\mu}) d\tau$$
(5.43)

with $d\tau \equiv \prod_{v=0}^{D} dx^v$ (Cartesian and temporal), and we extremize using an arbitrary $\eta(x^\mu)$ that vanishes on the boundaries, then expanding $\mathcal{L}$ to first order (as usual) gives:

$$S[\phi(x^\mu) + \eta(x^\mu)] = \int \mathcal{L}(\phi + \eta, \phi_{,\mu} + \eta_{,\mu}) \, d\tau$$

$$\approx \int \left(\mathcal{L}(\phi, \phi_{,\mu} + \eta_{,\mu}) + \frac{\partial \mathcal{L}(\phi, \phi_{,\mu} + \eta_{,\mu})}{\partial \phi} \eta \right) d\tau$$

$$\approx \underbrace{\int \mathcal{L}(\phi, \phi_{,\mu}) \, d\tau}_{= S[\phi(x^\mu)]} + \underbrace{\int \left(\frac{\partial \mathcal{L}(\phi, \phi_{,\mu})}{\partial \phi} \eta + \frac{\partial \mathcal{L}(\phi, \phi_{,\mu})}{\partial \phi_{,\mu}} \eta_{,\mu} \right) d\tau}_{\equiv \delta S}.$$

$$(5.44)$$

The δS term has a summation in it, $\frac{\partial \mathcal{L}}{\partial \phi_{,\mu}} \eta_{,\mu}$, and we need to use integration by parts on each term of the sum. The sum contains all derivatives of $\mathcal{L}$ w.r.t. each derivative of ϕ dotted into the corresponding derivative of η – we can flip the derivatives one by one, leaving us with boundary terms which vanish by assumption,[2] and an overall ∂_μ acting on the partials $\frac{\partial \mathcal{L}}{\partial \phi_{,\mu}}$. Simply put:

$$\int \frac{\partial \mathcal{L}(\phi, \phi_{,\mu})}{\partial \phi_{,\mu}} \eta_{,\mu} \, d\tau = - \int \eta \, \partial_\mu \left(\frac{\partial \mathcal{L}(\phi, \phi_{,\mu})}{\partial \phi_{,\mu}} \right) d\tau. \qquad (5.46)$$

To extremize the action, we enforce $\delta S = 0$ for arbitrary $\eta(x^\mu)$:

$$\delta S = \int \left(\frac{\partial \mathcal{L}}{\partial \phi} - \partial_\mu \left(\frac{\partial \mathcal{L}}{\partial \phi_{,\mu}} \right) \right) \eta \, d\tau = 0 \longrightarrow \boxed{\frac{\partial \mathcal{L}}{\partial \phi} - \partial_\mu \left(\frac{\partial \mathcal{L}}{\partial \phi_{,\mu}} \right) = 0,} \quad (5.47)$$

which is a natural generalization of the Euler–Lagrange equations of motion from classical mechanics.

As a check, let's see how these field equations apply to our scalar field Lagrangian $\mathcal{L} = \frac{1}{2} \phi_{,\alpha} g^{\alpha\beta} \phi_{,\beta}$ (dropping constants again):

$$\partial_\mu \left(\frac{\partial \mathcal{L}}{\partial \phi_{,\mu}} \right) = \partial_\mu \left(g^{\mu\beta} \phi_{,\beta} \right) = \partial^\beta \partial_\beta \phi$$

$$(5.48)$$

$$= -\frac{1}{c^2} \frac{\partial^2 \phi}{\partial t^2} + \frac{\partial^2 \phi}{\partial x^2} = 0,$$

which is just another equivalent form for the wave equation.

---

[2] This is a feature of the divergence theorem – for surface terms that vanish on the boundary of some domain:

$$0 = \int \partial_\mu \left(\frac{\partial \mathcal{L}}{\partial \phi_{,\mu}} \eta \right) d\tau$$

$$(5.45)$$

$$= \int \left(\eta \, \partial_\mu \left(\frac{\partial \mathcal{L}}{\partial \phi_{,\mu}} \right) + \frac{\partial \mathcal{L}}{\partial \phi_{,\mu}} \eta_{,\mu} \right) d\tau.$$

Under an integral, then, we can take $\frac{\partial \mathcal{L}}{\partial \phi_{,\mu}} \eta_{,\mu} \longrightarrow -\eta \, \partial_\mu \left(\frac{\partial \mathcal{L}}{\partial \phi_{,\mu}} \right)$.

So to sum up, for a metric representing $D + 1$-dimensional flat space with a temporal component and D Cartesian spatial components, the Euler–Lagrange equations for the Lagrange density $\mathcal{L}(\phi, \phi_{,\mu})$ corresponding to extremal action are:

$$\partial_\mu \left(\frac{\partial \mathcal{L}}{\partial \phi_{,\mu}} \right) - \frac{\partial \mathcal{L}}{\partial \phi} = 0. \tag{5.49}$$

We could go further, this equation could be modified to handle densities like $\mathcal{L}(\phi, \phi_{,\mu}, \phi_{,\mu\nu})$ or other high-derivative combinations, but we have no reason currently to do this. The situation is similar to classical mechanics, where we do not generally consider Lagrangians that depend on acceleration.

5.4 Integral transformations and the action

Just as the particle Lagrangian from classical mechanics can be used to deduce properties of the system (conserved quantities, momenta), the field Lagrange densities are useful even prior to finding the field equations. In particular, the invariance of a particle action under various transformations tells us about constants of the motion via Noether's theorem: if the action is unchanged by translation, momentum is conserved, if infinitesimal rotations leave the action fixed, angular momentum is conserved, and of course, time-translation insensitivity leads to energy conservation. The same sort of ideas hold for field actions – the invariance of the action under coordinate transformation, in particular, implies that the field equations we obtain will satisfy general covariance, i.e. that the field equations will be tensorial.

The details of integration in generic spaces, which will be used to ensure that the action is a scalar, take a little time to go through. We will present a few results that will be more useful later on, but fit nicely into our current discussion. In the back of our minds is the scalar Lagrange density, a simple model to which we can apply our ideas.

5.4.1 Coordinate transformation

There is an integration transformation problem in our classical field Lagrangian. We have seen that:

$$S = \int \mathcal{L}(\phi, \phi_{,\mu}) \, d\tau \tag{5.50}$$

is appropriate for Cartesian spatial coordinates and time, but what if we wanted to transform to some other coordinate system?

Suppose we are in a two-dimensional Euclidean space with Cartesian coordinates $x^j \doteq (x, y)^T$, so that the metric is just the identity matrix. Consider the "volume" integral, in this context:

$$I \equiv \int_\Omega f(x, y)\, dx\, dy,$$ (5.51)

for some parametrized volume Ω. How does I change if we go to some new coordinate system, $\bar{x}^j \doteq (\bar{x}, \bar{y})^T$? Well, we know from vector calculus that there is a factor of the determinant of the Jacobian of the transformation between the two coordinates involved so as to keep area elements equal. The integral written in terms of $\bar{x}$ and $\bar{y}$ is:

$$\bar{I} = \int_{\bar\Omega} f(\bar{x}, \bar{y})\, |\det(J)|\, d\bar{x}\, d\bar{y}.$$ (5.52)

Where the Jacobian (whose determinant appears in the above) is:

$$J^j_k = J(x^j, \bar{x}^k) = \frac{\partial x^j}{\partial \bar{x}^k} \doteq \begin{pmatrix} \frac{\partial x^1}{\partial \bar{x}^1} & \frac{\partial x^1}{\partial \bar{x}^2} \\ \frac{\partial x^2}{\partial \bar{x}^1} & \frac{\partial x^2}{\partial \bar{x}^2} \end{pmatrix},$$ (5.53)

and we view $x^1(\bar{x}^1, \bar{x}^2)$ and $x^2(\bar{x}^1, \bar{x}^2)$ as functions of the new coordinates. There is a connection between the Jacobian and the transformed metric – according to the covariant transformation law:

$$\bar{g}_{\mu\nu} = \frac{\partial x^\alpha}{\partial \bar{x}^\mu} \frac{\partial x^\beta}{\partial \bar{x}^\nu} g_{\alpha\beta}.$$ (5.54)

This means that the new metric can be viewed as a matrix product of the Jacobian of the transformation (just a new name for $\frac{\partial x^\alpha}{\partial \bar{x}^\mu}$) with the original metric:

$$\bar{g}_{\mu\nu} = J^\alpha_\mu g_{\alpha\beta} J^\beta_\nu.$$ (5.55)

The interesting feature here is that we can relate the determinant of the Jacobian matrix to the determinant of the metric – $\det(\bar{g}_{\mu\nu}) = \left(\det(J^\mu_\nu)\right)^2$ (I've used the tensor notation, but we are viewing these as matrices when we take the determinant). The determinant of the metric is generally denoted $g \equiv \det(g_{\mu\nu})$, and then the integral transformation law reads:

$$\bar{I} = \int_{\bar\Omega} f(\bar{x}, \bar{y}) \sqrt{\bar{g}}\, d\bar{\tau}.$$ (5.56)

Example

Take the transformation from Cartesian coordinates to polar, $x = r\cos\theta$, $y = r\sin\theta$ – the Jacobian is:

$$J^j_k \doteq \begin{pmatrix} \frac{\partial x}{\partial r} & \frac{\partial x}{\partial \theta} \\ \frac{\partial y}{\partial r} & \frac{\partial y}{\partial \theta} \end{pmatrix} = \begin{pmatrix} \cos\theta & -r\sin\theta \\ \sin\theta & r\cos\theta \end{pmatrix}$$ (5.57)

and the determinant is $\det(J_k^j) = r$. The metric for polar coordinates is:

$$\bar{g}_{\mu\nu} \doteq \begin{pmatrix} 1 & 0 \\ 0 & r^2 \end{pmatrix} \tag{5.58}$$

with determinant $\bar{g} = r^2$, so it is indeed the case that $\det\left(J_k^j\right) = \sqrt{\bar{g}}$.

If we think of the integral over a "volume" in this two-dimensional space:

$$\int dx dy = \int r \, dr d\phi, \tag{5.59}$$

then we can clearly see the transformation $d\tau = dx dy$ is related to $d\bar{\tau} = dr d\phi$ via the Jacobian:

$$d\tau = |J| \, d\bar{\tau}. \tag{5.60}$$

The metric determinant in Cartesian coordinates is $g = 1$, and in polar, $\bar{g} = r^2$, so we have:

$$\bar{g} = |J|^2 g \tag{5.61}$$

and it is clear from (5.60) and (5.61) that:

$$\sqrt{g} \, d\tau = \frac{|J|}{|J|} \sqrt{\bar{g}} \, d\bar{\tau} = \sqrt{\bar{g}} \, d\bar{\tau}, \tag{5.62}$$

a scalar.

The issue is that our action (5.50) does not transform as a scalar. The preceding example shows that to support general coordinate transformations, we should use $\sqrt{g} \, \mathcal{L}$ as our integrand for S.

5.4.2 Final form of field action

To make our action a scalar, we want to insert a factor like $\sqrt{g}$ into our integrals. But our "spaces" are really "space-times", we have a metric with a negative determinant– in Cartesian spatial coordinates, the determinant of $g_{\mu\nu}$ for the Minkowski metric is $g = -1$, so taking the square root introduces a factor of $i = \sqrt{-1}$. To keep our volume interpretation, then, we actually have $|\det(J_\nu^\mu)| = \sqrt{-\bar{g}}$ for our space-time metrics.

So we will introduce a factor of $\sqrt{-g}$ in all of our actions – this doesn't change anything we've done so far, since for us, $\sqrt{-g}$ has been one until now (i.e. for Minkowski), but in the interest of generality, the action for a scalar field theory is:

$$S[\phi(x^\mu)] = \int d\tau \underbrace{\left(\sqrt{-g} \, \bar{\mathcal{L}}(\phi, \phi_{,\mu})\right)}_{\equiv \mathcal{L}}, \tag{5.63}$$

where we have a scalar $\bar{\mathcal{L}}(\phi, \phi_{,\mu})$, and the Lagrange density $\mathcal{L}$ is just the product of $\bar{\mathcal{L}}$ and $\sqrt{-g}$, the Jacobian correction factor. This is an interesting shift, because while an expression like $\phi_{,\mu} g^{\mu\nu} \phi_{,\nu}$ is clearly a scalar, we suspect that $\sqrt{-g}$ is *not*, since $d\tau$ is not – the whole point of putting in the $\sqrt{-g}$ factor was to make the integration procedure itself a scalar operation. To sum up: $\bar{\mathcal{L}}$ is a scalar, $\sqrt{-g}\,d\tau$ is a scalar, and $\sqrt{-g}\,\bar{\mathcal{L}}$ is a "density" in the sense described below.

5.5 Transformation of the densities

Let's look at how the determinant of the metric transforms under arbitrary coordinate transformation. We have:

$$\bar{g}_{\mu\nu} = \frac{\partial x^\alpha}{\partial \bar{x}^\mu} \frac{\partial x^\beta}{\partial \bar{x}^\nu} g_{\alpha\beta} \tag{5.64}$$

and if we view the right-hand side as three matrices multiplied together, then using the multiplicative property of determinants, $\det(\mathbb{A}\,\mathbb{B}) = \det(\mathbb{A})\det(\mathbb{B})$, we have:

$$\bar{g} = \det\left(\frac{\partial x}{\partial \bar{x}}\right)^2 g \tag{5.65}$$

and the matrix $\frac{\partial x}{\partial \bar{x}}$ is the Jacobian for $x \longrightarrow \bar{x}$ defined in (5.53) (with size appropriate to the dimension of our space-time). The Jacobian for the inverse transformation, taking us from $\bar{x} \longrightarrow x$, is just the matrix inverse of $\frac{\partial x}{\partial \bar{x}}$, so that:

$$\bar{g} = \det\left(\frac{\partial \bar{x}}{\partial x}\right)^{-2} g \tag{5.66}$$

and a quantity that transforms in this manner, picking up a factor of the determinant of the coordinate transformation, is called a tensor *density*. The name density, again, connects the role of g to the transformation of volumes under integration. We say that a tensor density A has weight given by p in the general transformation:

$$\bar{A} = \det\left(\frac{\partial \bar{x}}{\partial x}\right)^p A \tag{5.67}$$

(for nonscalar A, we introduce the appropriate additional factors for the tensor indices) – so the determinant of the metric is a scalar density of weight $p = -2$, and $\sqrt{-g}$ is a scalar density of weight -1. The volume factor itself:

$$d\bar{\tau} = \det\left(\frac{\partial \bar{x}}{\partial x}\right) d\tau \tag{5.68}$$

is a scalar density of weight $p = 1$. Density weights add when multiplied together. For example:

$$\sqrt{-\bar{g}}\, d\bar{\tau} = \left(\sqrt{-g}\, \det\left(\frac{\partial \bar{x}}{\partial x}\right)^{-1}\right)\left(\det\left(\frac{\partial \bar{x}}{\partial x}\right) d\tau\right) = \sqrt{-g}\, d\tau \qquad (5.69)$$

and we see that $\sqrt{-g}\, d\tau$ is a density of weight 0, which is to say, a normal scalar.

Again, if we take our usual $\bar{\mathcal{L}}$ to be a scalar (like $\phi_{,\mu}\, g^{\mu\nu}\, \phi_{,\nu}$), then $\mathcal{L} = \bar{\mathcal{L}}\sqrt{-g}$ is a scalar density of weight -1, just right for forming an action since we must multiply by $d\tau$.

5.5.1 Tensor density derivatives

While we're at it, it's a good idea to set some of the notation for derivatives of densities, as these come up any time integration is involved. Recall the covariant derivative of a first-rank (zero-weight) tensor:

$$A^{\mu}_{\ ;\nu} = A^{\mu}_{\ ,\nu} + \Gamma^{\mu}_{\ \sigma\nu}\, A^{\sigma}. \qquad (5.70)$$

What if we had a tensor density of weight p: $\mathcal{A}^{\mu}$?[3] We can construct a tensor of weight zero (our "usual" ones) $A^{\mu} = (\sqrt{-g})^{p}\, \mathcal{A}^{\mu}$ from this, then apply the covariant derivative as above, and finally multiply by $\sqrt{-g}^{-p}$ to restore the proper weight, suggesting that:

$$\begin{aligned}
\mathcal{A}^{\mu}_{\ ;\nu} &= (\sqrt{-g})^{-p}\left((\sqrt{-g})^{p}\, \mathcal{A}^{\mu}\right)_{;\nu} \\
&= (\sqrt{-g})^{-p}\left(((\sqrt{-g})^{p}\, \mathcal{A}^{\mu})_{,\nu} + \Gamma^{\mu}_{\ \sigma\nu}(\sqrt{-g})^{p}\, \mathcal{A}^{\sigma}\right) \qquad (5.72) \\
&= \mathcal{A}^{\mu}_{\ ,\nu} + \Gamma^{\mu}_{\ \sigma\nu}\, \mathcal{A}^{\sigma} + \mathcal{A}^{\mu}\frac{p}{2g}\frac{\partial g}{\partial x^{\nu}}.
\end{aligned}$$

The advantage of $\sqrt{-g}$ in integration comes from the observation that, for a tensor A^{μ}:[4]

$$\boxed{\sqrt{-g}\, A^{\mu}_{\ ;\mu} = \left(\sqrt{-g}\, A^{\mu}\right)_{,\mu},} \qquad (5.73)$$

---

[3] Transforming as:

$$\bar{\mathcal{A}}^{\mu} = \det\left(\frac{\partial \bar{x}}{\partial x}\right)^{p}\frac{\partial \bar{x}^{\mu}}{\partial x^{\alpha}}\, \mathcal{A}^{\alpha}. \qquad (5.71)$$

[4] This relies on the two properties of derivatives: $\sqrt{-g}_{;\mu} = 0$, which is pretty clear via the chain rule, and $A^{\mu}_{\ ;\mu} = \frac{1}{\sqrt{-g}}\left(\sqrt{-g}\, A^{\mu}\right)_{,\mu}$.

so we can express the flat-space divergence theorem $\int_\Omega \nabla \cdot \mathbf{E}\, d\tau = \oint_{\partial\Omega} \mathbf{E} \cdot d\mathbf{a}$ in tensor form:

$$\int d\tau \sqrt{-g}\, A^\mu_{\ ;\mu} = \oint \sqrt{-g}\, A^\mu\, da_\mu. \tag{5.74}$$

Notice also that the covariant divergence of the integrand reduces to the "ordinary" divergence when $\sqrt{-g}$ is involved:

$$\boxed{(\sqrt{-g}\, A^\mu)_{;\mu} = (\sqrt{-g}\, A^\mu)_{,\mu}.} \tag{5.75}$$

Scalar fields in spherical coordinates

Let's see how all of this works for our scalar field. We have the general field equations:

$$\left(\partial_\mu \left(\frac{\partial \mathcal{L}}{\partial \phi_{,\mu}}\right) - \frac{\partial \mathcal{L}}{\partial \phi}\right) = 0. \tag{5.76}$$

If we take the usual four-dimensional $\{c\,t, x, y, z\}$ flat coordinates, then:

$$\mathcal{L} = \frac{1}{2} \phi_{,\mu}\, g^{\mu\nu}\, \phi_{,\nu} \sqrt{-g} = \frac{1}{2} \phi_{,\mu}\, g^{\mu\nu}\, \phi_{,\nu} \tag{5.77}$$

and we recover:

$$\partial_\mu \partial^\mu \phi = -\frac{1}{c^2}\frac{\partial^2 \phi}{\partial t^2} + \nabla^2 \phi = 0. \tag{5.78}$$

But suppose we have in mind spherical coordinates? Then the Minkowski metric takes the form:

$$g_{\mu\nu} \doteq \begin{pmatrix} -1 & 0 & 0 & 0 \\ 0 & 1 & 0 & 0 \\ 0 & 0 & r^2 & 0 \\ 0 & 0 & 0 & r^2 \sin^2\theta \end{pmatrix} \tag{5.79}$$

and $\sqrt{-g} = r^2 \sin\theta$. We have:

$$\partial_\mu \left(g^{\mu\beta} \sqrt{-g}\, \phi_{,\beta}\right) = 0. \tag{5.80}$$

Separating out the temporal and spatial portions, this gives:

$$-\frac{1}{c^2}\frac{\partial^2 \phi}{\partial t^2} + \nabla^2 \phi = 0, \tag{5.81}$$

where now the field $\phi = \phi(t, r, \theta, \phi)$ and the Laplacian refers to the spherical one.

This is an interesting development – we started with a Lagrange density that made the action a scalar, and now we discover that the field equations themselves behave as tensors. Appropriately, the field equations have the same form no matter what coordinate system we use.

Problem 5.1

Given a Lagrange density for a scalar field ϕ in Minkowski space-time, find the field equations if $\mathcal{L}$ takes the form:

$$\mathcal{L}(\phi, \partial\phi, \partial\partial\phi), \tag{5.82}$$

that is, the Lagrangian depends on the field, its derivatives, and second derivatives (assume, here, that $\delta\phi$ vanishes on the boundary region, and that $\delta\phi_{,\mu}$ also vanishes there).

Problem 5.2

Given the Lagrange density:

$$\mathcal{L} = \sqrt{-g}\left(\frac{1}{2}\phi_{,\mu}\, g^{\mu\nu}\, \phi_{,\nu} - \alpha\, \phi^4\right), \tag{5.83}$$

find the field equations governing ϕ.

Problem 5.3

The correct, scalar form of the action yields field equations for a scalar ψ:

$$\frac{\partial}{\partial x^\mu}\left(g^{\mu\beta}\sqrt{-g}\psi_{,\beta}\right) = 0, \tag{5.84}$$

and it is clear that for Minkowski space-time represented in Cartesian coordinates, this yields the D'Alembertian:

$$-\frac{1}{c^2}\frac{\partial^2\psi}{\partial t^2} + \nabla^2\psi = 0. \tag{5.85}$$

Show that using cylindrical coordinates in Minkowski gives back the same equation (i.e. write out (5.84) explicitly using the Minkowski metric written in cylindrical coordinates, and show that you can combine terms to form (5.85), where ∇^2 is the cylindrical Laplacian).

*Problem 5.4

Think of a string in three dimensions, the location of a segment of string is given by $\mathbf{u}(s, t)$, a vector that points from the origin to the piece of string located at s (along the unstretched string) at time t. At a specific location along the string, s_0, the kinetic energy in the neighborhood of s_0 is given by:

$$dT = \frac{1}{2}\mu(s_0)\, ds\, \frac{\partial\mathbf{u}(s_0, t)}{\partial t} \cdot \frac{\partial\mathbf{u}(s_0, t)}{\partial t}, \tag{5.86}$$

where $\mu(s_0)$ is the mass per unit length of the string at s_0 and ds is an infinitesimal length. Take the string to be locally connected by springs with constant k and equilibrium length ds, then the *potential* energy at s_0 is:

$$dU = \frac{1}{2} k \left(|\mathbf{u}(s_0, t) - \mathbf{u}(s_0 - ds, t)| - ds \right)^2 \approx \frac{1}{2} k \, ds^2 \left(\left| \frac{\partial \mathbf{u}(s, t)}{\partial s} \right| - 1 \right)^2 \quad (5.87)$$

(the limit as $ds \longrightarrow 0$ is what we are after). Thinking back to the addition of springs, define K via $k \, ds = K$, where K is a macroscopic description of the "springiness" of the string (like Young's modulus, for example). Using dots to represent time-derivatives, and primes to represent s-derivatives, we can write the action:

$$S = \int (dT - dU) \, dt = \int \left(\frac{1}{2} \mu \, (\dot{\mathbf{u}} \cdot \dot{\mathbf{u}}) - \frac{1}{2} K \left(|\mathbf{u}'| - 1 \right)^2 \right) ds \, dt. \quad (5.88)$$

(a) Taking μ constant, vary the action (or use the appropriate Euler–Lagrange equations) to find the equations of motion for $\mathbf{u}$ (we are in Cartesian space here).

(b) Write those equations in terms of the curvature $\mathbf{k}$ defined in (4.41) – this should allow you to make sense of the statement: "a nonlinear string is governed by a wave equation sourced by the local curvature of the string". This formulation of the nonlinear wave equation was suggested to me by Andrew Rhines.

5.6 Continuity equations

We have finished the definition of Lagrange density for a generic space-time described by a metric $g_{\mu\nu}$. The action is now, appropriately, a scalar, meaning that its value is insensitive to coordinate choice. The fact that the action cannot change under a change of variables will lead to some constraints on the field solutions (and their derivatives). This is an example of Noether's theorem, and we will recover a number of familiar results.

Keep in mind that all we have discussed so far are free (scalar) fields, that is, functions that are self-consistent but sourceless. We have not yet described the connection between sources and fields, nor the physical effect of fields on test particles. Both field sources and field "forces" can be put into our current Lagrange framework, but before we do that, we must find the explicit manifestation of coordinate invariance for the fields. For this, we do not need more than the free-field action.

While the following refers to general field actions (scalar, vector, and higher), it is useful to continue our connection to scalar fields. We will develop the energy–momentum tensor and its set of four continuity equations.

5.6.1 Coordinates and conservation

Returning to the coordinate invariance of the action, the idea is to make a formal, infinitesimal coordinate transformation, see what changes (to first order) that induces in the action $S \longrightarrow S + \delta S$ and require that $\delta S = 0$, enforcing the overall scalar nature of S. We begin with the action:

$$S = \int \mathcal{L}(\phi, \phi_{,\mu}, g_{\mu\nu}) \, d\tau, \tag{5.89}$$

where $\mathcal{L} = \sqrt{-g} \, \bar{\mathcal{L}}(\phi, \phi_{,\mu}, g_{\mu\nu})$ is the Lagrange density, and we indicate the metric dependence in $\mathcal{L}$ (since the metric itself depends on the coordinates). Now we introduce an infinitesimal coordinate transformation:

$$\bar{x}^\mu = x^\mu + \eta^\mu \tag{5.90}$$

with η^μ a small, but general perturbation, $\eta^\mu = \eta^\mu(x)$. The coordinate transformation generates a perturbation to both the field $\phi(x)$ and the metric $g_{\mu\nu}(x)$ – let's write these as $\bar{\phi}(\bar{x}) = \phi(x) + \delta\phi(x)$ and $\bar{g}_{\mu\nu}(\bar{x}) = g_{\mu\nu}(x) + \delta g_{\mu\nu}(x)$. We can connect these explicitly to the perturbation $\eta^\mu(x)$ by Taylor expansion – take $\phi(\bar{x})$:

$$\bar{\phi}(\bar{x}) = \phi(x(\bar{x})) = \phi(\bar{x} - \eta) = \phi(x) - \frac{\partial\phi}{\partial x^\mu} \eta^\mu + O(\eta^2), \tag{5.91}$$

so we would call $\delta\phi = -\frac{\partial\phi}{\partial x^\mu} \eta^\mu$. We'll return to the analogous procedure for $\bar{g}_{\mu\nu}(\bar{x})$ in a moment.

Viewing the Lagrange density as a function of ϕ, $\phi_{,\mu}$, and $g_{\mu\nu}$, we have:

$$\mathcal{L}(\phi + \delta\phi, \phi_{,\mu} + \delta\phi_{,\mu}, g_{\mu\nu} + \delta g_{\mu\nu})$$

$$\approx \mathcal{L}(\phi, \phi_{,\mu}, g_{\mu\nu}) + \frac{\partial\mathcal{L}}{\partial\phi} \delta\phi + \frac{\partial\mathcal{L}}{\partial\phi_{,\mu}} \delta\phi_{,\mu} + \frac{\partial\mathcal{L}}{\partial g_{\mu\nu}} \delta g_{\mu\nu}, \tag{5.92}$$

and it is pretty clear what will happen to the action when we put this in and integrate by parts – we'll get:

$$S[\phi + \delta\phi] = S[\phi] + \underbrace{\int \left(\frac{\partial\mathcal{L}}{\partial\phi} - \partial_\mu \left(\frac{\partial\mathcal{L}}{\partial\phi_{,\mu}} \right) \right) \delta\phi \, d\tau}_{=0} + \underbrace{\int \frac{\partial\mathcal{L}}{\partial g_{\mu\nu}} \delta g_{\mu\nu} d\tau}_{\equiv \delta S}. \tag{5.93}$$

Notice that the first perturbative term above vanishes by the field equations – indeed, this first term is just the result of an induced variation of ϕ (the field equation for ϕ comes from arbitrary variation, we are just picking a particular variation $\delta\phi$ associated with coordinate transformations, so the field equations still hold). The second term, labeled δS, must be zero for S to remain unchanged. That is what will, when combined with our assumed dependence on the metric explicitly through $\sqrt{-g}$ and implicitly through $\bar{\mathcal{L}}$, put a restriction on the fields themselves. It is in this sense that we have a form of Noether's theorem – our symmetry is coordinate-invariance, and the conservation law is precisely the stress-tensor conservation we are after.

We need to connect the change in the metric to the change in coordinates. It is tempting to simply set $\frac{\partial \mathcal{L}}{\partial g_{\mu\nu}} = 0$, but this is overly restrictive ($\delta g_{\mu\nu}$ here refers to a *specific* type of variation for the metric, not a generic one). Consider the general transformation of a second-rank covariant tensor:

$$\bar{g}_{\mu\nu}(\bar{x}) = \frac{\partial x^\alpha}{\partial \bar{x}^\mu} \frac{\partial x^\beta}{\partial \bar{x}^\nu} g_{\alpha\beta}(x). \tag{5.94}$$

If we rewrite the transformation in terms of the original variables (so that we can recover S in terms of the original coordinates) using $\bar{x}^\mu = x^\mu + \eta^\mu$, then:

$$\bar{g}_{\mu\nu}(x) = g_{\mu\nu} + \left(-\eta^\beta_{,\nu} g_{\mu\beta} - \eta^\alpha_{,\mu} g_{\alpha\nu} - g_{\mu\nu,\sigma} \eta^\sigma \right), \tag{5.95}$$

with both sides functions of x now.

A moral point: we rarely (if ever) leave partial derivatives alone – a comma (as in $g_{\mu\nu,\sigma}$) is, as we have learned, not a tensorial operation. The whole point of introducing covariant differentiation was to induce tensor character, so it is always a good idea to use it. We are in luck, the metric's covariant derivative vanishes (for our space-times, recall (3.97)), so that:

$$g_{\mu\nu;\sigma} = g_{\mu\nu,\sigma} - \Gamma^\alpha_{\mu\sigma} g_{\alpha\nu} - \Gamma^\alpha_{\sigma\nu} g_{\mu\alpha} = 0. \tag{5.96}$$

The partial derivative on $\eta^\beta_{,\nu}$ in (5.95) can also be replaced by the covariant derivative:

$$\eta^\beta_{;\nu} = \eta^\beta_{,\nu} + \Gamma^\beta_{\sigma\nu} \eta^\sigma \longrightarrow \eta^\beta_{,\nu} = \eta^\beta_{;\nu} - \Gamma^\beta_{\sigma\nu} \eta^\sigma, \tag{5.97}$$

and similarly for the $\eta^\alpha_{,\mu}$. Until, finally, we have whittled down the term in parentheses from (5.95):

$$\bar{g}_{\mu\nu}(x) = g_{\mu\nu} - \eta_{\mu;\nu} - \eta_{\nu;\mu}. \tag{5.98}$$

Evidently, the perturbation is $\delta g_{\mu\nu} = -(\eta_{\mu;\nu} + \eta_{\nu;\mu})$[5] and we are ready to return to δS.

Referring to (5.93), we define a new density built from a tensor $T^{\mu\nu}$:

$$\frac{1}{2} \sqrt{-g}\, T^{\mu\nu} \equiv -\frac{\partial \mathcal{L}}{\partial g_{\mu\nu}}, \tag{5.99}$$

---

[5] This combination of vector and derivatives should look familiar – when it is zero, we call η_μ a Killing vector.

which is symmetric, so that:

$$\delta S = -\frac{1}{2} \int d\tau \sqrt{-g}\, T^{\mu\nu}\, \delta g_{\mu\nu} = +\frac{1}{2} \int d\tau \sqrt{-g}\, T^{\mu\nu}\, \left(\eta_{\mu;\nu} + \eta_{\nu;\mu}\right)$$

$$= \int d\tau \sqrt{-g}\, T^{\mu\nu}\, \eta_{\mu;\nu}. \tag{5.100}$$

We have used the symmetric property of $T^{\mu\nu}$ to simplify (notice how the strange factor of $\frac{1}{2}$ and the minus sign from (5.99) have played a role). Now we can break the above into a total derivative, which will turn into a surface integral that vanishes (by assumption, on the boundary of the integration, η^μ goes away) and an additional term:

$$\int d\tau \sqrt{-g}\, T^{\mu\nu}\, \eta_{\mu;\nu} = \underbrace{\int d\tau \left(\sqrt{-g}\, T^{\mu\nu}\, \eta_\mu\right)_{;\nu}}_{=0} - \int d\tau \sqrt{-g}\, T^{\mu\nu}{}_{;\nu}\, \eta_\mu. \tag{5.101}$$

For arbitrary, infinitesimal $\eta_\mu(x)$, then, we have:

$$\boxed{T^{\mu\nu}{}_{;\nu} = 0.} \tag{5.102}$$

The tensor $T^{\mu\nu}$ is called the "energy–momentum" tensor (it is also known as the "stress–energy" tensor, or "stress tensor"), and the statement (5.102) is a continuity equation. We have not done anything new, just exposed a property of solutions to the scalar field equations by appealing to a property of the action. We will think about (5.102) first by analogy with the more familiar charge conservation from E&M.

Think of a four-divergence in Minkowski space – an expression of the form $A^\mu{}_{;\mu} = 0$ reduces to (in flat space, "; $\longrightarrow$,", where covariant derivatives become "normal" partial derivatives, and we take Minkowski in its usual Cartesian form) $A^\mu{}_{,\mu} = 0$. Written out explicitly, we have:

$$\frac{1}{c}\frac{\partial A^0}{\partial t} + \nabla \cdot \mathbf{A} = 0 \longrightarrow \frac{\partial A^0}{\partial t} = -c\, \nabla \cdot \mathbf{A}, \tag{5.103}$$

with $\mathbf{A}$ the spatial components of the four-vector A^μ. This is a continuity statement – in generic language, we say that the zero component is the conserved "charge", and the vector portion the "current", terms coming from one of the most common examples: charge conservation, where $J^\mu \doteq (c\,\rho, J_x, J_y, J_z)^T = (c\,\rho, \mathbf{J})^T$, and then:

$$\frac{\partial \rho}{\partial t} = -\nabla \cdot \mathbf{J}. \tag{5.104}$$

The Lorentz gauge condition from E&M can also be expressed this way, with $A^\mu \doteq (\frac{V}{c}, A_x, A_y, A_z)^T$, leading to:

$$\frac{1}{c^2} \frac{\partial V}{\partial t} = -\nabla \cdot \mathbf{A}. \tag{5.105}$$

5.7 The stress–energy tensor

For a second-rank tensor in this flat-space $(; \longrightarrow,)$ setting, the statement (5.102) becomes four equations:

$$0 = \frac{1}{c} \frac{\partial T^{00}}{\partial t} + \frac{\partial T^{0j}}{\partial x^j}$$

$$0 = \frac{1}{c} \frac{\partial T^{j0}}{\partial t} + \frac{\partial T^{jk}}{\partial x^k} \qquad j = 1, 2, 3. \tag{5.106}$$

If we integrate over an arbitrary spatial volume, and use the usual form of the divergence theorem, we can interpret these four equations as continuity equations as well:

$$\frac{\partial}{\partial t} \int \frac{1}{c} T^{00} \, d\tau = - \oint T^{0j} \, da_j$$

$$\frac{\partial}{\partial t} \int \frac{1}{c} T^{0j} \, d\tau = - \oint T^{jk} \, da_k, \tag{5.107}$$

with the obvious identification of a scalar and three-vector in flat space.

In order to understand the actual physics of this $T^{\mu\nu}$ tensor, we will find the explicit form in terms of the Lagrange density and see what this form implies about our scalar field.

5.7.1 The tensor $T^{\mu\nu}$ for fields

From its definition in (5.99):

$$T^{\mu\nu} = -\frac{2}{\sqrt{-g}} \frac{\partial \mathcal{L}}{\partial g_{\mu\nu}}, \tag{5.108}$$

we need the derivative of the density $\mathcal{L}$ with respect to $g_{\mu\nu}$. A typical density, like our scalar field, will depend on $g_{\mu\nu}$ through g, the determinant, and potentially, a metric that is used to contract terms as in $\phi_{,\mu} g^{\mu\nu} \phi_{,\nu}$. Let's assume the form is:

$$\mathcal{L}(\phi, \phi_{,\mu}, g_{\mu\nu}) = \sqrt{-g} \, \bar{\mathcal{L}}(\phi, \phi_{,\mu}, g_{\mu\nu}) \tag{5.109}$$

and note that $\frac{\partial \sqrt{-g}}{\partial g_{\mu\nu}} = \frac{1}{2} \sqrt{-g} \, g^{\mu\nu}$, so:

$$\frac{\partial \mathcal{L}}{\partial g_{\mu\nu}} = \frac{1}{2} \sqrt{-g} \, g^{\mu\nu} \, \bar{\mathcal{L}} + \sqrt{-g} \, \frac{\partial \bar{\mathcal{L}}}{\partial g_{\mu\nu}}. \tag{5.110}$$

The tensor of interest can be written as:

$$T^{\mu\nu} = -\left(g^{\mu\nu}\,\bar{\mathcal{L}} + 2\,\frac{\partial\bar{\mathcal{L}}}{\partial g_{\mu\nu}}\right). \tag{5.111}$$

Free scalar field (Minkowski)

As an example, we'll construct the stress–energy tensor for a simple, free scalar field in Minkowski space-time. This familiar example will allow us to interpret the T^{00} component of field stress tensors as an energy density.

For the free scalar field, the Lagrange scalar is $\bar{\mathcal{L}} = \frac{1}{2}\phi_{,\alpha}\,g^{\alpha\beta}\,\phi_{,\beta}$. If we input this Lagrange scalar in (5.111), we need to evaluate the term:[6]

$$\frac{\partial\bar{\mathcal{L}}}{\partial g_{\mu\nu}} = -\frac{1}{2}\phi_{,\alpha}\,\phi_{,\beta}\,g^{\alpha\mu}\,g^{\beta\nu}, \tag{5.112}$$

and the $T^{\mu\nu}$ tensor is:

$$T^{\mu\nu} = \phi_{,\alpha}\,\phi_{,\beta}\,g^{\alpha\mu}\,g^{\beta\nu} - \frac{1}{2}g^{\mu\nu}\,\phi_{,\alpha}\,g^{\alpha\beta}\,\phi_{,\beta}. \tag{5.113}$$

Again, we have the question of interpretation here – referring to our original discrete Lagrangian, from whence all of this came, we can transform to a Hamiltonian, and this will give us an expression (upon taking the continuum limit) for what might reasonably be called the energy density of the field. The procedure is motivated by its classical analogue – where the temporal derivative is privileged – this seems strange in our current homogenous treatment, but lends itself to an obvious energetic interpretation.

In the particle case (ball and spring model), the canonical momentum associated with the motion of an individual portion of the rod (for example) would be $\frac{\partial L}{\partial\dot{\phi}(x,t)} \equiv \pi$, suggesting that we take, as the canonical momentum for the field ϕ, $\frac{\partial\bar{\mathcal{L}}}{\partial\dot{\phi}} \equiv \pi$. Starting from, in two dimensions:

$$\bar{\mathcal{L}} = \frac{1}{2}\phi_{,\mu}\,g^{\mu\nu}\,\phi_{,\nu} = \frac{1}{2}\left(-\frac{1}{v^2}\dot{\phi}^2 + \phi'^2\right), \tag{5.114}$$

we identify $\pi = -\frac{1}{v^2}\dot{\phi}$, and then:

$$\begin{aligned}
\bar{\mathcal{H}} = \dot{\phi}\,\pi - \bar{\mathcal{L}} &= -\frac{1}{v^2}\dot{\phi}^2 - \left(-\frac{1}{2\,v^2}\dot{\phi}^2 + \frac{1}{2}\phi'^2\right)\\
&= -\frac{1}{2}\left(\frac{1}{v^2}\dot{\phi}^2 + \phi'^2\right) \equiv -\mathcal{E},
\end{aligned} \tag{5.115}$$

---

[6] Using $\frac{\partial g^{\alpha\beta}}{\partial g_{\mu\nu}} = -g^{\alpha\mu}\,g^{\beta\nu}$, where explicit symmetrization with respect to $\mu \leftrightarrow \nu$ is unnecessary because of subsequent contraction, as in (5.112).

where we define the energy density $\mathcal{E}$ to be the negative of $\bar{\mathcal{H}}$ (this is just a matter of the metric signature, nothing deep). Compare this with T^{00} from (5.113) – that component is:

$$T^{00} = \phi_{,0}\,\phi_{,0} + \frac{1}{2}\left(-\phi_{,0}^2 + \phi_{,1}^2\right) = \frac{1}{2}\left(\frac{1}{v^2}\,\dot{\phi}^2 + \phi'^2\right)$$

$$= \mathcal{E},$$

(5.116)

so that the zero–zero (pure temporal) component of $T^{\mu\nu}$ is naturally identified with the energy density of the system.

If we introduce our overall factor (multiply by $\mu\,v^2$ in $\bar{\mathcal{L}}$ to recover the form from Section 5.3) to make contact with real longitudinal oscillations, then the energy density is (with units, and referring again to the physical rod, not vacuum):

$$\mathcal{E} = \frac{1}{2}\,\mu\,\dot{\phi}^2 + \frac{1}{2}\,\mu\,v^2\,\phi'^2,$$

(5.117)

which is pretty clearly the kinetic and potential energy per unit length (think of what we would have got out of the Hamiltonian for balls and springs).

Now we want to find the natural momentum density here, and we can work directly from the energy for a short segment of the rod $E = \int_a^b \mathcal{E}\,dx$ – as time goes on, energy will flow into and out of this segment, and we can calculate the temporal dependence of that flow. If we Taylor expand the difference in energy at t and $t + dt$:

$$E(t + dt) - E(t) = \frac{dE}{dt}\,dt + O(dt^2),$$

(5.118)

and compare with the Taylor expansion of the integral of energy density:

$$E(t+dt) = \int\left[\frac{1}{2}\,\mu\,(\dot{\phi}(t+dt))^2 + \frac{1}{2}\,\mu\,v^2\,(\phi'(t+dt))^2\right]dx$$

$$= \int\left[\frac{1}{2}\,\mu\,\dot{\phi}(t)^2 + \frac{1}{2}\,\mu\,v^2\,\phi'(t)^2 + dt\left(\mu\,\ddot{\phi}\,\dot{\phi} + \mu\,v^2\,\phi'\,\dot{\phi}'\right)\right]dx,$$

(5.119)

then it is clear we should associate the derivative of energy with the integral:

$$\frac{dE}{dt} = \int\left(\mu\,\ddot{\phi}\,\dot{\phi} + \mu\,v^2\,\phi'\,\dot{\phi}'\right)dx.$$

(5.120)

Using the field equation $\ddot{\phi} = v^2\,\phi''$, we can write the integral on the right as a total derivative:

$$\frac{dE}{dt} = \int\frac{\partial}{\partial x}\left(\mu\,v^2\,\dot{\phi}\,\phi'\right)dx$$

(5.121)

which, as a total derivative, can be evaluated at the endpoints as usual. In terms of the local statement, we have:

$$\frac{\partial \mathcal{E}}{\partial t} = -\frac{\partial}{\partial x} \left(-\mu v^2 \phi \phi' \right).$$
(5.122)

From (5.106), and the association of energy density with $\mathcal{E}$, the derivative on the right should be $v T^{0x}$ (we are using a generic speed v, which is c for vacuum). Computing T^{0x} directly from (5.113), with the factor of μv^2 in place gives:

$$
\begin{aligned}
T^{0x} &= \mu v^2 \left(\phi_{,\alpha} \phi_{,\beta} g^{\alpha 0} g^{\beta x} - g^{0x} \phi_{,\alpha} g^{\alpha\beta} \phi_{,\beta} \right) \\
&= -\mu v^2 \phi_{,0} \phi_{,x} \\
&= -\mu v \phi \phi'.
\end{aligned}
$$
(5.123)

Finally, we see that $\frac{\partial}{\partial x} \left(v T^{0x} \right)$ is precisely the right-hand side of (5.122).

In addition, we can interpret this component in terms of momentum density – for a patch of field between x and $x + dx$, we have mass displaced from x into the interval $\mu \phi(x)$, and mass displaced out of the interval on the right $\mu \phi(x + dx)$. Then the momentum in this interval is:

$$\mathrm{p}\, dx = \mu \dot{\phi} \left(\phi(x) - \phi(x + dx) \right) \approx -\mu \dot{\phi} \phi'\, dx,$$
(5.124)

and this is related to T^{0x} by one factor of v: $\mathrm{p} = \frac{1}{v} T^{0x}$, where p is the momentum density (momentum per unit length).

5.7.2 Conservation of angular momentum

Given the conservation of the energy–momentum tensor itself, we can also construct the angular momentum analogue. Define:

$$M^{\alpha\beta\gamma} \equiv x^\beta T^{\gamma\alpha} - x^\gamma T^{\beta\alpha},$$
(5.125)

working in four-dimensional Minkowski space-time. We have the derivative:

$$\frac{\partial M^{\alpha\beta\gamma}}{\partial x^\alpha} = T^{\gamma\beta} - T^{\beta\gamma} = 0,$$
(5.126)

from the symmetry of the stress tensor, and conservation: $T^{\mu\nu}{}_{,\nu} = 0$. This gives us a set of conserved quantities in the usual way:

$$\frac{1}{c} \frac{\partial M^{0\beta\gamma}}{\partial t} = -\frac{\partial M^{i\beta\gamma}}{\partial x^i}$$
(5.127)

or, in integral form:

$$\frac{d}{dt} \int \frac{1}{c} M^{0\beta\gamma}\, d\tau = -\oint M^{i\beta\gamma}\, da_i.$$
(5.128)

The left-hand side represents six independent quantities – of most interest are the spatial components. Let $J^{\alpha\beta} = \frac{1}{c}M^{0\alpha\beta}$, then the densities are just as we expect:

$$J^{ij} = \frac{1}{c}M^{0ij} = \frac{1}{c}[x^i\, T^{j0} - x^j\, T^{i0}], \tag{5.129}$$

so that, in terms of the spatial components of momentum density, we have:

$$J^{ij} = x^i\, \mathfrak{p}^j - x^j\, \mathfrak{p}^i, \tag{5.130}$$

precisely the angular momentum (density).

Problem 5.5

A "massive" scalar field has action (Klein–Gordon):

$$S[\phi] = \int d\tau \sqrt{-g}\left[\frac{1}{2}\left(\phi_{,\alpha}g^{\alpha\beta}\phi_{,\beta} + m^2\phi^2\right)\right]. \tag{5.131}$$

Working in $D = 3 + 1$ Minkowski with Cartesian spatial coordinates:
(a) Find the field equations and stress tensor $T^{\mu\nu}$ associated with this action.
(b) Verify that $T^{\mu\nu}{}_{,\nu} = 0$ as a consequence of the field equations.
(c) Write out the T^{00} component of the stress tensor, this is the energy density associated with the field.
(d) Take a plane-wave ansatz:

$$\phi(t, \mathbf{x}) = A\, e^{i\, p_\mu\, x^\mu} \tag{5.132}$$

for four-vector p_μ. What constraint does the field equation put on p_μ? Think of p_μ as the four-momentum of a particle, and explain why ϕ coming from (5.131) is called a "massive" scalar field.

Problem 5.6

For a massless Klein–Gordon field ϕ (set $m = 0$ in (5.131)), take a plane-wave solution $\phi = A\, e^{i\, x^\mu\, p_\mu}$ (in Minkowski space-time written in Cartesian spatial coordinates).
(a) Find the constraint on p_μ placed on the solution by the field equations.
(b) Construct the stress–energy tensor – write $T^{\mu\nu}$ in terms of p_μ.
(c) What is the M^{00j} component(s) of $M^{\alpha\beta\gamma}$ from (5.125)? How about the M^{i0j} component(s) (i and j here are spatial indices)? Do these satisfy (5.127)?

Problem 5.7

The general scalar field Lagrangian (in Minkowski space-time with Cartesian spatial coordinates) looks like:

$$\mathcal{L} = \frac{1}{2}\phi_{,\alpha}g^{\alpha\beta}\phi_{,\beta} - V(\phi). \tag{5.133}$$

(a) Compute the general field equations for arbitrary $V(\phi)$. Write these explicitly for $D = 1 + 1$ Minkowski space-time in Cartesian coordinates and the potential $V(\phi) = -\frac{m^2}{4v^2}(\phi^2 - v^2)^2$ (note that this potential is a natural nonlinear extension of the Klein–Gordon field, for small v, we recover a mass term and a ϕ^4 term).

(b) Our goal will be to generate the full solution from the stationary form. So set $\phi(x, t) = \phi(x)$ and solve the above field equations, assume $\phi'(x) = 0$ as $x \longrightarrow \pm\infty$ and $\phi = \pm v$ at infinity. Finally, you can take $\phi(x = 0) = 0$.

(c) The full field equations are manifestly Lorentz invariant – what we'll do is put the Lorentz invariance back into our purely spatial solution by boosting out of a frame instantaneously at rest. Write x with $t = 0$ as a combination of $\bar{x}$ and $\bar{t}$ boosted along a shared x-axis moving with velocity u. The resulting function $\phi(x(\bar{x}, \bar{t})) \equiv \phi(\bar{x}, \bar{t})$ *is* a solution to the full field equations.

Problem 5.8

For the time-independent scalar field equation $\nabla^2\phi(x, y, z) = 0$, write down a spherically symmetric ansatz for ϕ and solve for ϕ for all points except the origin.

Problem 5.9

What is $F^{\mu}_{;\mu}$ (the four-divergence) of a vector field in Euclidean three-dimensional space written in spherical coordinates? Did you get the expression you expect (from, for example, [17])? If not, explain why and show how to recover the familiar expression.

5.8 Stress tensors, particles, and fields

We continue looking at the energy–momentum tensor, first establishing the usual interpretations for a single free particle, and then making the connection to the field tensors we began to develop in Section 5.7.1. Along the way, we introduce the Lagrangian density for point particles, and this will prove useful when we actually attempt to couple theories to matter.

5.8.1 Energy–momentum tensor for particles

Because it is the fundamental object of interest for matter coupling, and also because it sheds some light on the field energy–momentum tensor, we want to connect the $T^{\mu\nu}$ that comes from particles themselves to the field version. Consider the usual action for free particles in special relativity from Section 2.2.2:

$$S = -m\,c\int \sqrt{-\dot{x}^\mu\, g_{\mu\nu}\, \dot{x}^\nu}\, ds, \tag{5.134}$$

where dots refer to differentiation by s. If we take s to be the proper time, then this reduces to $S = -m \, c^2 \int ds$, and we can expand ds in the low-velocity limit to recover our three-dimensional $L = \frac{1}{2} m \, v^2$. Since we are talking about free particles in proper time parametrization, the quantity $\sqrt{-\dot{x}^\mu \, g_{\mu\nu} \, \dot{x}^\nu}$ is actually constant (with value c for Minkowski).

We have an action, so we should be able to develop an energy–momentum tensor by finding the appropriate $\mathcal{L}$. The problem is, we do not have a four-volume integral in the action (5.134) from which to extract $\mathcal{L}$. We do know that the particle is moving along some trajectory, $x^\mu(s)$, and then we can view $\rho \longrightarrow m \, \delta^4(x^\mu - x^\mu(s))$ – the four-dimensional delta function is itself a density, so there is a scalar here:

$$S_p \equiv -m \, c \int \int ds d\tau \, \sqrt{-\dot{x}^\mu \, g_{\mu\nu} \, \dot{x}^\nu} \, \delta^4(x^\mu - x^\mu(s)). \tag{5.135}$$

The energy–momentum tensor *density* $\mathcal{T}^{\mu\nu} = \sqrt{-g} \, T^{\mu\nu}$ can be obtained from our general form:

$$\frac{1}{2} \sqrt{-g} \, T^{\mu\nu} = -\frac{\partial \mathcal{L}}{\partial g_{\mu\nu}} \longrightarrow T^{\mu\nu} = -2 \frac{\partial \mathcal{L}}{\partial g_{\mu\nu}}, \tag{5.136}$$

with $\mathcal{L}$ the Lagrange density, given for the particle by the integrand of (5.135):

$$\mathcal{L} = -\int ds \, m \, c \, \sqrt{-\dot{x}^\mu \, g_{\mu\nu} \, \dot{x}^\nu} \, \delta^4(x^\mu - x^\mu(s)). \tag{5.137}$$

The tensor density is:

$$\mathcal{T}^{\mu\nu} = \int ds \, \frac{m \, c \, \dot{x}^\mu \, \dot{x}^\nu}{\sqrt{-\dot{x}^\mu \, g_{\mu\nu} \, \dot{x}^\nu}} \delta(t - t(s)) \, \delta^3(\mathbf{x} - \mathbf{x}(t))$$

$$= m \, \dot{x}^\mu \, \dot{x}^\nu \, \frac{ds}{dt} \, \delta^3(\mathbf{x} - \mathbf{x}(s(t))), \tag{5.138}$$

where we have transformed the integral over s to one over t, used the $\delta(t - t(s))$ to perform the integration, and then transformed back to the s parametrization. We have also used the invariance of the interval (in proper time parametrization) $\sqrt{-\dot{x}^\mu \, g_{\mu\nu} \, \dot{x}^\nu} = c$. The three-dimensional delta function just says that we take the stress tensor for the particle and evaluate it along the particle trajectory in three dimensions, with appropriate Lorentz factor. Note that for Minkowksi space, $\frac{ds}{dt}$ is given by:

$$c^2 \, ds^2 = c^2 \, dt^2 - dx^2 - dy^2 - dz^2 \longrightarrow \frac{ds}{dt} = \sqrt{1 - \frac{v^2}{c^2}}, \tag{5.139}$$

where v is the time-parametrized particle speed.

If we are in the rest-frame of the particle, $\frac{ds}{dt} = 1$, and $\frac{dx^0}{ds} = \frac{c \, dt}{ds} = c$, so the energy–momentum tensor density has only one term, $T^{00} = m \, c^2 \, \delta^3(\mathbf{x} - \mathbf{x}(t))$

(along its trajectory), which is the energy density of the particle. Take the temporal–spatial three-vector (not necessarily in the rest-frame), we have, mixing notation a bit in the second equality to make the point:

$$T^{0j} = m c \frac{dt}{ds} \frac{dx^j}{ds} \frac{ds}{dt} \doteq \frac{m c \mathbf{v}}{\sqrt{1 - \frac{v^2}{c^2}}} \delta^3(\mathbf{x} - \mathbf{x}(t)), \qquad (5.140)$$

which is, modulo the factor of c in the numerator, the three-velocity portion of the four-velocity (with v written in "usual" $\frac{d}{dt}$ form). Indeed, the full four-vector is:

$$\boxed{T^{0\mu} \doteq \left(\begin{array}{c} \frac{m c^2}{\sqrt{1 - \frac{v^2}{c^2}}} \\ \frac{m c \mathbf{v}}{\sqrt{1 - \frac{v^2}{c^2}}} \end{array} \right) \delta^3(\mathbf{x} - \mathbf{x}(t)) = c \, p^\mu \, \delta^3(\mathbf{x} - \mathbf{x}(t)),} \qquad (5.141)$$

with p^μ the four-momentum vector of special relativity. The spatial delta function ensures we are evaluating the particle along its trajectory.

5.8.2 Energy and momenta for fields

If we view the $T^{00} = \mathcal{E}$ entry as energy density for the fields (as we saw for the scalar field in Section 5.7.1), then the integrals associated with $T^{\mu\nu}{}_{;\nu} = 0$ (precisely (5.107)) give a natural interpretation to the other elements:

$$\frac{\partial}{\partial t} \int \mathcal{E} \, d\tau = - \oint c \, T^{0j} \, da_j \qquad (5.142)$$

represents the time-derivative (on the left) of the energy enclosed in a volume, the total change of energy in the volume in time. On the right is energy passing out through a surface (i.e. $c \, T^{0j}$ is energy per unit area per unit time) – the energy flux density. If we compare with the particle case, we see, from (5.141), that $\frac{1}{c} T^{0j} \sim \mathbf{p}$ is the momentum density, so that it is reasonable to associate the (physical) field spatial momenta with:

$$P^j = \frac{1}{c} \int T^{0j} \, d\tau, \qquad (5.143)$$

as we did with the scalar field.

The purely spatial components give another continuity statement:

$$\frac{\partial}{\partial t} \underbrace{\int \frac{1}{c} T^{0j} \, d\tau}_{= \frac{\partial}{\partial t} P^j} = - \oint T^{jk} \, da_k, \qquad (5.144)$$

so that the spatial components of $T^{0\mu}$ also change with time, with flux density given by T^{jk}, which has units of force per unit area. Thinking back to E&M, we associate the T^{00} component with energy density and the T^{0j} component with the Poynting vector (or more generally, momentum density for the fields). Similarly, the change in momentum density comes from the Maxwell stress tensor, which forms a momentum flux density. We will see all of this explicitly later on, but the point is – the energy and momentum conservation statements for a field theory are entirely encapsulated in the vanishing of the divergence of the energy–momentum tensor $T^{\mu\nu}$.

5.9 First-order action

We now switch gears a bit, and return to the treatment of actions, this time by way of a "Hamiltonian". I use this term loosely here, what we will really be doing is a series of Legendre transforms. Hamiltonians, in field theory, still represent a specific breakup of time and space, so the procedure below is more of a Hamiltonian-ization of the action, and not what most people would call the Hamiltonian of a field theory which we actually constructed in Section 5.7.1.

The Lagrange density is the most natural way to define the action for a field theory, coming to us, as it does, directly from a discrete multi-particle Lagrangian. But $\mathcal{L}$ does not necessarily provide the most useful description of the fields, it leads to PDEs that are second order in the "coordinates" (time and space, now).

In our usual classical mechanics setting, it is possible to turn a single second-order equation of motion into two first-order equations through the Hamiltonian. The same can be done on the field side. Consider the free scalar field action:

$$S = \int d\tau \sqrt{-g} \underbrace{\left(\frac{1}{2} \phi_{,\mu} g^{\mu\nu} \phi_{,\nu} \right)}_{=\bar{\mathcal{L}}}, \qquad (5.145)$$

and notice the correspondence between this and a free particle action $S = \int ds \left(\frac{1}{2} \dot{x}^\mu g_{\mu\nu} \dot{x}^\nu \right)$ – what if we imagine the replacement $\dot{x}^\mu \longrightarrow \phi_{,\mu}$? That would make the "canonical momenta" for the field:

$$\pi^\alpha \equiv \frac{\partial \bar{\mathcal{L}}}{\partial \phi_{,\alpha}}, \qquad (5.146)$$

and we could get a Hamiltonian just by taking the Legendre transform:

$$\bar{\mathcal{H}} = \phi_{,\alpha} \pi^\alpha - \bar{\mathcal{L}}. \qquad (5.147)$$

Then, just as in classical mechanics, we can form the so-called "first-order action" (because it leads to first-order field equations) by inverting the Legendre transform:

$$\boxed{S = \int d\tau \sqrt{-g} \left(\phi_{,\alpha} \pi^\alpha - \bar{\mathcal{H}} \right).} \qquad (5.148)$$

Aside from a slight change of notation, there is a relatively large shift in focus here – a Hamiltonian is expressed entirely in terms of the momenta and coordinates (field momenta π^α and field ϕ here), and the action written in the above form has an explicit term depending on $\phi_{,\alpha}$ but no other hidden $\phi_{,\alpha}$ dependence.

For our scalar action, we have:

$$\frac{\partial \bar{\mathcal{L}}}{\partial \phi_{,\alpha}} = g^{\alpha \nu} \phi_{,\nu} \qquad \bar{\mathcal{H}} = \frac{1}{2} \pi^\alpha \, g_{\alpha\beta} \, \pi^\beta, \tag{5.149}$$

so that the first-order action is:

$$S[\phi, \pi] = \int d\tau \sqrt{-g} \left(\phi_{,\alpha} \pi^\alpha - \frac{1}{2} \pi^\alpha \, g_{\alpha\beta} \, \pi^\beta \right). \tag{5.150}$$

Now we can vary w.r.t. ϕ and π independently – the ϕ variation gives the usual:

$$\delta S = \int d\tau \sqrt{-g} \, \pi^\alpha \, \delta\phi_{,\alpha}$$
$$= \underbrace{\int d\tau \left(\sqrt{-g} \, \pi^\alpha \, \delta\phi \right)_{;\alpha}}_{=0} - \int d\tau \left(\sqrt{-g} \, \pi^\alpha \right)_{;\alpha} \delta\phi, \tag{5.151}$$

and the second term tells us that $\pi^\alpha_{\;;\alpha} = 0$ (remember that $(\sqrt{-g})_{;\alpha} = 0$).

Varying the action w.r.t. π^α gives:

$$\delta S = \int d\tau \sqrt{-g} \left(\phi_{,\alpha} - g_{\alpha\beta} \, \pi^\beta \right) \delta\pi^\alpha, \tag{5.152}$$

and for this to be zero for arbitrary $\delta\pi^\alpha$, we must have:

$$\phi_{,\alpha} - g_{\alpha\beta} \, \pi^\beta = 0 \longrightarrow \pi^\beta = g^{\alpha\beta} \, \phi_{,\alpha}, \tag{5.153}$$

which reproduces the definition of π^β. Putting these two together, our two first-order field equations are:

$$\left. \begin{array}{l} \pi^\beta = g^{\alpha\beta} \, \phi_{,\alpha} \\ \pi^\alpha_{\;;\alpha} = 0 \end{array} \right\} \longrightarrow \phi^{;\alpha}_{\;;\alpha} = 0. \tag{5.154}$$

So what? Well, the advantage here is that the action for a free scalar field is, at this point, easy to guess. Suppose we knew from the start that a free field has $\bar{\mathcal{H}} = \frac{1}{2} \pi^\alpha \pi_\alpha$, which is a pretty reasonable form. Then we are told that, via Legendre, we must introduce derivatives of a scalar field (those are first-rank tensors), the resulting action is then fixed. It's clear from $\pi^\alpha \pi_\alpha$ that we are dealing with scalar fields, and then a term like $\phi_{,\alpha} \pi^\alpha$ is obvious.

Massive scalar fields

We can use first-order form to introduce potentials (although we have no immediate reason to do so). If we want to include a potential for ϕ, there are a few terms we might consider, $\sim \phi$, $\sim \phi^2$, etc. The simplest would be $\alpha \phi$, but this gives a relatively uninteresting Poisson equation with constant source. A term quadratic in ϕ changes the differential equation. For the general case:

$$S_m = \int d\tau \sqrt{-g} \left(\pi^\alpha \phi_{,\alpha} - \frac{1}{2} \pi^\alpha \pi^\beta g_{\alpha\beta} - V(\phi) \right) \qquad (5.155)$$

we have, upon variation of π^α, the usual $\pi^\beta g_{\alpha\beta} = \phi_{,\alpha}$, and under $\delta\phi$:

$$\delta S_m = \int d\tau \sqrt{-g} \left(-\pi^\alpha{}_{;\alpha} - \frac{\partial V}{\partial \phi} \right) \delta\phi = 0 \qquad (5.156)$$

which, in flat Minkowksi space-time, becomes:

$$\Box^2 \phi + \frac{\partial V}{\partial \phi} = 0. \qquad (5.157)$$

For a quadratic "mass" term, $V(\phi) = -\frac{1}{2} m^2 \phi^2$, we get the Klein–Gordon equation:

$$\Box^2 \phi - m^2 \phi = 0. \qquad (5.158)$$

This is an interesting theory on its own, and you have worked out some of the details and interpretation (especially associated with its stress–energy tensor) in the problems. In particular, the use of the word "mass" to describe the theory is provocative.

Structurally, first-order form has the same advantages we saw in the particle case. We move from second-order field equations (like the wave equation) to a pair of first-order equations. In addition, one of those tells us how to define the canonical field momentum. The other tells us the dynamical content of the theory, and potentially couples to sources – this logical breakup is important. Think, for example, of Maxwell's equations – two of those tell us that we can define a four-potential, while the other two relate the potential to sources. As written, Maxwell's equations mix these two statements. Using first-order form for the action, we recover a natural structure–sources pair. We shall see how this works for E&M in the next chapter.

Problem 5.10

We have defined the momenta as $\pi_\alpha = \frac{\partial \mathcal{L}}{\partial \phi_{,\alpha}}$, which is sensible from a Legendre transform point of view. But for quantum mechanics, the "Hamiltonian" for a theory is usually defined by taking the momentum to be $\pi \equiv \frac{\partial \mathcal{L}}{\partial \dot\phi}$, by analogy with the preferred role of the time-derivative in defining the momentum in classical mechanics as discussed in Section 5.7.1. Take the full four-dimensional, massive Klein–Gordon field Lagrangian:

$$\mathcal{L} = \frac{1}{2} \phi_{,\mu} \eta^{\mu\nu} \phi_{,\nu} + \frac{1}{2} m^2 \phi^2 \qquad (5.159)$$

(for $\eta^{\mu\nu}$ the Minkowski metric written in Cartesian coordinates), and form the momentum $\pi \equiv \frac{\partial \mathcal{L}}{\partial \dot{\phi}}$. Construct $\mathcal{H} = \dot{\phi}\,\pi - \mathcal{L}$, and show that it is the negative of the energy density T^{00} appropriate for this field. Use the Hamiltonian equations of motion to recover the massive Klein–Gordon field equations.

Problem 5.11

Here we examine a reformulation of massive scalar fields (work in Minkowski space-time in Cartesian coordinates).
(a) Write the scalar Lagrangian (in second-order form) for two uncoupled, real, massive scalar fields, ψ and η, with identical mass m.
(b) Define $\phi = \psi + i\,\eta$, and rewrite the Lagrangian in terms of ϕ and ϕ^*.
(c) Vary the Lagrangian with respect to both ϕ and ϕ^* separately, verify that the resulting field equations reduce to the correct ones for ψ and η.

Problem 5.12

Find the spherically symmetric, time-independent, solution to the massive scalar field equation $\Box^2 \phi = m^2\,\phi$ in $D = 3 + 1$ that vanishes as $r \longrightarrow \infty$. Does it reduce as expected when $m = 0$?

Problem 5.13

(a) In cylindrical coordinates, solve $\Box^2 \phi = 0$ for $\phi(t, x, y, z) = \phi(s)$ with $s = \sqrt{x^2 + y^2}$, that is, the axially symmetric solution.
(b) Using the same symmetries, find $\Box^2 \phi = m^2\,\phi$.

Problem 5.14

To make the role of the metric determinant in integration clear (and separate it from, for example, the reparametrization of volumes that must also occur when changing variables in integration), let's work out the area of a circle in two dimensions using Cartesian and polar coordinates. That is, we'll compute:

$$I = \int_\Omega \sqrt{g}\, d\tau \qquad\qquad (5.160)$$

for Ω a circle in two dimensions, using two different coordinate systems.
(a) A circle of radius R is centered on the origin. Write the two-dimensional integral that gives its area using Cartesian coordinates (remember that the boundary of a circle provides a relation between y and x: $y = \pm\sqrt{R^2 - x^2}$, use this in the limits of the y integration).
(b) Evaluate I for the same circle using polar coordinates: $x = r\,\cos\phi$, $y = r\,\sin\phi$ (what is the metric in this coordinate system? Use that to compute $\sqrt{g}$ appearing in the integral).

Problem 5.15

The stress tensor naturally arose as a contravariant object, defined via:

$$T^{\mu\nu} = -\frac{2}{\sqrt{-g}} \frac{\partial \mathcal{L}}{\partial g_{\mu\nu}}. \tag{5.161}$$

Show that the covariant form is:

$$T_{\mu\nu} = \frac{2}{\sqrt{-g}} \frac{\partial \mathcal{L}}{\partial g^{\mu\nu}}. \tag{5.162}$$

Hint: lower indices carefully, then use the chain rule, and evaluate $\frac{\partial g_{\mu\nu}}{\partial g^{\alpha\beta}}$.

Problem 5.16

The Dirac delta is a scalar density. Show that $\delta^3(x, y, z) = \delta(x)\,\delta(y)\,\delta(z)$ has weight negative one.

6

Tensor field theory

The scalar field example gives us a number of tools that can be applied to any relativistic field theory. We started with a physical model for longitudinal "density" waves (balls and springs) and shifted from a material (with intrinsic wave speed v) to the vacuum (with intrinsic wave speed c). That shift took us from real fields describing explicit physics for our model system to fields that do not have explicit physical manifestation – the scalar field ϕ satisfying $\Box^2 \phi = 0$ is not necessarily the density or pressure of anything. The theory of electricity and magnetism has, at its core, a vector field (the combined electric and magnetic potentials) that is an object in its own right, and does not correspond to a macroscopic property of materials – of course, the potential can be used to find the forces on objects, and it is these force-mediating fields that form the bulk of classical field theory. Electricity and magnetism provides a vehicle for discussing the appropriate form of field equations (and their precursors, the action and Lagrange densities) that support specific physical ideas – superposition, and special relativity, for example. Complementing these physically inspired properties, E&M is a good place to think about gauge freedom and the role that gauge-fixing can play in revealing the underlying physics while simplifying (in some cases) the field equations. We can use these ideas to "go backwards", that is, to start with an action that yields linear field equations with finite propagation speed for waves (written in some gauge), and see what physical interpretation we might give to the resulting fields. This program will give precisely E&M (and its massive version, the Proca field equations) when applied to vectors, and general relativity when applied to second-rank, symmetric, tensors.

We go from first-order scalar field Lagrange densities to the equivalent form for a vector field. Our ultimate goal, for vectors, is a description of vacuum electrodynamics. From the natural action, we derive the field equations – Maxwell's equations for potentials in the absence of sources. Sourceless field theories tell us

205

nothing about the constants that must appear in a general action, coupling sources
to the field theory density. By appealing to our E&M background, we can set
the various constants appropriate to source coupling. So we begin with the action
and field equations, then form the energy–momentum tensor for the fields alone,
and by setting our one overall constant so that the T^{00} component corresponds
to energy density, we recover the full (familiar) energy–momentum tensor for
E&M. After all of this, we are in good position to introduce charge and current
sources.

In preparation for the tensor case, we will also look at the inversion of the above
program, namely, we will start with field equations for scalars and vectors that have
desirable properties (like, for example, superposition) and generate the Lagrangians
and actions that, when varied, produce the target field equations. This will give us
a place to discuss the role of various terms in the field equations, and their physical
significance (or, in the case of gauge freedom, their lack of significance).

Finally, we move to second-rank tensor fields. The culmination of our work on
first-order actions will be the development of a compelling, relativistic, second-rank
tensor field theory. When such a theory is appropriately tuned for self-consistency,
we will find that we have come back to general relativity. This approach makes the
interpretation of the field as a metric, and the connection of that metric to geometry
a matter of destiny. In addition, the universal coupling of all other theories, both
particle and field, to gravity is transparent and becomes a requirement of the
mathematical process of variation.

6.1 Vector action

Our objective is E&M, and our starting point will be an action. The big change
comes in our Lagrange density – rather than $\mathcal{L}(\phi, \phi_{,\mu}, g_{\mu\nu})$, we have a set of fields,
call them A_μ, so the action can depend on scalars made out of the fields A_μ and their
derivatives $A_{\mu,\nu}$: $\mathcal{L}(A_\mu, A_{\mu,\nu}, g_{\mu\nu})$. Think of Minkowski space-time, and our usual
treatment of electricity and magnetism. We know that the "field strength tensor"
$F^{\mu\nu} = \partial^\mu A^\nu - \partial^\nu A^\mu$ is a tensor with respect to Lorentz transformations, but why
stop there? Why not make a full scalar action out of the field strength tensor, and
ignore its usual definition (we should recover this anyway, if our theory is properly
defined)? To that end, the only important feature of $F^{\mu\nu}$ is its antisymmetry:
$F^{\mu\nu} = -F^{\nu\mu}$.

Since we are shooting for a vector field, a singly indexed set of quantities, the
second-rank tensor $F^{\mu\nu}$ *must* be associated, in first-order form, with the momenta
of the field A_μ. The only scalar we can make comes from double-contraction,
$F^{\mu\nu} F_{\mu\nu}$. So taking this as the "Hamiltonian", our first-order action has to look

like:

$$S_V = \int d\tau \sqrt{-g} \left(F^{\mu\nu} (\text{derivatives of } A_\mu) - \frac{1}{2} F^{\mu\nu} g_{\alpha\mu} g_{\beta\nu} F^{\alpha\beta} \right). \quad (6.1)$$

Compare this with (5.148) – we needed a second-rank object, $F_{\mu\nu}$ to construct the $\bar{\mathcal{H}}$ portion, and then we contract with derivatives of the target field A_μ.

Because $F^{\mu\nu}$ is antisymmetric, we can form an antisymmetrization of $A_{\nu,\mu}$, i.e. $(A_{\nu,\mu} - A_{\mu,\nu})$. We are implicitly assuming that the metric here is just the usual Minkowski, $g_{\mu\nu} = \eta_{\mu\nu}$, and the presence of the ordinary partial derivatives appearing in the $A_{\mu,\nu}$ terms would appear to rule them out as suitable building blocks for scalars. This is actually not a problem for the antisymmetrized form, since:

$$A_{\nu;\mu} - A_{\mu;\nu} = \left(A_{\nu,\mu} - \Gamma^\sigma_{\nu\mu} A_\sigma \right) - \left(A_{\mu,\nu} - \Gamma^\sigma_{\mu\nu} A_\sigma \right)$$
$$= A_{\nu,\mu} - A_{\mu,\nu}, \quad (6.2)$$

by the symmetry (torsion-free) of the connection: $\Gamma^\sigma_{\mu\nu} = \Gamma^\sigma_{\nu\mu}$.

Our proposed action is:

$$\boxed{S_V = \int d\tau \sqrt{-g} \left(F^{\mu\nu} \left(A_{\nu,\mu} - A_{\mu,\nu} \right) - \frac{1}{2} F^{\mu\nu} g_{\alpha\mu} g_{\beta\nu} F^{\alpha\beta} \right).} \quad (6.3)$$

Again, since this is first-order form, we vary w.r.t. $F^{\mu\nu}$ and A_μ separately. Incidentally, the antisymmetry of $F^{\mu\nu}$ is now reinforced by the requirement that we build a tensor out of $A_{\mu,\nu}$ – because it is contracted with the antisymmetrization of $A_{\mu,\nu}$, only the antisymmetric portion of $F^{\mu\nu}$ will contribute, so we can vary w.r.t. all 16 components independently:[1]

$$\frac{\delta S_V}{\delta F^{\mu\nu}} = \sqrt{-g} \left[\left(A_{\nu,\mu} - A_{\mu,\nu} \right) - g_{\alpha\mu} g_{\beta\nu} F^{\alpha\beta} \right] = 0, \quad (6.4)$$

and this gives a relation between F and ∂A:

$$\boxed{F_{\mu\nu} = A_{\nu,\mu} - A_{\mu,\nu}.} \quad (6.5)$$

The variation of the A_μ is a little more difficult (we can play fewer indexing games by working with $\delta A_{\alpha,\beta}$, but it's good to get used to simplifying tricks now

---

[1] We use the standard notation here: $\frac{\delta S}{\delta F^{\mu\nu}}$ stands for the integrand of δS varied w.r.t. $F^{\mu\nu}$ – we perform all integration by parts, and are left with something "dotted into" $\delta F^{\mu\nu}$, that is the portion denoted $\frac{\delta S}{\delta F^{\mu\nu}}$. For a Lagrange density depending only on $F^{\mu\nu}$ and its first derivatives, for example, we have:

$$\delta S = \int d\tau \underbrace{\left(\frac{\partial \mathcal{L}}{\partial F^{\mu\nu}} - \frac{\partial}{\partial x^\gamma} \frac{\partial \mathcal{L}}{\partial F^{\mu\nu}_{,\gamma}} \right)}_{= \frac{\delta S}{\delta F^{\mu\nu}}} \delta F^{\mu\nu}.$$

rather than later):

$$\delta S_V = \int d\tau \sqrt{-g} \, \frac{\partial}{\partial A_{\mu,\nu}} \left(F^{\mu\nu} \left(A_{\nu,\mu} - A_{\mu,\nu} \right) \right) \delta A_{\mu,\nu}$$

$$= \int d\tau \sqrt{-g} \, \frac{\partial}{\partial A_{\mu,\nu}} \left(A_{\mu,\nu} \left(F^{\nu\mu} - F^{\mu\nu} \right) \right) \delta A_{\mu,\nu} \qquad (6.6)$$

$$= \int d\tau \sqrt{-g} \left(F^{\nu\mu} - F^{\mu\nu} \right) \delta A_{\mu,\nu},$$

and we really have in mind the covariant derivative (the $\sqrt{-g}$ ensures that we mean what we say).

We can use integration by parts to get a total divergence term, which as usual can be converted to a boundary integral where it must vanish:

$$\delta S_V = \underbrace{\int d\tau \left[\sqrt{-g} \left(F^{\nu\mu} - F^{\mu\nu} \right) \delta A_\mu \right]_{;\nu}}_{=0} - \int d\tau \left[\sqrt{-g} \left(F^{\nu\mu} - F^{\mu\nu} \right) \right]_{;\nu} \delta A_\mu.$$

$$(6.7)$$

The second term's integrand must then vanish for arbitrary δA_μ. Using $(\sqrt{-g})_{;\nu} = 0$, the above gives the second set of field equations:

$$\left(F^{\nu\mu} - F^{\mu\nu} \right)_{;\nu} = 0 \qquad (6.8)$$

and because we know that $F_{\mu\nu}$ is antisymmetric (either *a priori* or from (6.5)), this reduces to:

$$\boxed{F^{\mu\nu}{}_{;\nu} = 0.} \qquad (6.9)$$

Combining this with the other first-order field equation (6.5), and writing in terms of A_μ, we find that the fields A_μ satisfy (written in Cartesian coordinates):

$$\partial^\mu \partial_\nu A^\nu - \partial^\nu \partial_\nu A^\mu = 0. \qquad (6.10)$$

If we are really in Minkowski space-time with Cartesian spatial coordinates, this is a set of four independent equations which we may write as:

$$0 = -\frac{1}{c} \frac{\partial}{\partial t} \left[\frac{1}{c} \frac{\partial A^0}{\partial t} + \nabla \cdot \mathbf{A} \right] - \left[-\frac{1}{c^2} \frac{\partial^2 A^0}{\partial t^2} + \nabla^2 A^0 \right]$$

$$0 = \nabla \left[\frac{1}{c} \frac{\partial A^0}{\partial t} + \nabla \cdot \mathbf{A} \right] - \left[-\frac{1}{c^2} \frac{\partial^2 \mathbf{A}}{\partial t^2} + \nabla^2 \mathbf{A} \right]. \qquad (6.11)$$

Note the similarity with equations (10.4) and (10.5) from [17]:

$$\nabla^2 V + \frac{\partial}{\partial t} (\nabla \cdot \mathbf{A}) = -\frac{\rho}{\epsilon_0}$$

$$\left(\nabla^2 \mathbf{A} - \mu_0 \epsilon_0 \frac{\partial^2 \mathbf{A}}{\partial t^2} \right) - \nabla \left(\nabla \cdot \mathbf{A} + \mu_0 \epsilon_0 \frac{\partial V}{\partial t} \right) = -\mu_0 \mathbf{J}. \tag{6.12}$$

It appears reasonable to associate the four-vector A^μ with $A^0 = \frac{V}{c}$ and $A^j \doteq \mathbf{A}$, the electric and magnetic vector potentials respectively. To proceed, we know that there must be gauge freedom in A_μ – the physical effects of the fields are transmitted through $F^{\mu\nu}$. There is gauge freedom, although we have not yet done anything to establish it here, there being no sources in our theory. The gauge freedom is expressed in the connection between the field strength tensor and the potential:

$$F_{\mu\nu} = A_{\nu;\mu} - A_{\mu;\nu} = A_{\nu,\mu} - A_{\mu,\nu}. \tag{6.13}$$

If we take $A_\mu \longrightarrow A'_\mu = A_\mu + \psi_{,\mu}$, then:

$$F'_{\mu\nu} = A'_{\nu,\mu} - A'_{\mu,\nu} = \left(A_{\nu,\mu} + \psi_{,\nu\mu} \right) - \left(A_{\mu,\nu} + \psi_{,\mu\nu} \right) = A_{\nu,\mu} - A_{\mu,\nu}, \tag{6.14}$$

and nothing changes. We can exploit this to set the total divergence of A^μ. Suppose we start with A_μ such that $\partial_\mu A^\mu = F(x)$, some arbitrary function of the coordinates. Then we introduce a four-gradient of a scalar ψ: $A^\mu \longrightarrow A^\mu + \partial^\mu \psi$, and the new divergence is:

$$\partial_\mu (A^\mu + \partial^\mu \psi) = F + \partial_\mu \partial^\mu \psi. \tag{6.15}$$

If we like, this new divergence can be set to zero by appropriate choice of ψ (i.e. $\partial_\mu \partial^\mu \psi = -F$, Poisson's equation for the D'Alembertian operator). Then we can *a priori* set $\partial_\mu A^\mu = 0$ (Lorentz gauge) and the field equations can be simplified (from (6.10)):

$$\partial^\mu \partial_\nu A^\nu - \partial^\nu \partial_\nu A^\mu = -\partial^\nu \partial_\nu A^\mu, \tag{6.16}$$

which becomes, under our identification $A^0 = \frac{V}{c}$:

$$0 = -\frac{1}{c^2} \frac{\partial^2 V}{\partial t^2} + \nabla^2 V$$

$$0 = -\frac{1}{c^2} \frac{\partial^2 \mathbf{A}}{\partial t^2} + \nabla^2 \mathbf{A}, \tag{6.17}$$

appropriate to the source-free potential formulation of E&M in Lorentz gauge.

6.1.1 The field strength tensor

We can go on to define the components of the field strength tensor, but we are still lacking an interpretation in terms of **E** and **B**. That's fine, for now, we can use our previous experience as a crutch, but to reiterate – we cannot say exactly what the fields are (the independent elements of $F^{\mu\nu}$) until we have a notion of force.

From $F^{\mu\nu} = \partial^\nu A^\mu - \partial^\mu A^\nu$, we have:[2]

$$F^{\mu\nu} \doteq \begin{pmatrix} 0 & -\frac{1}{c}\left(\frac{\partial A^x}{\partial t} + \frac{\partial V}{\partial x}\right) & -\frac{1}{c}\left(\frac{\partial A^y}{\partial t} + \frac{\partial V}{\partial y}\right) & -\frac{1}{c}\left(\frac{\partial A^z}{\partial t} + \frac{\partial V}{\partial z}\right) \\ \circ & 0 & \frac{\partial A^y}{\partial x} - \frac{\partial A^x}{\partial y} & \frac{\partial A^z}{\partial x} - \frac{\partial A^x}{\partial z} \\ \circ & \circ & 0 & \frac{\partial A^z}{\partial y} - \frac{\partial A^y}{\partial z} \\ \circ & \circ & \circ & 0 \end{pmatrix}, \tag{6.18}$$

with $\circ$ representing the antisymmetric entry.

If we take the typical definition $\mathbf{E} = -\nabla V - \frac{\partial \mathbf{A}}{\partial t}$ and $\mathbf{B} = \nabla \times \mathbf{A}$, then the field strength tensor takes its usual form:

$$F^{\mu\nu} \doteq \begin{pmatrix} 0 & \frac{E^x}{c} & \frac{E^y}{c} & \frac{E^z}{c} \\ -\frac{E^x}{c} & 0 & B^z & -B^y \\ -\frac{E^y}{c} & -B^z & 0 & B^x \\ -\frac{E^z}{c} & B^y & -B^x & 0 \end{pmatrix}. \tag{6.19}$$

Problem 6.1

For the stress tensor of E&M:

$$T^{\mu\nu} = \alpha\left[F^{\mu\sigma} F^\nu{}_\sigma - \frac{1}{4} g^{\mu\nu} F^{\alpha\beta} F_{\alpha\beta}\right], \tag{6.20}$$

(a) Establish conservation using the source-free field equations (and any other identities you need).

(b) Show that the stress tensor is traceless in four dimensions (a hallmark of E&M).

(c) What are the restrictions on a plane-wave ansatz:

$$A_\mu = K_\mu \, e^{i p_\nu x^\nu} \tag{6.21}$$

if this is to be a solution to the vacuum field equations in Lorentz gauge?

(d) For the plane-wave solution, compute $T^{\mu\nu}$.

---

[2] Again using Minkowski, the "usual" gradient is a covariant tensor, while the four-potential A^μ is naturally contravariant, so it is A^μ that is represented by $(\frac{V}{c}, A^x, A^y, A^z)^{\mathrm{T}}$ for example. Raising and lowering changes the sign of the zero component only.

Problem 6.2

In (6.6), we varied quickly by using symmetries and some clever names for the dummy (summed) indices. Try varying:

$$S_V = \int d\tau \sqrt{-g} \left[F^{\mu\nu} \left(A_{\nu,\mu} - A_{\mu,\nu} \right) - \frac{1}{2} F^{\mu\nu} g_{\alpha\mu} g_{\beta\nu} F^{\alpha\beta} \right] \qquad (6.22)$$

with respect to $\delta A_{\alpha,\beta}$ without using any shortcuts (you may assume you already know, from $F^{\mu\nu}$ variation, that $F_{\mu\nu} = A_{\nu,\mu} - A_{\mu,\nu}$) – work in Cartesian spatial coordinates.

6.2 Energy–momentum tensor for E&M

We now understand the Lagrangian density (6.3) as the appropriate integrand of the action for the electromagnetic potential, or at least, the source-free E&M form. Our next immediate task is to identify the energy–momentum tensor for this field theory, and we will do this via the definition of the $T^{\mu\nu}$ tensor (from (5.111)):

$$T^{\mu\nu} = -\left(g^{\mu\nu} \bar{\mathcal{L}} + 2 \frac{\partial \bar{\mathcal{L}}}{\partial g_{\mu\nu}} \right), \qquad (6.23)$$

with

$$\bar{\mathcal{L}} = F^{\mu\nu} \left(A_{\nu,\mu} - A_{\mu,\nu} \right) - \frac{1}{2} F^{\sigma\rho} g_{\alpha\sigma} g_{\beta\rho} F^{\alpha\beta}. \qquad (6.24)$$

The most important term is the derivative w.r.t. $g_{\mu\nu}$, appearing in the second term above – we can simplify life by rewriting one set of dummy indices so as to get a $g_{\mu\nu}$ in between the two field strengths. Remember that we could do this on either of the two metrics, so we will pick up an overall factor of two:

$$\frac{\partial}{\partial g_{\mu\nu}} \left(\frac{1}{2} F^{\sigma\rho} g_{\alpha\sigma} g_{\beta\rho} F^{\alpha\beta} \right) = \frac{\partial}{\partial g_{\mu\nu}} \left(\frac{1}{2} F^{\nu\rho} g_{\mu\nu} g_{\beta\rho} F^{\mu\beta} \right)$$

$$= F^{\nu\rho} g_{\beta\rho} F^{\mu\beta}. \qquad (6.25)$$

Putting this together with the field relation between $F^{\mu\nu}$ and the derivatives of A_μ, we have:

$$T^{\mu\nu} = -\left(g^{\mu\nu} \left(\frac{1}{2} F^{\alpha\beta} F_{\alpha\beta} \right) - 2 F^{\nu\rho} g_{\beta\rho} F^{\mu\beta} \right)$$

$$= 2 \left(F^{\mu\beta} F^{\nu}_{\ \beta} - \frac{1}{4} g^{\mu\nu} F^{\alpha\beta} F_{\alpha\beta} \right). \qquad (6.26)$$

Overall factors in a free Lagrange density do not matter (field equations equal zero, so there is no need to worry about constants). For coupling to matter, however, we must put in the appropriate units. One way to do this is to connect the stress tensor to its known form. Suppose we started from the action $\tilde{S}_V \equiv \alpha \, S_V$, then the energy–momentum tensor is:

$$T^{\mu\nu} = 2\alpha \left(F^{\mu\beta} \, F^\nu{}_\beta - \frac{1}{4} g^{\mu\nu} \, F^{\alpha\beta} \, F_{\alpha\beta} \right). \qquad (6.27)$$

Take the zero component of this:

$$T^{00} = \alpha \left(\mathbf{B} \cdot \mathbf{B} + \frac{1}{c^2} \mathbf{E} \cdot \mathbf{E} \right). \qquad (6.28)$$

The usual expression for energy density is given by (see [17]):

$$u = \frac{1}{2} \left(\epsilon_0 \, E^2 + \frac{1}{\mu_0} B^2 \right) = \frac{\epsilon_0}{2} \left(E^2 + c^2 \, B^2 \right) \qquad (6.29)$$

so evidently, we can identify:

$$T^{00} = \frac{2\alpha}{\epsilon_0 \, c^2} u \longrightarrow T^{00} = u \quad \text{for } \alpha = \frac{\epsilon_0 \, c^2}{2}. \qquad (6.30)$$

With this factor in place, we have the final form:

$$T^{\mu\nu} \doteq \begin{pmatrix} \frac{1}{2} \epsilon_0 \left(E^2 + c^2 \, B^2 \right) & c \, \epsilon_0 \, \mathbf{E} \times \mathbf{B} \\ c \, \epsilon_0 \, \mathbf{E} \times \mathbf{B} & -T^{ij} \end{pmatrix}, \qquad (6.31)$$

with T^{ij} the usual Maxwell stress tensor:

$$T^{ij} = \epsilon_0 \left(E^i \, E^j - \frac{1}{2} \delta^{ij} \, E^2 \right) + \epsilon_0 \, c^2 \left(B^i \, B^j - \frac{1}{2} \delta^{ij} \, B^2 \right). \qquad (6.32)$$

We also have the association $\frac{1}{c} T^{0j} = \epsilon_0 \, \mathbf{E} \times \mathbf{B}$, the familiar momentum density from E&M. The energy density is indeed the T^{00} component, and the spatial portion is (the negative of) Maxwell's stress tensor.

6.2.1 Units

The above allowed us to set SI units for the action via the prefactor $\frac{\epsilon_0 c^2}{2} = \frac{1}{2\mu_0}$, the electromagnetic action reads:

$$S_V = \frac{1}{2\mu_0} \int d\tau \, \sqrt{-g} \left(F^{\mu\nu} \left(A_{\nu,\mu} - A_{\mu,\nu} \right) - \frac{1}{2} F^{\mu\nu} \, g_{\alpha\mu} \, g_{\beta\nu} \, F^{\alpha\beta} \right), \qquad (6.33)$$

but this is not the most obvious, certainly not the usual, form for the action formulation of E&M. Most common are Gaussian units, where E and B have the same units. The basic rule taking us from SI to Gaussian is: $\epsilon_0 \longrightarrow \frac{1}{4\pi}$, $\mu_0 \longrightarrow \frac{4\pi}{c^2}$ (ensuring that $\frac{1}{\epsilon_0 \mu_0} = c^2$), and $c \, \mathbf{B} \longrightarrow \mathbf{B}$. For example, the electromagnetic field energy in a volume in SI units is:

$$U = \frac{1}{2} \int \left(\epsilon_0 E_{SI}^2 + \frac{1}{\mu_0} B_{SI}^2 \right) d\tau. \tag{6.34}$$

Using our conversion, we have:

$$U = \frac{1}{2} \int \left(\frac{1}{4\pi} E^2 + \frac{c^2}{4\pi\mu_0} B_{SI}^2 \right) d\tau$$

$$= \frac{1}{8\pi} \int \left(E^2 + B^2 \right) d\tau. \tag{6.35}$$

We can convert the action itself – in its usual form (so-called "second-order form"), we write the action entirely in terms of the field strength tensor $F^{\mu\nu}$, so we lose the connection between the potential and the fields.

$$S_V = \frac{1}{16\pi} \int d\tau\, F^{\mu\nu} F_{\mu\nu}, \tag{6.36}$$

where the c^2 that was in the numerator from $\mu_0 \longrightarrow \frac{4\pi}{c^2}$ got soaked into the field strength tensor – because it is quadratic, the integrand depends on components like $\sim \frac{E^i E^j}{c^2}$ and the c^2 out front kills that, and in addition, $\sim B_{SI}^i B_{SI}^j$ which becomes, upon multiplication by c^2, just $\sim B^i B^j$. In other words, the field strength tensor in Gaussian units looks like (6.19) with $c \longrightarrow 1$. As a final note, you will generally see this action written with an overall minus sign (as in, for example, [27]) – that is convention, and leads to a compensating sign in the stress tensor definition. Regardless, keep the physics in mind, and you will be safe.

6.3 E&M and sources

E&M is a good place to begin talking about how sources should be introduced to free field theories, since we already know the answer from Maxwell's equations. We will look at how a source J^μ can be correctly coupled to a free vector field. By introducing the simplest possible term in the action, it will be clear that the coupling is correct, but will lead to inconsistencies since there is no field theory for J^μ – we imagine a set of charges and currents specified in the usual way, and the action will be incomplete until the terms associated with the free field theory for J^μ are introduced.

In addition to giving us a model for how to couple to a field theory, we expose a new symmetry of the action in the new source term. The gauge invariance of $F^{\mu\nu}$ (changing the four-potential A^μ by a total derivative) implies that J^μ, whatever it may be, is conserved.[3] This conservation leads to an interest in more general J^μ, not just those coming from macroscopic E&M. So we will look at the coupling of *fields* to E&M, and for this, we need some massive scalar field results.

6.3.1 Introducing sources

We are looking for a connection between source charges ρ and currents, and the electric and magnetic fields themselves. In this setting, the target is a relation between $F^{\mu\nu}$ (or A^μ) and $J^\mu \doteq (\rho\, c, \mathbf{J})^T$. We have not yet attempted to introduce sources, and indeed it is unclear in general how sources should relate to $F^{\mu\nu}$. Fortunately, we are familiar with the structure of electrodynamics, and can use this to guide our approach.

For the four-potential, in Lorentz gauge, we know that Maxwell's equations reduce to:

$$-\frac{\rho}{\epsilon_0} = \nabla^2 V - \frac{1}{c^2}\frac{\partial^2 V}{\partial t^2}$$
$$-\mu_0 \mathbf{J} = \nabla^2 \mathbf{A} - \frac{1}{c^2}\frac{\partial^2 \mathbf{A}}{\partial t^2},$$
(6.37)

or, in four-vector language (with $A^0 = \frac{V}{c}$):

$$\left(-\frac{1}{c^2}\frac{\partial^2}{\partial t^2} + \nabla^2\right)\begin{pmatrix}\frac{V}{c}\\ \mathbf{A}\end{pmatrix} = -\begin{pmatrix}\frac{\rho c}{\epsilon_0 c^2}\\ \frac{\mathbf{J}}{\epsilon_0 c^2}\end{pmatrix},$$
(6.38)

which implies the usual identification:

$$J^\mu \doteq \begin{pmatrix}\rho\, c\\ \mathbf{J}\end{pmatrix}$$
(6.39)

and the equations become:

$$\partial^\nu \partial_\nu A^\mu = -\frac{1}{\epsilon_0\, c^2} J^\mu.$$
(6.40)

These are to be introduced via the $F^{\mu\nu}$ equation – we have:

$$\partial_\nu F^{\mu\nu} = \partial_\nu (\partial^\mu A^\nu - \partial^\nu A^\mu) \underbrace{= -\partial_\nu \partial^\nu A^\mu}_{\text{in Lorentz gauge}} = \frac{1}{\epsilon_0\, c^2} J^\mu.$$
(6.41)

Remember that the equation $\partial_\nu F^{\mu\nu} = 0$, in the source-free case, came from variation of A^μ. So we want to introduce a term in the action that gives, upon

---

[3] The gauge choice in general relativity will correspond to coordinate freedom, and we already know that coordinate freedom implies stress–energy conservation: $T^{\mu\nu}{}_{;\nu} = 0$.

variation of A^μ, the source $\frac{1}{\epsilon_0 c^2} J^\mu$. The most obvious such term (that is still a scalar) is $A_\mu J^\mu$. Suppose we make the natural extension:

$$S = \frac{\epsilon_0 c^2}{2} \int d\tau \sqrt{-g} \left(F^{\mu\nu} (A_{\nu,\mu} - A_{\mu,\nu}) - \frac{1}{2} F^{\mu\nu} F_{\mu\nu} \right) + \beta \int d\tau \sqrt{-g} J^\mu A_\mu,$$

(6.42)

then variation of $F^{\mu\nu}$ gives, as before:

$$\frac{\epsilon_0 c^2}{2} \left(A_{\nu,\mu} - A_{\mu,\nu} - F_{\mu\nu} \right) = 0$$

(6.43)

and we recover the description of field strength in terms of the potential fields. The variation w.r.t. A_μ is different – we get:

$$\int d\tau \sqrt{-g} \left[\frac{\epsilon_0 c^2}{2} (F^{\nu\mu} - F^{\mu\nu})_{;\mu} + \beta J^\nu \right] \delta A_\mu = 0,$$

(6.44)

and using the antisymmetry of $F^{\mu\nu}$, this gives us the field equation:

$$\epsilon_0 c^2 F^{\nu\mu}{}_{;\mu} = -\beta J^\nu,$$

(6.45)

which is supposed to read $F^{\nu\mu}{}_{;\mu} = \mu_0 J^\nu \longrightarrow \beta = -1$.

So the coupling of E&M to sources gives an action of the form:

$$S = \frac{\epsilon_0 c^2}{2} \int d\tau \sqrt{-g} \left(F^{\mu\nu} (A_{\nu,\mu} - A_{\mu,\nu}) - \frac{1}{2} F^{\mu\nu} F_{\mu\nu} - \frac{2}{\epsilon_0 c^2} J^\mu A_\mu \right).$$

(6.46)

Notice that, from the field equation itself $F^{\mu\nu}{}_{;\nu} = \mu_0 J^\mu$, we have:

$$F^{\mu\nu}{}_{;\nu\mu} = \mu_0 J^\mu{}_{;\mu} = 0$$

(6.47)

because we are contracting an antisymmetric tensor $F^{\mu\nu}$ with the symmetric $\partial_\mu \partial_\nu$. The antisymmetry of $F^{\mu\nu}$ is a manifestation of the gauge freedom of the potentials. As we shall see later on, the gauge freedom is enough to enforce charge conservation (in Minkowski space-time):

$$J^\mu{}_{;\mu} = \frac{1}{c} \frac{\partial(c\,\rho)}{\partial t} + \nabla \cdot \mathbf{J} = 0.$$

(6.48)

We have defined J^μ in its usual E&M way, but that means that J^μ is itself a field. Where are the terms corresponding to the dynamical properties of J^μ? Presumably, this four-current is generated by some distribution of charge and current, specified by ρ and $\mathbf{v}$. It's all true, of course, and what we are apparently lacking is an additional term in the action corresponding to the free field action for J^μ. We will return to this issue after we see how gauge freedom implies source conservation at the level of the action, as opposed to the discussion above, which applies to the field equations.

6.3.2 Kernel of variation

Irrespective of the lack of a complete description of the system $\{J^\mu, F^{\mu\nu}\}$, we can nevertheless make some progress just by knowing the form of the source. That there should be *some* source and that it should be describable by J^μ is already a new situation.

This new $A_\mu J^\mu$ term in the action provides a new insight into symmetry. We know that arbitrary variation of $A_\mu \longrightarrow A_\mu + \delta A_\mu$ leads to the equations of motion. The free field action for E&M is actually less restrictive – consider the explicit variation:

$$
\begin{aligned}
S[A_\mu + \delta A_\mu] = \int d\tau \sqrt{-g} \left(F^{\mu\nu} (A_{\nu,\mu} - A_{\mu,\nu}) - \frac{1}{2} F^{\mu\nu} F_{\mu\nu} \right) \\
+ \int d\tau \sqrt{-g} \left(F^{\mu\nu} (\delta A_{\nu,\mu} - \delta A_{\mu,\nu}) \right),
\end{aligned}
\tag{6.49}
$$

so we have:

$$
\delta S = \int d\tau \sqrt{-g} (F^{\mu\nu} - F^{\nu\mu}) \, \delta A_{\nu,\mu}
\tag{6.50}
$$

and it was the integration by parts that gave us the statement $F^{\mu\nu}{}_{;\nu} = 0$. But we see here that there is an entire class of possibilities that are completely missed in the above – forgetting for the moment that $F^{\mu\nu}$ is antisymmetric, it certainly appears in antisymmetric form here, and so if it were the case that $\delta A_{\nu,\mu} = \delta A_{\mu,\nu}$, then $\delta S = 0$ automatically and we have not constrained the fields at all. So while it is true that the integration-by-parts argument will work for arbitrary variation of A_μ, we have apparently missed the class of variations $\delta A_{\nu,\mu} = \eta_{,\nu\mu}$ for arbitrary scalar functions η. Well, to say that they are "missing" is a little much, rather, they don't matter. The physically relevant field ($F^{\mu\nu}$, since it forces test particles) is left unchanged by a transformation of A_μ of the form:

$$
A_\mu \longrightarrow A'_\mu = A_\mu + \eta_{,\mu},
\tag{6.51}
$$

in which case $\delta S = 0$ automatically. This is precisely the gauge choice that underlies E&M.

What is interesting is that when we introduce a source of the form $A_\mu J^\mu$ in the action, there enters a naked δA_μ upon variation, and so it is no longer the case that we can free ourselves of a potential restriction. In other words, for all variations δA_μ *but* $\delta A_\mu = \eta_{,\mu}$, we know the field equations. In that special case, we have a contribution that comes from:

$$\delta S_J = \int d\tau \sqrt{-g} \, J^\mu \, \delta A_\mu = \int d\tau \sqrt{-g} \, J^\mu \, \eta_{,\mu} \qquad (6.52)$$

and we must have this term vanish separately – integrating by parts gives:

$$\delta S_J = \underbrace{\int d\tau \left(\sqrt{-g} \, J^\mu \eta \right)_{,\mu}}_{=0} - \int d\tau \sqrt{-g} \, J^\mu_{;\mu} \, \eta \qquad (6.53)$$

so that $J^\mu_{;\mu} = 0$ is a *requirement* if we are to obtain $\delta S_J = 0$. Then gauge freedom has actually imposed continuity on the sources, apart from our original E&M concerns. This tells us that if we were looking for more exotic sources, they would all need a notion of continuity, or would violate the most minimal coupling.

Evidently, this particular $\eta_{,\mu}$ variation is, in a sense, in the null space of the variation of the (uncoupled) action, once we have specified the antisymmetric term in (6.49), we automatically miss any variation of this form – with the additional structure of the current–vector coupling, we have revealed the deficiency.

Note the similarity with the other "invariance-leads-to-conservation" argument we have seen recently – that of coordinate invariance in an action and energy–momentum tensor conservation. It's a similar situation here with the gauge choice for potentials.

Scalar fields

This same issue, of "missing" variation, occurs in massless scalar fields, and there is a "gauge freedom" there, too. Recall the free field Lagrangian for a (massless) scalar field:

$$S = \int d\tau \sqrt{-g} \left(\phi_{,\mu} \pi^\mu - \frac{1}{2} \pi_\mu \pi^\mu \right). \qquad (6.54)$$

When we vary w.r.t. π^μ, we find $\phi_{,\mu} = \pi_\mu$, which is reasonable. But think about the $\phi_{,\mu}$ variation. As with E&M, only the derivatives of ϕ show up, so we expect trouble. Indeed, for generic $\delta\phi$:

$$\delta S = \int d\tau \sqrt{-g} \, \pi^\mu \, \delta\phi_{,\mu} \qquad (6.55)$$

and once again, we see that for variation of the form $\delta\phi = $const. the action automatically vanishes; for all the rest, we obtain $\pi^\mu_{;\mu} = 0$. This is the usual sort of idea for scalars (think of the electrostatic potential), that fundamentally, a constant can be added to the definition without changing the equations of motion.

What we need now is a valid source J^μ (we do not yet have one) with $J^\mu_{;\mu} = 0$, and a full Lagrangian contribution for the dynamics of the fields comprising J^μ (and ensuring its conservation). Actually generating the appropriate, consistent source is not easy, and brings up the same sort of self-consistency issues that arise in the development of general relativity. In the next section, we will construct a toy theory for electromagnetic sources out of uncoupled, massless scalar fields to show how the full action can be made self-consistent in the presence of sources *and* their governing field Lagrangians.

6.4 Scalar fields and gauge

We will discuss the use of multiple fields to expand our notion of symmetries and conservation. Using a natural "current" that comes from a complex scalar field theory, we have a candidate source for coupling to E&M. We go through the usual procedure of minimal coupling, followed by a process of consistency that we will use to introduce the idea of local gauge invariance. The approach can be thought of in terms of augmenting a Lagrangian until a consistent theory is obtained, and displays the pattern of guess–check–reguess–recheck that in this case ends after one iteration. The nice text [34] discusses gauge freedom and the use of local gauge invariance in this classical field setting.

6.4.1 Two scalar fields

Two non-interacting scalar fields can be developed from our current considerations – the simplest representation would be:

$$\bar{\mathcal{L}} = \left(\frac{1}{2}\psi_{,\mu}\,\psi_,{}^\mu\right) + \left(\frac{1}{2}\eta_{,\mu}\,\eta_,{}^\mu\right). \tag{6.56}$$

As a sum of two independent Lagrangians, it is easy to see that the variations of ψ and η do not talk to each other, so we will get a pair of independent massless Klein–Gordon fields.

Given that there are two fields, we "immediately" think of the real and imaginary parts of a complex number, and define the *independent* fields ϕ and ϕ^* by taking $\phi \equiv \psi + i\,\eta$, $\phi^* = \psi - i\,\eta$. Then in these variables:

$$\bar{\mathcal{L}} = \phi^*_{,\mu}\,g^{\mu\nu}\,\phi_{,\nu} \tag{6.57}$$

and to obtain the field equations, we need to vary w.r.t ϕ and ϕ^* independently (it's a two-field theory, after all). The two field equations are identical:

$$0 = \Box^2 \phi$$
$$0 = \Box^2 \phi^*. \qquad (6.58)$$

It is pretty clear from the action that our variation, w.r.t. $\delta\phi$ and $\delta\phi^*$, which constrains the fields themselves, says nothing at all about invariance of the Lagrangian under variations that change the phase of the fields. For example, if we take $\phi \longrightarrow e^{i\alpha}\phi$ for constant, real α, then $\phi^* \longrightarrow e^{-i\alpha}\phi^*$ and the Lagrangian has not changed.[4] Since it is a pure phase, and the complex Lagrangian is real, we know that any value for α leaves the action unchanged, and so it is reasonable to ask what this implies about conservation (a Noetherian question).[5]

Given that we cannot directly vary w.r.t. the parameter α, we need a new way to think about this type of symmetry. Suppose we take the infinitesimal form of $e^{i\alpha} \approx (1 + i\alpha)$ – then we are looking at the transformation:

$$\phi \longrightarrow \phi(1 + i\alpha)$$
$$\phi^* \longrightarrow \phi^*(1 - i\alpha), \qquad (6.59)$$

so we can write the perturbed Lagrangian:

$$\bar{\mathcal{L}}\left((\phi + i\alpha\phi)_{,\mu}, (\phi^* - i\alpha\phi^*)_{,\mu}\right) \qquad (6.60)$$

(noting that the Lagrangian depends only on the derivatives) and Taylor expansion gives the difference between the perturbed and unperturbed value of $\bar{\mathcal{L}}$:

$$\Delta\bar{\mathcal{L}} \approx \frac{\partial\bar{\mathcal{L}}}{\partial\phi_{,\mu}}(i\alpha)\phi_{,\mu} + \frac{\partial\bar{\mathcal{L}}}{\partial\phi^*_{,\mu}}(-i\alpha)\phi^*_{,\mu}. \qquad (6.61)$$

We know that the Lagrangian is invariant for any value of α, so $\Delta\bar{\mathcal{L}} = 0$, automatically. From the field equations for ϕ and ϕ^* (precursors to (6.58)):

$$0 = \partial_\mu\left(\frac{\partial\bar{\mathcal{L}}}{\partial\phi_{,\mu}}\right)$$
$$0 = \partial_\mu\left(\frac{\partial\bar{\mathcal{L}}}{\partial\phi^*_{,\mu}}\right) \qquad (6.62)$$

---

[4] In terms of the real fields ψ and η from (6.56), this phase corresponds to taking linear combinations of the two fields.

[5] Note that the introduction of a mass term does not spoil this phase invariance.

(with $\frac{\partial \bar{\mathcal{L}}}{\partial \phi} = \frac{\partial \bar{\mathcal{L}}}{\partial \phi^*} = 0$), we can augment the two terms comprising $\Delta \bar{\mathcal{L}}$ from (6.61) with zeros in order to write a total derivative:

$$
\begin{aligned}
0 &= \frac{\partial \bar{\mathcal{L}}}{\partial \phi_{,\mu}} (i\,\alpha)\, \phi_{,\mu} + \frac{\partial \bar{\mathcal{L}}}{\partial \phi^*_{,\mu}} (-i\,\alpha)\, \phi^*_{,\mu} \\[2mm]
&\quad + \underbrace{\partial_\mu \left(\frac{\partial \bar{\mathcal{L}}}{\partial \phi_{,\mu}} \right) (i\,\alpha)\, \phi}_{=0} + \underbrace{\partial_\mu \left(\frac{\partial \bar{\mathcal{L}}}{\partial \phi^*_{,\mu}} \right) (-i\,\alpha)\phi^*}_{=0} \\[2mm]
&= \alpha\, \partial_\mu \underbrace{i \left(\frac{\partial \bar{\mathcal{L}}}{\partial \phi_{,\mu}} \phi - \frac{\partial \bar{\mathcal{L}}}{\partial \phi^*_{,\mu}} \phi^* \right)}_{\equiv j^\mu}.
\end{aligned}
\tag{6.63}
$$

Explicitly, this global (constant α) phase transformation leads us to the conserved "current" (so named because $\partial_\mu j^\mu = 0$ just as with the physical current in E&M):

$$
j^\mu = i \left((\partial^\mu \phi^*)\, \phi - (\partial^\mu \phi)\, \phi^* \right).
\tag{6.64}
$$

The current j^μ has $\partial_\mu j^\mu = 0$ irrespective of α – phase-invariance was just a vehicle for uncovering the conserved quantity. Because of the association with E&M four-currents, the zero component is sometimes called the conserved "charge", and the spatial components the "current". Integrating the conservation statement $\partial_\mu j^\mu = 0$ gives the usual:

$$
\frac{1}{c} \frac{d}{dt} \int j^0\, d\tau = - \oint \mathbf{j} \cdot d\mathbf{a},
\tag{6.65}
$$

using the Minkowski metric (then $d\tau = dx\,dy\,dz$ is the spatial three-volume).

6.4.2 Current and coupling

That's all very nice, but what does it do for us? Thinking about E&M, we know that any four-vector J^μ that is supposed to provide a source for the electric and magnetic fields must be conserved, $\partial_\mu J^\mu = 0$. That's always true for the physical currents that we use in E&M, swarms of charges, for example. But it makes more sense, in a way, to try to couple the vector field theory for A_μ to *other field theories*. In other words, we take two free field theories, E&M and a complex scalar field, and combine them. In order to do this, we need to have in place a conserved four-current, and we now know that our complex scalar fields have one built-in.

Consider the full action (dispensing with the units for now):

$$S = \int d\tau \sqrt{-g} \left[\left(F^{\mu\nu} \left(A_{\nu,\mu} - A_{\mu,\nu} \right) - \frac{1}{2} F^2 \right) \right.$$

$$+ \phi^*_{,\mu} \, g^{\mu\nu} \, \phi_{,\nu} \qquad\qquad (6.66)$$

$$\left. - \gamma \, j^\mu \, A_\mu \right],$$

where γ is the coupling strength between the ϕ and ϕ^* fields and A_μ, and $j^\mu \equiv i((\partial^\mu \phi^*) \phi - (\partial^\mu \phi) \phi^*)$ from (6.64). This action is straightforward, but there is a potential problem – we know that $\partial_\mu j^\mu = 0$ for the free fields ϕ and ϕ^*, but by introducing the coupling, it is not clear that j^μ *remains* conserved. This is a persistent issue in classical field theory – we make the simplest possible theory, but we have not yet established that this simple theory is consistent with itself. We see that variation w.r.t. A_μ will couple the scalar fields to E&M, but now there will be A_μ terms in the ϕ and ϕ^* field equations because j^μ sits next to the four-potential in the coupling term $\gamma \, j^\mu \, A_\mu$. So we need to check that the whole theory is consistent. That involves finding the field equations and showing, explicitly in this expanded setting, that $\partial_\mu j^\mu = 0$ (it does not).

The field equations that come from varying A_μ, $F^{\mu\nu}$, ϕ, and ϕ^* are:

$$F_{\mu\nu} = A_{\nu,\mu} - A_{\mu,\nu}$$

$$2 F^{\mu\nu}{}_{,\nu} = \gamma \, j^\mu$$

$$(6.67)$$

$$-\partial^\mu \partial_\mu \phi + i \gamma \, \partial^\mu \left(\phi \, A_\mu \right) + i \gamma \, \partial^\mu \phi \, A_\mu = 0$$

$$-\partial^\mu \partial_\mu \phi^* - i \gamma \, \partial^\mu \left(\phi^* \, A_\mu \right) - i \gamma \, \partial^\mu \phi^* \, A_\mu = 0.$$

Again, by symmetry, we see the need for the conservation of j^μ from the second equation above – but:

$$\partial_\mu j^\mu = i \left(\phi \, \partial^\mu \partial_\mu \phi^* + \partial^\mu \phi^* \, \partial_\mu \phi - \phi^* \, \partial^\mu \partial_\mu \phi - \partial^\mu \phi \, \partial_\mu \phi^* \right)$$

$$= i \left(\phi \, \partial^\mu \partial_\mu \phi^* - \phi^* \, \partial^\mu \partial_\mu \phi \right)$$

$$(6.68)$$

which, using the field equations for ϕ and ϕ^*, becomes:

$$\partial_\mu j^\mu = 2 \gamma \, \partial_\mu \left(\phi \, \phi^* \, A^\mu \right) \neq 0. \qquad (6.69)$$

We started out by finding a conserved current for the free ϕ field(s), preparatory to coupling the ϕ fields to E&M. But, when we made the full, combined theory of the scalar fields and E&M, we found that the current was no longer conserved – now we have the task of fixing it. It is pretty clear how to do this – we basically want to introduce a term that will kill off the offending term in (6.69), a term in the Lagrangian whose variation looks like $-2\gamma\,\phi\,\phi^*\,A^\mu$. Instead, we will impose local gauge invariance to find the corrected Lagrangian.

6.4.3 Local gauge invariance

We know that the electromagnetic field (and its action) is unchanged under $A_\mu \longrightarrow A_\mu + \psi_{,\mu}$, and that there is an internal phase invariance for the free scalar fields $\phi \longrightarrow e^{i\,\alpha}\,\phi$. How can we combine these two ideas? It is not particularly clear that we *should* combine them. Why should the coupled system exhibit the same symmetries as the individual ones? As it turns out, this is a fundamental (and new, from our point of view) guiding principle for generating "good" field theories – that somehow, merging two field theories should have as much gauge structure as the free theories did. That in itself is not unreasonable – after all, if the coupling between the theories is weak enough, then experiments must return results from each separately as if the coupling was not there at all. So we expect a certain retention of gauge freedom. The really surprising element is that by introducing *more* symmetry than existed in either theory, we will recover precisely the correct terms in our action to counterbalance the loss of self-consistency that resulted from our naive mashing together of the two theories in (6.66).

We have basically one option available to us if we want to combine the gauge transformations for A_μ and ϕ – because ψ is an arbitrary function of x, we cannot obtain a relation between ψ and α (the phase constant) that holds everywhere. Solution: make α a function of position, and set it equal to ψ. Now the gauge function $\psi(x)$ is itself a field, and it is clear from the derivatives in the complex scalar Lagrangian that $\phi \longrightarrow \phi' = e^{i\,\psi(x)}\,\phi$ is not a symmetry of the free field theory[6] – in particular:

$$\partial_\mu\left(e^{i\,\psi}\,\phi\right) = e^{i\,\psi}\left(\phi_{,\mu} + i\,\psi_{,\mu}\,\phi\right) \tag{6.70}$$

shows us that the derivative term will not lose all ψ dependence. Here is the unusual question we now ask (and a similar issue comes up in GR, in a very similar setting) – is it possible to redefine the partial derivative, $\partial_\mu \longrightarrow D_\mu$, such that $D_\mu\left(e^{i\,\psi}\,\phi\right) = e^{i\,\psi}\,D_\mu\,\phi$? Then forming the action out of this modified differential operator will automatically enforce local gauge invariance.

---

[6] A mass term in the scalar portion of the Lagrangian would be fine, even for $\psi(x)$: $m^2\,\phi\,\phi^* \longrightarrow m^2\,\phi'\,\phi'^* = m^2\,\phi\,\phi^*$, indicating that this story holds more or less unchanged if our fields ϕ and ϕ^* were massive Klein–Gordon fields.

If we define the primed fields $\phi' \equiv e^{i\psi} \phi$ and $A'_\mu \equiv A_\mu + \gamma^{-1} \psi_{,\mu}$, then solving for $\psi_{,\mu}$ in terms of A_μ and A'_μ, we can take the derivative of ϕ':

$$\partial_\mu \phi' = e^{i\psi} \left(\partial_\mu \phi + i\gamma \left(A'_\mu - A_\mu \right) \phi \right)$$

$$\downarrow$$

$$\partial_\mu \phi' - i\gamma \phi' A'_\mu = e^{i\psi} \left(\partial_\mu \phi - i\gamma A_\mu \phi \right),$$

(6.71)

which immediately suggests that $D_\mu \equiv \partial_\mu - i\gamma A_\mu$ is the most likely candidate, so that the above reads $D_\mu \phi' = e^{i\psi} D_\mu \phi$ – this operator is called the "covariant derivative".[7] If we write the new Lagrangian:

$$\bar{\mathcal{L}} = F^{\mu\nu} \left(A_{\nu,\mu} - A_{\mu,\nu} \right) - \frac{1}{2} F^2 + D_\mu \phi \, g^{\mu\nu} \left(D_\nu \phi \right)^*,$$

(6.72)

then we will get a consistent field theory for E&M, our scalar field, and their coupling. The coupling itself is now hidden inside the redefined derivative operator, D_μ, so we have what looks, upon a quick scan of the Lagrangian, like two free field theories. The electromagnetic portion of the Lagrangian is being treated somewhat preferentially here – we don't introduce our new derivative in the first term of $\bar{\mathcal{L}}$ – that's because it was really the scalar part that was causing problems, a lack of conservation or, from our current point of view, incomplete gauge freedom.

Our new derivative is more notation than anything else, and provides a compact form for the Lagrangian. But let's be clear, when you get down to the physics of the theory, like the continuity equation, it is our familiar ∂_μ that is useful – that's the one which allows us to integrate, so conservation must still be expressed as $\partial_\mu J^\mu = 0$. The current J^μ will not be the j^μ we started with, of course – the final form of the current will be dictated by the field equations, although you can probably guess what replacement will take us from $j^\mu = i \left((\partial^\mu \phi^*) \phi - (\partial^\mu \phi) \phi^* \right)$ to the final form.

Field equations of the sourced system

If we write out all the terms in (6.72) in preparation for variation, we have:

$$\bar{\mathcal{L}} = F^{\mu\nu} \left(A_{\nu,\mu} - A_{\mu,\nu} \right) - \frac{1}{2} F^2 + \phi_{,\mu} \, g^{\mu\nu} \, \phi^*_{,\nu}$$
$$+ \underbrace{i\gamma \, \phi_{,\mu} \, g^{\mu\nu} A_\nu \phi^* - i\gamma A_\mu \phi \, g^{\mu\nu} \, \phi^*_{,\nu}}_{=-\gamma j^\mu A_\mu} + \gamma^2 A_\mu \phi \, g^{\mu\nu} A_\nu \phi^*.$$

(6.73)

---

[7] This is apparently a new use of the term "covariant derivative", although as we shall see, there are connections to the program of derivative modification found in general relativity.

Our enforcement of local gauge invariance has yielded the predictable corrective term. From the point of view of conservation, what we lacked was a term that looked like $2\gamma^2 \phi\phi^* A^\mu$ under A^μ variation – the final term in our expression for $\bar{\mathcal{L}}$ above is the new piece, and it is quadratic in both the vector potential, and the fields, as it had to be.

Nothing has changed for the $F^{\mu\nu}$ variation, and we recover the usual $F_{\mu\nu} = A_{\nu,\mu} - A_{\mu,\nu}$. The variation with respect to A_μ is more complicated:

$$\delta S = \int d\tau \sqrt{-g} \left[\delta A_{\mu,\nu} \left(F^{\nu\mu} - F^{\mu\nu} \right) \right.$$
$$\left. - i\gamma \left(\phi\, \partial^\mu \phi^* - \phi^* \, \partial^\mu \phi \right) \delta A_\mu + 2\gamma^2 A^\mu \phi\phi^* \delta A_\mu \right]. \tag{6.74}$$

Flipping the derivative on the first term, to turn $\delta A_{\mu,\nu} \longrightarrow \delta A_\mu$, we get the field equation:

$$2\, F^{\mu\nu}{}_{,\nu} = i\gamma \left(\phi\, \partial^\mu \phi^* - \phi^* \, \partial^\mu \phi \right) - 2\gamma^2 \phi\phi^* A^\mu, \tag{6.75}$$

and clearly, we now have the natural current definition:

$$J^\mu \equiv i \left(\phi\, \partial^\mu \phi^* - \phi^* \, \partial^\mu \phi \right) - 2\gamma^2 \phi\phi^* A^\mu$$
$$= i \left(\phi\, (D^\mu \phi)^* - \phi^* \, D^\mu (\phi) \right). \tag{6.76}$$

This conserved current can be "obtained" from the original j^μ by replacing ∂^μ with D^μ.[8]

For the ϕ^* variation, we have (referring to (6.72) for simplicity):

$$- D^\mu D_\mu \phi = 0, \tag{6.77}$$

and we get the same thing for ϕ^* via ϕ variation (namely, $(D_\mu D^\mu \phi)^* = 0$). It is interesting that the effect of the move $\partial \longrightarrow D$ is just replacement, we can treat it just like ∂_μ in the variation of ϕ. This is no accident, it is the source that gets modified, and those modifications do not depend on the source field ϕ – instead, D_μ is dependent on the electromagnetic potential.[9]

---

[8] Again, to highlight the similarities with GR to come, we might call this replacement a "minimal substitution".

[9] The same sort of argument will be made when we get to general relativity. There, there is a similar consistency problem, and the solution is to redefine the derivative for all other theories to depend on the natural second-rank field (which turns out to be the metric). That process ends in the assignment of covariant (in the metric-connection sense) derivatives for partial derivatives.

Finally, we want to check that conservation holds for this new J^μ, so let's verify that $\partial_\mu J^\mu = 0$:

$$
\begin{aligned}
\partial_\mu J^\mu &= i \left(\partial_\mu \phi \, \partial^\mu \phi^* + \phi \, \partial_\mu \partial^\mu \phi^* - \partial_\mu \phi^* \, \partial^\mu \phi - \phi^* \, \partial_\mu \partial^\mu \phi \right) \\
&\quad - 2\alpha \, \phi_{,\mu} \, \phi^* A^\mu - 2\alpha \, \phi \, \phi^*_{,\mu} A^\mu - 2\alpha \, \phi \, \phi^* A^\mu_{,\mu} \\
&= i \left(\phi \, (D^\mu D_\mu \phi)^* - \phi^* \, (D^\mu D_\mu \phi) \right) \\
&= 0,
\end{aligned}
\tag{6.78}
$$

where the final equality holds by virtue of the field equations.

To sum up – we started with an action that had E&M, and two scalar fields that did not interact. We included an interaction term that correctly coupled a natural (conserved, in the free scalar field case) current to E&M. This led to the usual Maxwell equations with source provided from ϕ and ϕ^*, but the action coupled the other direction as well, so ϕ and ϕ^* were themselves sourced by elements of the electromagnetic field A^μ. In order to restore conservation of the source, we imposed local gauge invariance, and this made our entire, coupled system, self-consistent.[10] We could have started with the uncoupled action, coupled the two theories together, and then, with no guiding principle whatsoever except our wits, generated the correct form. The point of local gauge invariance is to provide a recipe that mechanizes this process.

While we won't have much explicit use for this technique in what follows, our discussion highlights a few elements that will be familiar as we move on to general field theories. First, that considering a field theory in isolation does not always directly inform the content of an action in the presence of other fields. Second, that there exist very specific terms in an action that can be used to produce field equations with desired properties. If we look back at (6.73), we notice that all terms are scalar, and there are no terms higher than quadratic in any field. Each component of (6.73) is required to produce the overall, self-consistent, coupled theory. However strange those terms may look at first glance, I hope it is clear, upon reflection, that what we have are *all* possible scalar combinations with at most quadratic field dependence (for ϕ and A_μ separately), excluding the mass term, and this observation provides some indication of how we might have guessed at the structure of that final form from the start.

6.5 Construction of field theories

We are beginning our final descent, and I'll take the opportunity to look at the freedom we have in designing rational (by which I mean relatively sane) field

[10] It also made the field equations more difficult to solve, but that's a different problem.

theories out of Lagrangians.[11] Starting with the scalar fields, there is not much to say: at quadratic order we have a mass term and quadratic derivative term that recovers massive Klein–Gordon scalars. For vector theories like E&M, there are two distinct options – massive (with Lorentz gauge fixed) and massless (with gauge freedom). We will count the number of valid (slash useful) quadratic terms for vector fields, leading to a tabulation of "all" free vector fields that support superposition (and have at most second derivatives in the field equations).

By "free", here, we mean that the theory is unsourced – that is a natural starting point, and basically means that we will not consider terms linear in the field (remember that, for example, $A^\mu J_\mu$ is the source for E&M, variation w.r.t. A^μ is what gives the naked J_μ source term in Maxwell's equations). We will work with the Minkowski metric in Cartesian coordinates for simplicity in what follows (you can generalize as you go).

6.5.1 Available terms for scalar fields

We want to create a free field, what constructs are available for an action? If we demand the field equations support superposition (linearity) so that the sum of two free fields is also a field, then we must have no more than quadratic terms in the Lagrangian. If we want wave-like solutions, then there should be second derivatives in the field equations and hence at most quadratic first derivatives in the Lagrangian. Wave solutions are not required, but in terms of the structure of provide-able physical data, they are natural – two spatial boundaries, and two temporal initial conditions are a typical set. In addition, wave-like equations have local propagation speeds that can be set to c.

Finally, to support general covariance, our $\bar{\mathcal{L}}$ must be a scalar, then $\mathcal{L} \equiv \sqrt{-g}\,\bar{\mathcal{L}}$ is, appropriately, a density. For a scalar $\phi(x, y, z, t)$, the *only* option is:

$$\bar{\mathcal{L}}_s = \alpha\,\phi_{,\mu}\,g^{\mu\nu}\,\phi_{,\nu} + \beta\,\phi^2, \tag{6.79}$$

for arbitrary α and β. The creation of scalar $\bar{\mathcal{L}}$ implies the metric as a fundamental building block. From here, first-order form follows naturally, and the above is clearly just the Klein–Gordon equation for massive scalar fields.

6.5.2 Available terms for vector fields

Vector fields allow more options. Our two available terms from the scalar case become, for vector A_μ: $A_\mu A^\mu$ (mass term), $A_{\mu,\nu} A^{\mu\,\nu}$, $A_{\mu,\nu} A^{\nu\,\mu}$ and $(A^\alpha_{\,,\alpha})^2$. So

---

[11] For a comprehensive discussion of terms and model-building for relativistic field theory, see [41]. The particular massless E&M setting is described in [34].

we can imagine a general Lagrangian of the form:

$$\bar{\mathcal{L}}_v = \alpha \, A_\mu \, A^\mu + \beta \, A_{\mu,\nu} \, A^{\mu\,,\nu} + \gamma \, A_{\mu,\nu} \, A^{\nu\,,\mu} + \delta \, (A^\mu_{\,,\mu})^2, \qquad (6.80)$$

with arbitrary constants α, β, γ, and δ. The non-derivative term is pretty clearly a "mass" term, of the usual $m^2 \, \phi^2$ variety (for scalars), and we know how this works in the field equations (from Section 5.9) – so set $\alpha = 0$ for now, and take the rest. The field equations, after variation, are:

$$2 \, (\gamma + \delta) \, \partial^\nu \partial^\mu \, A_\mu + 2 \, \beta \, \partial^\mu \partial_\mu \, A^\nu = 0, \qquad (6.81)$$

which seems reasonable.

There is a slight problem, or rather, we will *create* a slight problem for ourselves. At issue is the nature of objects of the form A_μ – we want to talk about "real" vectors, here, and not derivatives of scalars. It is clear that there is a "derivative" (in both senses of the word) vector we can make from a scalar ψ, namely $\psi_{,\mu}$. This is really just a scalar, providing no new information, but our goal is a true vector field theory. So what we'd like to do is get rid of any potential (no deep significance intended) scalar derivative polluting our new theory – we already dealt with scalar fields above. Our target is a set of field equations that make no reference to the scalar portion of a vector at all.

Recall from three-dimensional vector calculus that a vector, like $\mathbf{E}$, is determined from its divergence and curl – generically $\mathbf{E} = \nabla \times \mathbf{F} + \nabla G$, and then $\nabla \cdot \mathbf{E} = \nabla^2 G$ so that $\nabla^2 G$ is the divergence of $\mathbf{E}$ and $\nabla \times \mathbf{E} = \nabla (\nabla \cdot \mathbf{F}) - \nabla^2 \mathbf{F}$ is the curl of $\mathbf{E}$. This breakup persists in four-dimensional vector fields, and in particular, we are currently interested in the $\mathbf{F}$ part rather than the G part. Just as $\nabla \cdot \mathbf{E} = 0$ implies a purely solenoidal potential in three dimensions, the requirement $\partial_\mu A^\mu = 0$ (Lorentz gauge) indicates that there is no generic scalar gradient lurking in A^μ.

We will take a projective approach – breaking the full A^μ into a portion that has no $\psi_{,\mu}$ and a pure $\psi_{,\mu}$ part, then require that the field equation be insensitive to the scalar, and work backwards to the form of the action that generates this field equation. Start by assuming we can write $A_\mu = \bar{A}_\mu + \psi_{,\mu}$ (in general, we can) – then the field equation (6.81) becomes:

$$2 \, (\gamma + \delta) \left[\partial^\nu \partial^\mu \, \bar{A}_\mu + \partial^\nu \partial^\mu \partial_\mu \, \psi \right] + 2 \, \beta \, (\partial_\mu \partial^\mu \, \bar{A}^\nu + \partial^\mu \partial_\mu \partial^\nu \, \psi) = 0, \qquad (6.82)$$

and if we want no reference to ψ, then the requirement is $\beta + \gamma = -\delta$. Again, the logic here is that we have no scalar field information in our initial Lagrangian, so we have not coupled (accurately or not) the scalar ψ to the vector $\bar{A}_\mu$, and we want to make this lack of information explicit in the field equations.

Then the original Lagrangian reads (again, ignoring the mass term):

$$\bar{\mathcal{L}}_v = \beta \, A_{\mu,\nu} \, A^{\mu\,,\nu} + \gamma \, A_{\mu,\nu} \, A^{\nu\,,\mu} - (\beta + \gamma) \, (A^\mu_{\,,\mu})^2. \qquad (6.83)$$

Remember that $\bar{\mathcal{L}}_v$ is part of the larger action structure:

$$S = \int d\tau \sqrt{-g}\,\bar{\mathcal{L}}_v. \tag{6.84}$$

For any F^μ depending on A_μ, we can add a term $\partial_\mu F^\mu$ with $F^\mu \longrightarrow 0$ at spatial infinity (so that the surface term vanishes), which will not contribute to the field equations – i.e. the following gives the same field equations as $\mathcal{L}_v$ by itself:

$$S = \int d\tau \sqrt{-g}\,(\bar{\mathcal{L}}_v + \partial_\mu F^\mu). \tag{6.85}$$

As a particular function, take $F^\mu = (A^\mu{}_{,v} A^v - A^\mu A^v{}_{,v})$. Then from the point of view of $\bar{\mathcal{L}}_v$ variation, $\partial_\mu F^\mu\text{"="}0$. Formally, this divergence is:

$$\begin{aligned}
\partial_\mu F^\mu &= (\partial_\mu \partial_v A^\mu)\,A^v + \partial_v A^\mu\,\partial_\mu A^v - \partial_\mu A^\mu\,\partial_v A^v - A^\mu\,\partial_\mu \partial_v A^v \\
&= \partial_v A^\mu\,\partial_\mu A^v - \partial_\mu A^\mu\,\partial_v A^v,
\end{aligned} \tag{6.86}$$

and we can add this with impunity to the Lagrangian – suppose we add $-(\beta + \gamma)(A_{\mu,v} A^{v,\mu} - A^\mu{}_{,\mu} A^v{}_{,v})$ which is of the above form, to the revised $\bar{\mathcal{L}}$ in (6.83), the clear choice for killing off the $(A^\alpha{}_{,\alpha})^2$ term:

$$\bar{\mathcal{L}}_v = \beta \left(A_{\mu,v} A^{\mu,v} - A_{\mu,v} A^{v,\mu} \right) \tag{6.87}$$

and we have finally arrived, uniquely, at the antisymmetric derivative Lagrangian we began with in our study of E&M (think of the first-order form of (6.87)).

Evidently, the antisymmetry of the field-strength tensor is somehow related to its "pure vector" character. This procedure also introduces, explicitly, the gauge invariance of A_μ – its field equations and Lagrangian are totally independent of scalar gradients, and only the vector part contributes to the physics of electricity and magnetism.

Once we have (6.87), the conjugate momenta are clear – take $F^{\mu v} \equiv \frac{\partial \bar{\mathcal{L}}}{\partial A_{\mu,v}} = \beta (A^{\mu,v} - A^{v,\mu})$, as usual. First-order form follows from the Legendre transform, as always. The introduction of a mass term is separate, and provides a theory distinguishable from E&M, but it is interesting that aside from this, the physically motivated requirements of superposition of solutions, second-order field equations, and Lorentz covariance have provided the full structure of Maxwell's equations as a matter of course.

Mass term

One can, with the above Lagrangian in place, introduce a mass term in the usual way – take:

$$\bar{\mathcal{L}}_{vm} = \frac{1}{2}\alpha\, F^{\mu v}\,(A_{v,\mu} - A_{\mu,v}) - \beta\, F^2 - \frac{1}{2}m^2\, A^\mu A_\mu. \tag{6.88}$$

Then variation gives:

$$\beta F_{\mu\nu} = \alpha (A_{\nu,\mu} - A_{\mu,\nu})$$
$$-\alpha F^{\mu\nu}{}_{,\nu} - m^2 A^\mu = 0. \tag{6.89}$$

The first of these field equations tells us that the field strength tensor can be developed from a combination of the derivatives of the vector potential (as in E&M), while the second equation, known as Proca's equation, differs from the electromagnetic case. It is interesting that, by taking the divergence again, we get:

$$\partial_\mu \partial_\nu F^{\mu\nu} + m^2 \partial_\mu A^\mu = m^2 \partial_\mu A^\mu = 0, \tag{6.90}$$

which is precisely the Lorentz gauge condition. For massive A^μ, we are required, for internal consistency, to have a divergence-less field. This is equivalent to the condition that there can be *no* scalar divergence associated with A_μ – just as $\nabla \cdot \mathbf{B} = 0$ implies that $\mathbf{B}$ is a pure curl with no $\nabla\psi$ admixture. So, far from releasing us from scalar dependence (by making no reference to it at all), the massive vector theory requires that the gradient be omitted.

It was precisely this gauge freedom for A_μ, we remember, that led to charge conservation when we coupled to a source J^μ – we have now lost that freedom, and the relevant field equations in the sourced setting will be:

$$F^{\mu\nu}{}_{,\nu} - (m^2 A^\mu + \gamma J^\mu) = 0, \tag{6.91}$$

from which we learn only that the combination $m^2 A^\mu + \gamma J^\mu$ must be divergence-less.

All of this is to say that massive vector fields are quite different from massless vector fields. In the language of field theory, we call massless A^μ a "gauge field", and it is responsible for light, for example. The massive vector A^μ is *not* a gauge field, as it lacks the gauge-independence, and does not lead to the theory of E&M.

6.5.3 One last ingredient

There is a final building block, that I hesitate to mention. We know that for vector fields, for example, A_μ, $A_{\mu,\nu}$ and $g_{\mu\nu}$ are the major players, but there is a further element. The Levi–Civita tensor density can also be used to develop Lagrangians. We define, *in any coordinate system*, the numerical symbol:

$$\epsilon^{\alpha\beta\gamma\delta} \equiv \begin{cases} 1 & \text{for indices in "even" order} \\ -1 & \text{for indices in "odd" order} \\ 0 & \text{for any repeated index,} \end{cases} \tag{6.92}$$

where we take some canonical ordering (if $\{\alpha, \beta, \gamma, \delta\}$ are in $0 \longrightarrow 3$, for example, we take $\epsilon^{0123} = 1$) as the base ordering and count even and odd permutations (number of flips) to get any given ordering. This is the antisymmetric "symbol", it doesn't transform as a fourth-rank contravariant tensor because we require it to be numerically identical in all coordinate systems. What, then, is it?

Following [40], suppose we treat $\epsilon^{\alpha\beta\gamma\delta}$ as a fourth-rank contravariant tensor density with weight p, then:

$$\bar{\epsilon}^{\alpha\beta\gamma\delta} = \det\left(\frac{\partial\bar{x}}{\partial x}\right)^p \frac{\partial\bar{x}^\alpha}{\partial x^\rho}\frac{\partial\bar{x}^\beta}{\partial x^\sigma}\frac{\partial\bar{x}^\gamma}{\partial x^\mu}\frac{\partial\bar{x}^\delta}{\partial x^\nu}\epsilon^{\rho\sigma\mu\nu}, \tag{6.93}$$

and consider the object:

$$\frac{\partial\bar{x}^\alpha}{\partial x^\rho}\frac{\partial\bar{x}^\beta}{\partial x^\sigma}\frac{\partial\bar{x}^\gamma}{\partial x^\mu}\frac{\partial\bar{x}^\delta}{\partial x^\nu}\epsilon^{\rho\sigma\mu\nu}. \tag{6.94}$$

This, as an expression, is totally antisymmetric in $\{\alpha, \beta, \gamma, \delta\}$, and so must be proportional to $\epsilon^{\alpha\beta\gamma\delta}$. Let the proportionality constant be A, then to find the value for A, note that the determinant of a matrix M with elements M_{ij} can be written as:

$$\det M = \epsilon^{i_1 i_2 i_3 i_4} M_{1i_1} M_{2i_2} M_{3i_3} M_{4i_4} \tag{6.95}$$

in four dimensions. We only need one value for the object in (6.94) to set the constant A, so take:

$$\frac{\partial\bar{x}^0}{\partial x^\rho}\frac{\partial\bar{x}^1}{\partial x^\sigma}\frac{\partial\bar{x}^2}{\partial x^\mu}\frac{\partial\bar{x}^3}{\partial x^\nu}\epsilon^{\rho\sigma\mu\nu} = \det\left(\frac{\partial\bar{x}}{\partial x}\right)\epsilon^{0123} = \det\left(\frac{\partial\bar{x}}{\partial x}\right), \tag{6.96}$$

and the transformation for ϵ reads:

$$\bar{\epsilon}^{\alpha\beta\gamma\delta} = \det\left(\frac{\partial\bar{x}}{\partial x}\right)^p \left(\det\left(\frac{\partial\bar{x}}{\partial x}\right)\epsilon^{\alpha\beta\gamma\delta}\right). \tag{6.97}$$

If we insist on $\bar{\epsilon}^{\alpha\beta\gamma\delta} = \epsilon^{\alpha\beta\gamma\delta}$ (from its definition), then the Levi–Civita symbol must be a tensor density of weight $p = -1$.

I mention all of this because it is possible, on the vector side, to introduce a fourth invariant – the scalar density $\epsilon^{\alpha\beta\gamma\delta} A_{\beta,\alpha} A_{\delta,\gamma}$ combined with the density $d\tau$ – and we are honor bound to find its role in the Lagrangian and field equations. It suffices to look again to the massless theory.

$$\mathcal{L}_v = \alpha\sqrt{-g}\left(A_{\mu,\nu}(A^{\mu\,\nu} - A^{\nu\,\mu})\right) + \beta\,\epsilon^{\alpha\beta\gamma\delta} A_{\beta,\alpha} A_{\delta,\gamma}, \tag{6.98}$$

but the Levi–Civita term in this setting can be written as a total divergence – namely:

$$\partial_\mu\left(\epsilon^{\mu\nu\alpha\beta}(A_\nu A_{\beta,\alpha})\right) = \epsilon^{\mu\nu\alpha\beta}(A_{\nu,\mu} A_{\beta,\alpha} + A_\nu A_{\beta,\alpha\nu})$$
$$= \epsilon^{\mu\nu\alpha\beta}(A_{\nu,\mu} A_{\beta,\alpha}), \tag{6.99}$$

where the second term in the first line dies under $\alpha \leftrightarrow \nu$ symmetry of partials hitting the antisymmetry of $\epsilon^{\mu\nu\alpha\beta}$.

The Levi–Civita contribution, then, is not a contribution at all. We have gained nothing by its introduction in $D = 4$ – this is not always the case (see Problem 6.7).

6.6 Second-rank tensor field theory

We have done the warm-up: E&M. The first-rank tensor field theory we developed in Section 6.5.2 gives us the guiding principles for generating higher-rank theories. Second-rank tensors present a proliferation of possibilities, and for that reason, we will take a slightly different approach to the theory. Instead of tabulating the (as it turns out 13) available combinations of quadratics, we will begin at the field equation itself. Our guiding principles are: superposition, second order, and no vector pollution. That is, we want the most general linear combination of second derivatives of the field $h_{\mu\nu}$, where vector contributions (like $h_{\mu\nu} = A_{\mu,\nu} + A_{\nu,\mu}$) are ignored. This is akin to the removal of scalars from E&M, leading to its gauge structure.

Once we have the free field equations, we will work backwards toward an action. Our target will be first-order form from the start, so we expect the introduction of an auxiliary field $\pi^{\mu\nu\alpha}$, playing the role of canonical momentum. Once we have the appropriate action, we will, in the next section, see that it requires nonlinear modification in order to couple to its own stress tensor, and that will finish the argument, giving us the first-order action form for general relativity.

We begin in flat, Minkowski space-time, and we'll take our metric to be in Cartesian coordinates, generalization will follow.

6.6.1 General, symmetric free field equations

The most general linear combination of second derivatives of a symmetric field $h_{\mu\nu} = h_{\nu\mu}$ that can appear in a field equation governing its vacuum dynamics is:

$$\alpha_1 h_{\mu\nu,}{}^{\rho}{}_{\rho} + \alpha_2 h_{\mu\rho,}{}^{\rho}{}_{\nu} + \alpha_3 h_{\nu\rho,}{}^{\rho}{}_{\mu} + g_{\mu\nu}\left(\alpha_4 h^{\rho}{}_{\rho,}{}^{\gamma}{}_{\gamma} + \alpha_5 h^{\rho\gamma}{}_{,\rho\gamma}\right) + \alpha_6 h^{\rho}{}_{\rho,\mu\nu} = 0. \tag{6.100}$$

If we introduce the ansatz $h_{\mu\nu} = A_{\mu,\nu} + A_{\nu,\mu}$, a pure vector contribution, and require that the left-hand side of (6.100) be identically zero, effectively projecting out vector contributions, then we have the following four linear equations:

$$\alpha_1 + \alpha_2 = \alpha_1 + \alpha_3 = 0 \quad \alpha_2 + \alpha_3 + 2\alpha_6 = 0 \quad \alpha_4 + \alpha_5 = 0, \tag{6.101}$$

and the solution to these is $\alpha_2 = \alpha_3 = -\alpha_1$, $\alpha_6 = \alpha_1$ and $\alpha_5 = -\alpha_4$. Defining the lone relevant constant $\alpha \equiv \frac{\alpha_4}{\alpha_1}$, our free field equation reads:

$$h_{\mu\nu},^{\rho}{}_{\rho} - h_{\mu\rho},^{\rho}{}_{\nu} - h_{\nu\rho},^{\rho}{}_{\mu} + h^{\rho}{}_{\rho,\mu\nu} + \alpha \, g_{\mu\nu} \left(h^{\rho}{}_{\rho,}{}^{\gamma}{}_{\gamma} - h^{\rho\gamma}{}_{,\rho\gamma} \right) = 0. \quad (6.102)$$

Note that in addition to vector contributions, the above is also free of scalars – setting $h_{\mu\nu} = \phi_{,\mu\nu}$ gives zero for the left-hand side as well. So our field equations refer only to the pure second-rank (symmetric) portion of the tensor $h_{\mu\nu}$.

We can obtain a simple relation between $h^{\rho}{}_{\rho,}{}^{\gamma}{}_{\gamma}$ and $h^{\rho\gamma}{}_{,\rho\gamma}$ by taking the $\mu - \nu$ trace of (6.102):

$$2 \, h^{\gamma}{}_{\gamma,}{}^{\rho}{}_{\rho} - 2 h^{\gamma}{}_{\rho,}{}^{\rho}{}_{\gamma} + \alpha \, D \left(h^{\rho}{}_{\rho,}{}^{\gamma}{}_{\gamma} - h^{\rho\gamma}{}_{,\rho\gamma} \right) = 0 \quad (6.103)$$

(D is the dimension of the space-time and comes from $g^{\mu}{}_{\mu} = D$) from which we learn that:

$$h^{\gamma}{}_{\gamma,}{}^{\rho}{}_{\rho} = h^{\rho\gamma}{}_{,\rho\gamma}. \quad (6.104)$$

The final form for the vacuum field equation is:

$$\boxed{h_{\mu\nu},^{\rho}{}_{\rho} - h_{\mu\rho},^{\rho}{}_{\nu} - h_{\nu\rho},^{\rho}{}_{\mu} + h^{\rho}{}_{\rho,\mu\nu} = 0.} \quad (6.105)$$

Now we just need an action that returns this. Notice the interesting feature of the above – it can be written as a total divergence:

$$\boxed{\partial^{\rho} \left(h_{\mu\nu,\rho} - h_{\mu\rho,\nu} - h_{\nu\rho,\mu} + \frac{1}{2} g_{\nu\rho} \, h^{\gamma}{}_{\gamma,\mu} + \frac{1}{2} g_{\mu\rho} \, h^{\gamma}{}_{\gamma,\nu} \right) = 0,} \quad (6.106)$$

and this suggests that we define:

$$\pi_{\rho\mu\nu} = \frac{1}{2} \left(h_{\rho\mu,\nu} + h_{\rho\nu,\mu} - h_{\mu\nu,\rho} \right), \quad (6.107)$$

then we could write the field equation for $h_{\mu\nu}$ in terms of a divergence of $\pi_{\rho\mu\nu}$ and its traces. As it turns out, the development of the action benefits from a slightly modified form of the field $h_{\mu\nu}$ – this doesn't change the content of the field equations, but makes keeping track of the relevant terms a little easier.

Trace-reversed field equations

We will work with the "trace-reversed" form of the field $h_{\mu\nu}$ – let $H_{\mu\nu} \equiv h_{\mu\nu} - \frac{1}{2} g_{\mu\nu} h$,[12] where $h \equiv h^{\alpha}{}_{\alpha}$, the trace of $h_{\mu\nu}$. The field equation (6.105) becomes (with $H \equiv H^{\alpha}{}_{\alpha}$):

$$H_{\mu\nu},^{\rho}{}_{\rho} - H_{\mu\rho},^{\rho}{}_{\nu} - H_{\nu\rho},^{\rho}{}_{\mu} - \frac{1}{2} g_{\mu\nu} H^{\rho}{}_{,\rho} = 0. \quad (6.108)$$

---

[12] In $D = 4$, then, the inversion is $h_{\mu\nu} = H_{\mu\nu} - \frac{1}{2} g_{\mu\nu} H$.

We can play the same game, writing the entire equation in terms of a divergence:

$$\partial^\rho \left(H_{\mu\nu,\rho} - H_{\mu\rho,\nu} - H_{\nu\rho,\mu} - \frac{1}{2} g_{\mu\nu} H^\gamma{}_{\gamma,\rho} \right) = 0. \qquad (6.109)$$

Now, using the auxiliary field from (6.107), written in terms of $H_{\mu\nu}$, we have:

$$\pi_{\rho\mu\nu} = \frac{1}{2} \left(h_{\rho\mu,\nu} + h_{\rho\nu,\mu} - h_{\mu\nu,\rho} \right)$$

$$= \frac{1}{2} \left(H_{\rho\mu,\nu} + H_{\rho\nu,\mu} - H_{\mu\nu,\rho} - \frac{1}{2} g_{\rho\mu} H_{,\nu} - \frac{1}{2} g_{\rho\nu} H_{,\mu} + \frac{1}{2} g_{\mu\nu} H_{,\rho} \right). \qquad (6.110)$$

The definition of canonical momentum is, in a sense, up to us – we just need a triply indexed object, related to the field H via a derivative. Taking the original π field and rewriting it in terms of H has the advantage of decoupling its traces (if you write out the traces of π in terms of $h_{\mu\nu}$, you get $\pi^\mu{}_{\mu\gamma} = \frac{1}{2} h^\mu{}_{\mu,\gamma}$, but $\pi_\gamma{}^\mu{}_\mu = h_{\gamma}{}^\mu{}_\mu - \frac{1}{2} h^\mu{}_{\mu,\gamma}$):

$$\pi^\mu{}_{\mu\nu} = -\frac{1}{2} H_{,\nu} \qquad (6.111)$$

$$\pi_\rho{}^\mu{}_\mu = H_\rho{}^\mu{}_{,\mu}.$$

Now, we are going to rewrite the divergence in (6.109) in terms of π using (6.111):

$$\partial^\rho \left(-2\pi_{\rho\mu\nu} - \frac{1}{2} g_{\rho\mu} H_{,\nu} - \frac{1}{2} g_{\rho\nu} H_{,\mu} \right) = 0 \qquad (6.112)$$

$$\partial^\rho \left(-2\pi_{\rho\mu\nu} + g_{\rho\mu} \pi^\gamma{}_{\gamma\nu} + g_{\rho\nu} \pi^\gamma{}_{\gamma\mu} \right) = 0.$$

Our base Lagrangian, returning the second of these under $H^{\mu\nu}$ variation, is:

$$\tilde{\mathcal{L}} = H^{\mu\nu,\rho} \left(2\pi_{\rho\mu\nu} - 2 g_{\rho\nu} \pi^\gamma{}_{\gamma\mu} \right), \qquad (6.113)$$

where we have used the symmetry $H^{\mu\nu} = H^{\nu\mu}$ to combine the second and third terms in (6.112). This Lagrangian, by construction, is guaranteed to return the field equations upon $H^{\mu\nu}$ variation. But the variation with respect to the separate, independent field $\pi_{\alpha\beta\gamma}$ will not necessarily enforce the defining relation between π and H from (6.110). In order to recover this necessary constraint, we must augment the current action. The logic, now, is to vary (6.113) with respect to $\pi_{\alpha\beta\gamma}$, see what we get in terms of π, and then subtract from this base action terms that reproduce the detritis upon variation by π. Notice that this process will not alter the field equation itself, since the term we will add will be quadratic in π (as we shall see), and not depend on $H^{\mu\nu}$.

Example: scalar fields

Because the process for the second-rank tensor field is a little bit different than our work on scalars and vectors, I want to quickly run the current argument for scalars.

We started with the most general field equation – for a scalar field, the most general, linear, second-order differential equation is (ignoring mass term):

$$\phi^{\alpha}_{,\alpha} = 0. \tag{6.114}$$

From this equation, we are motivated to introduce the auxiliary field $\pi_{\alpha} \equiv \phi_{,\alpha}$, at which point (6.114) becomes:

$$\partial^{\alpha} \pi_{\alpha} = 0. \tag{6.115}$$

The Lagrangian returning the above field equation, and analogous to (6.113) (in that it forces the divergence-less character of the momentum under ϕ variation), is:

$$\tilde{\mathcal{L}} = \phi_{,\alpha} \pi^{\alpha}. \tag{6.116}$$

In developing the Lagrangian, we effectively defined $\pi_{\alpha} \equiv \phi_{,\alpha}$ – but is this information contained in the Lagrangian itself? If we vary with respect to π^{α}, we get:

$$\frac{\partial \tilde{\mathcal{L}}}{\partial \pi^{\alpha}} = \phi_{,\alpha}. \tag{6.117}$$

As a field equation, we would normally set that to zero, from which we would conclude that $\phi_{,\alpha} = 0$. This statement is not useful, it gives us no relation connecting ϕ and π, and a first-order action must contain such a relation. So, our question: what additional term could we introduce that would, upon π^{α} variation, return $\pi_{\alpha} = \phi_{,\alpha}$? We want to generate an $\hat{\mathcal{L}}$ with:

$$\frac{\partial \hat{\mathcal{L}}}{\partial \pi^{\alpha}} = \pi_{\alpha}, \tag{6.118}$$

and $\frac{\partial \hat{\mathcal{L}}}{\partial \phi} = 0$, since we already have the correct field equations – all we are trying to do is enforce the relation between ϕ and its canonical momentum.

Suppose we could construct such a $\hat{\mathcal{L}}$, then our full Lagrangian will be:

$$\mathcal{L} = \tilde{\mathcal{L}} - \hat{\mathcal{L}}, \tag{6.119}$$

with the correct field equations:

$$\frac{\partial \mathcal{L}}{\partial \phi} = \frac{\partial \tilde{\mathcal{L}}}{\partial \phi} - \frac{\partial \hat{\mathcal{L}}}{\partial \phi} = \pi^{\alpha}_{,\alpha} - 0 = 0$$

$$\frac{\partial \mathcal{L}}{\partial \pi^{\alpha}} = \frac{\partial \tilde{\mathcal{L}}}{\partial \pi^{\alpha}} - \frac{\partial \hat{\mathcal{L}}}{\partial \pi^{\alpha}} = \phi_{,\alpha} - \pi_{\alpha} = 0. \tag{6.120}$$

This should be a familiar first-order form story, by now. One field equation constrains the momentum, the other connects the momentum to the field.

The final move is to construct $\hat{\mathcal{L}}$ satisfying $\frac{\delta \mathcal{L}}{\delta \pi^\alpha} = \pi_\alpha$ and $\frac{\delta \mathcal{L}}{\delta \phi} = 0$ – we know it must be independent of ϕ, and in order for its variation to be linear in π_α, $\hat{\mathcal{L}}$ must be quadratic in the momenta. The obvious choice works here – $\hat{\mathcal{L}} = \frac{1}{2} \pi^\alpha \pi_\alpha$ satisfies both requirements, and gives back the first-order form for a massless scalar field:

$$\mathcal{L} = \tilde{\mathcal{L}} - \hat{\mathcal{L}} = \pi^\alpha \phi_{,\alpha} - \frac{1}{2} \pi^\alpha \pi_\alpha. \tag{6.121}$$

We will use this same logic in what follows.

Variation of $\tilde{\mathcal{L}}$

We know that varying $\tilde{\mathcal{L}}$ with respect to $H^{\mu\nu}$, and using the relation between $H^{\mu\nu}{}_{,\rho}$ and $\pi_{\rho\mu\nu}$ from (6.110), will recover the field equation (6.109). But what happens if we vary $\tilde{\mathcal{L}}$ with respect to $\pi^{\alpha\beta\gamma}$? In preparation, rewrite:

$$\tilde{\mathcal{L}} = H^{\mu\nu}{}_{,\rho} \left[2 \pi_{\alpha\beta\gamma} \, \delta^\alpha_\rho \, \delta^\beta_\mu \, \delta^\gamma_\nu - 2 g_{\nu\rho} \, \pi_{\alpha\beta\gamma} \, g^{\alpha\beta} \, \delta^\gamma_\mu \right], \tag{6.122}$$

now the variation is:

$$\frac{\delta \tilde{\mathcal{L}}}{\delta \pi_{\alpha\beta\gamma}} = 2 \, H^{\beta\gamma}{}^{,\alpha} - 2 \, g^{\alpha\beta} \, H^{\gamma\rho}{}_{,\rho}. \tag{6.123}$$

From (6.110), we can write $H_{\beta\gamma,\alpha}$ in terms of $\pi_{\gamma\alpha\beta}$ and $\pi_{\beta\gamma\alpha}$:

$$\pi_{\gamma\alpha\beta} + \pi_{\beta\gamma\alpha} = H_{\beta\gamma,\alpha} - \frac{1}{2} g_{\beta\gamma} \, H_{,\alpha}, \tag{6.124}$$

and the traces of H appearing in the variation can be rewritten in terms of the traces of π from (6.111) – the end result is that the right-hand side of (6.123) can be expressed as:

$$2 \, H^{\beta\gamma}{}^{,\alpha} - 2 \, g^{\alpha\beta} \, H^{\gamma\rho}{}_{,\rho} = 2 \left(\pi^{\gamma\alpha\beta} + \pi^{\beta\gamma\alpha} - g^{\beta\gamma} \, \pi^{\mu\alpha}{}_\mu - g^{\alpha\beta} \, \pi^{\gamma\mu}{}_\mu \right). \tag{6.125}$$

The right-hand side of (6.125) is the result of applying (6.110) (together with the derived information (6.111)) to the left-hand side. It therefore represents the target form required for an additional component, the analogue to the $\hat{\mathcal{L}}(\pi)$ from (6.118). In order to construct the full Lagrangian, we just need to subtract a piece $\hat{\mathcal{L}}(\pi)$, quadratic in the momenta, whose variation with respect to $\pi_{\alpha\beta\gamma}$ is precisely the right-hand side of (6.125) – that is, we want to concoct an additional piece $\hat{\mathcal{L}}$ with:

$$\frac{\partial \hat{\mathcal{L}}}{\partial \pi_{\alpha\beta\gamma}} = 2 \left(\pi^{\gamma\alpha\beta} + \pi^{\beta\gamma\alpha} - g^{\beta\gamma} \, \pi^{\mu\alpha}{}_\mu - g^{\alpha\beta} \, \pi^{\gamma\mu}{}_\mu \right). \tag{6.126}$$

To find $\hat{\mathcal{L}}$, note that:

$$\frac{\partial}{\partial \pi_{\alpha\beta\gamma}} \left(\pi_{\alpha\beta\gamma} \, \pi^{\gamma\alpha\beta} \right) = \pi^{\gamma\alpha\beta} + \pi^{\beta\gamma\alpha} \tag{6.127}$$

(the field $\pi_{\alpha\beta\gamma}$ is symmetric in its second two indices, inherited from the symmetry $h_{\mu\nu} = h_{\nu\mu}$ from (6.107)). In addition, the derivative:

$$\frac{\partial}{\partial \pi_{\alpha\beta\gamma}} \left(\pi^{\gamma}_{\ \gamma\alpha} \pi^{\alpha\rho}_{\ \ \rho} \right) = g^{\beta\gamma} \pi^{\mu\alpha}_{\ \ \mu} + g^{\alpha\beta} \pi^{\gamma\mu}_{\ \ \mu} \qquad (6.128)$$

gives the remaining two terms in (6.125). So set:

$$\hat{\mathcal{L}} = 2 \pi_{\alpha\beta\gamma} \pi^{\gamma\alpha\beta} - 2 \pi^{\gamma}_{\ \gamma\alpha} \pi^{\alpha\rho}_{\ \ \rho}, \qquad (6.129)$$

satisfying (6.126). Our final Lagrangian reads (getting rid of the irrelevant overall factor of 2):

$$\bar{\mathcal{L}} = \tilde{\mathcal{L}} - \hat{\mathcal{L}} = H^{\mu\nu\ \rho}_{\ \ ,} \left(\pi_{\rho\mu\nu} - g_{\rho\nu} \pi^{\gamma}_{\ \gamma\mu} \right) + \pi^{\gamma}_{\ \gamma\alpha} \pi^{\alpha\rho}_{\ \ \rho} - \pi_{\alpha\beta\gamma} \pi^{\gamma\alpha\beta}. \qquad (6.130)$$

If we take the contravariant field $H^{\mu\nu}$ as fundamental, and treat the partial derivative as it naturally appears, lowered, then the momentum is most naturally expressed as a mixed tensor $\pi^{\alpha}_{\ \beta\gamma}$, and respecting this convention, we can write the equivalent Lagrangian:

$$\boxed{\mathcal{L}_H = H^{\mu\nu} \left(\pi^{\rho}_{\ \mu\nu,\rho} - \pi^{\gamma}_{\ \gamma\mu,\nu} \right) + g^{\mu\nu} \left(\pi^{\gamma}_{\ \gamma\alpha} \pi^{\alpha}_{\ \mu\nu} - \pi^{\alpha}_{\ \mu\gamma} \pi^{\gamma}_{\ \alpha\nu} \right),} \qquad (6.131)$$

where I have also flipped the derivative of the field $H^{\mu\nu}$ onto the π terms. Finally, in order to avoid confusion later on, let's take the provocative re-labeling of the canonical momentum here – the $\pi^{\rho}_{\ \mu\nu}$ will now be called $\Gamma^{\rho}_{\ \mu\nu}$. This term will, in the end, turn out to be the Christoffel connection, its interpretation enforced by the final field equations.

6.7 Second-rank consistency

There are two remaining issues that need to be discussed en route to the final form of the symmetric second-rank tensor field Lagrangian. They are both tied to the process of introducing sources. The most natural source for the right-hand side of the vacuum field equations is a stress tensor, this is (appropriately) guaranteed to be symmetric and conserved. The first issue, then, is that our current Lagrangian (6.131) itself has a stress tensor (note the explicit dependence on $g^{\mu\nu}$, that's a dead giveaway). We will address this explicitly by calculating the stress tensor for (6.131), and coupling the field to it. This process will lead us, pretty directly, to the Einstein–Hilbert action which yields, upon variation, Einstein's equation (in vacuum), and the interpretation of the second-rank field as a metric. We will thus complete the process of the second-rank tensor field development: start with a second-rank, symmetric field on a flat background, couple it to itself appropriately, and you get general relativity.[13] That means we have a theory in which the *metric* itself is the field.

The second issue is the coupling of external ("matter") fields to the second-rank tensor one. Since every field theory we have experience with has a stress tensor, each of these will contribute as sources for the metric field. As with E&M, the coupling of the external field to this second-rank theory changes the field equations of the external field, and we will have to compute the modification induced by the coupling. What will end up happening, basically via the requirement that all external fields have Lagrange densities (and so depend on the metric through $\sqrt{-g}$), is that any partial derivatives from the theory written on a flat background become covariant derivatives (and these will also provide metric dependence). Because all field theories have stress tensors, everything couples to $g^{\mu\nu}$, and we have an explicit, mathematical process for enforcing the strong equivalence principle. In addition, the "principle of minimal substitution" (which is used to turn Lorentz-covariant field equations into generally covariant ones by the replacement of commas with semicolons) is motivated, and indeed, required.

We'll start by varying (6.131) in order to find the current content of our field equations. Then, we'll couple $H^{\mu\nu}$ to its own stress tensor, and develop the final first-order form for the Lagrange density governing second-rank, symmetric tensor fields. Variation of that final form will give us back a relation between the field and its momentum that is precisely that of the metric to the connection. In addition, we will recover the source-free Einstein equation. Finally, we will couple an external field to the full GR action, and see how strong equivalence is enforced.

6.7.1 Modified action variation

When we were done, the final action form read:

$$\bar{\mathcal{L}} = H^{\mu\nu}\left(\Gamma^{\alpha}_{\mu\nu,\alpha} - \Gamma^{\alpha}_{\mu\alpha,\nu}\right) + g^{\mu\nu}\left(\Gamma^{\alpha}_{\mu\nu}\,\Gamma^{\gamma}_{\alpha\gamma} - \Gamma^{\alpha}_{\beta\mu}\,\Gamma^{\beta}_{\alpha\nu}\right), \tag{6.132}$$

which will give two sets of "first-order" equations, corresponding to variation w.r.t. $H^{\mu\nu}$ and $\Gamma^{\alpha}_{\beta\gamma}$. We can combine the two to give one second-order equation, the same one we've always had, (6.108). But it is interesting to look at the first-order results alone. Varying w.r.t. $H^{\mu\nu}$, we find that:

$$\boxed{\Gamma^{\alpha}_{\mu\nu,\alpha} - \Gamma^{\alpha}_{\mu\alpha,\nu} = 0,} \tag{6.133}$$

---

[13] This is all nicely laid out in its original form, see [7].

and varying w.r.t. $\Gamma^{\alpha}_{\beta\gamma}$ effectively gives us the relation between derivatives of $H_{\mu\nu}$ and the momenta – we have:

$$0 = -H^{\rho\delta}_{\ ,\sigma} + \frac{1}{2}\left(H^{\rho\nu}_{\ ,\nu}\,\delta^{\delta}_{\sigma} + H^{\delta\nu}_{\ ,\nu}\,\delta^{\rho}_{\sigma}\right) + g^{\rho\delta}\,\Gamma^{\gamma}_{\sigma\gamma} + \frac{1}{2}\,g^{\mu\nu}\left(\Gamma^{\rho}_{\mu\nu}\delta^{\delta}_{\sigma} + \Gamma^{\delta}_{\mu\nu}\delta^{\rho}_{\sigma}\right)$$
$$- \left(g^{\delta\nu}\,\Gamma^{\rho}_{\sigma\nu} + g^{\rho\nu}\,\Gamma^{\delta}_{\sigma\nu}\right).$$

$$(6.134)$$

If we input the definition $H_{\mu\nu} = h_{\mu\nu} - \frac{1}{2}\,g_{\mu\nu}h$, then the above boils down to:

$$\boxed{\Gamma_{\alpha\mu\nu} = \frac{1}{2}\left(h_{\alpha\mu,\nu} + h_{\alpha\nu,\mu} - h_{\mu\nu,\alpha}\right),} \qquad (6.135)$$

where we are just recovering (6.107). At this point, I want to make sure that the importance of this *as a field equation* does not escape us. We are being told that our theory, a fairly general free second-rank tensor field has, built in somehow, the idea that the field itself defines what amounts to the usual metric connection we get out of curvilinear coordinates (for example). Somehow, making a relativistic symmetric second-rank field theory is *already* tied to some geometrical notions.

The object appearing in (6.133) depends on the field momenta exactly as the (linearized) Ricci tensor, defined as:

$$\tilde{R}_{\mu\nu} \equiv \Gamma^{\alpha}_{\mu\nu,\alpha} - \Gamma^{\alpha}_{\mu\alpha,\nu} \qquad (6.136)$$

depends on the connection. The linearized Ricci tensor comes from (4.46) with terms quadratic in Γ dropped from the Riemann tensor (the "linearization" is thus with respect to perturbations of the flat metric of Minkowski space-time). With our relation between Γ and h in place, we can write the linearized Ricci tensor as a function of the field $h_{\mu\nu}$:

$$\tilde{R}_{\mu\nu} = \frac{1}{2}\,g^{\alpha\rho}\left(h_{\rho\nu,\mu\alpha} - h_{\mu\nu,\rho\alpha} - h_{\rho\alpha,\mu\nu} + h_{\mu\alpha,\rho\nu}\right). \qquad (6.137)$$

Clearly, once the field equation (6.133) is satisfied, this is zero, that's just like saying $\partial_{\mu}F^{\mu\nu} = 0$ in electrodynamics. But the linearized Ricci tensor has a symmetry relation for its derivatives much like $\partial_{\mu}\partial_{\nu}F^{\mu\nu} = 0$. Note the divergence of this tensor:

$$\partial^{\mu}\tilde{R}_{\mu\nu} = \frac{1}{2}\left(h^{\mu\alpha}_{\ ,\mu\alpha\nu} - h^{\ ,\mu}_{,\ \mu\nu}\right), \qquad (6.138)$$

not clearly zero. This is an issue when we attempt to couple our $H^{\mu\nu}$ field to other fields, and we'll return to this point later on.

6.7.2 Stress tensor for $H^{\mu\nu}$

Recall the situation in electricity and magnetism – we had a conserved current J^μ ($\partial_\mu J^\mu = 0$), and we used this as a source, introducing it into the vacuum Lagrangian by the addition of the term $A_\mu J^\mu$. We'd like to develop a notion of source for our $H^{\mu\nu}$ field, and here, we'll need a symmetric, second-rank tensor to go on the right-hand side of (6.133). In addition, if only by analogy with E&M (or from physical considerations of continuity), we'd like the source tensor to be conserved, so that its divergence vanishes. What is available? In our study of field theory so far, there is a natural candidate second-rank symmetric tensor that is *always* conserved, the stress tensor for a field theory: $T^{\mu\nu}$. Then what we will do is introduce a term like $H_{\mu\nu} T^{\mu\nu}$ to the free Lagrangian, and achieve coupling precisely as in the vector case. We developed the stress tensor, a by-product of coordinate invariance, by taking the variational derivative of a Lagrange density with respect to the metric. If we are to take the stress tensor seriously as a source, then, we see that the first one we must consider is the $T^{\mu\nu}$ associated with the Lagrangian (6.132) itself. This is a new situation, where the field itself provides a source. Our first job is to compute and introduce this source, so that we have a final theory that does not refer to any external source, and is complete (in the sense that the implicit source is already accounted for).

Let's go back to the stress tensor that would arise from a theory for $H_{\mu\nu}$. We have not made a scalar action for curvilinear coordinates, starting as we did from the assumption that $g^{\mu\nu} = \eta^{\mu\nu}$ for flat Cartesian coordinates – but we can fix this. Notice also that this is the first action we have considered that *needs* fixing – the scalar theory used "normal" partials since $\phi_{,\mu} = \phi_{;\mu}$, and the vector theory used $A_{\mu,\nu} - A_{\nu,\mu} = A_{\mu;\nu} - A_{\nu;\mu}$.

Since we would like, ultimately, a Lagrange density, so that the action is a scalar, we need to take commas to semicolons in (6.132). In addition, we take the field $H^{\mu\nu}$ to be a density, call it $\mathfrak{H}^{\mu\nu}$. All we have done is take a Lagrangian written in flat space-time, and produced the correct form for it in curvilinear coordinates. Our starting Lagrangian density, then, is:

$$\mathcal{L} = \mathfrak{H}^{\mu\nu} \left(\Gamma^\alpha_{\mu\nu;\alpha} - \Gamma^\alpha_{\mu\alpha;\nu} \right) + \sqrt{-g}\, g^{\mu\nu} \left(\Gamma^\alpha_{\mu\nu}\, \Gamma^\gamma_{\alpha\gamma} - \Gamma^\alpha_{\beta\mu}\, \Gamma^\beta_{\alpha\nu} \right). \qquad (6.139)$$

Remember that stress tensor conservation comes from coordinate-invariance, our definition in (5.99) comes from the assumption that the associated action is unchanged by (infinitesimal) coordinate transformation, and is meant to simplify the statement of conservation. By making some observations about the relation of derivatives of $\mathcal{L}$ with respect to the contravariant or covariant form of the metric (see, for example, Problem 5.15), we can write the analogous expression for the derivative with respect to the contravariant form of the metric:

$$\frac{1}{2}\sqrt{-g}\,T_{\mu\nu} = \frac{\partial\mathcal{L}}{\partial g^{\mu\nu}}. \tag{6.140}$$

Our field equations represent the linearized Ricci tensor (6.133), which does not have zero divergence (as is evidenced by (6.138)). By introducing a coupling like $H^{\mu\nu}\,T_{\mu\nu}$, we will end up with field equations, under $H^{\mu\nu}$ variation that look like $\tilde{R}_{\mu\nu} \sim T_{\mu\nu}$, which is inconsistent in the sense that $\partial^\mu T_{\mu\nu} = 0$ (since the stress tensor is conserved), while $\partial^\mu \tilde{R}_{\mu\nu} \neq 0$. To account for this, we'll use the trace-reversed form of the stress tensor, so that our coupling will proceed from:

$$\bar{T}_{\mu\nu} \equiv T_{\mu\nu} - \frac{1}{2}g_{\mu\nu}\,T^{\sigma}_{\ \alpha}. \tag{6.141}$$

Note that the linearized Bianchi identity (from (4.85), (4.112)) holds, and we could use that to restore consistency by taking the trace-reversed form of the linearized Ricci tensor, and sourcing with a $T_{\mu\nu}$ right-hand side. This approach, trace-reversing the source tensor, amounts to the same thing (the content is identical), and is easier to compute.

We want to now calculate the derivative of $\mathcal{L}$ w.r.t. $g^{\mu\nu}$, that will provide the form of the stress tensor associated with the field (on flat space) $H^{\mu\nu}$. For notation, write the connection (associated with the metric $g_{\mu\nu}$) in our curvilinear coordinate system as $C^{\alpha}_{\ \beta\gamma}$, just to distinguish it from the $\Gamma^{\alpha}_{\ \beta\gamma}$. Then the derivative w.r.t. $g^{\mu\nu}$ is:

$$\frac{\partial\mathcal{L}}{\partial g^{\mu\nu}} = \left(\mathfrak{H}^{\alpha\beta}\,\Gamma^{\lambda}_{\ \alpha\beta}\,\delta^{\tau}_{\sigma} - 2\,\mathfrak{H}^{\lambda\gamma}\,\Gamma^{\tau}_{\ \sigma\gamma} + \mathfrak{H}^{\lambda\tau}\,\Gamma^{\alpha}_{\ \sigma\alpha}\right)\frac{\partial C^{\sigma}_{\ \lambda\tau}}{\partial g^{\mu\nu}}$$

$$+ \sqrt{-g}\left(\Gamma^{\alpha}_{\ \mu\nu}\,\Gamma^{\gamma}_{\ \alpha\gamma} - \Gamma^{\alpha}_{\ \beta\mu}\,\Gamma^{\beta}_{\ \alpha\nu}\right) + \frac{1}{2}\sqrt{-g}\,g_{\mu\nu}\left(-\Gamma^{\rho\sigma}_{\ \ \sigma}\,\Gamma^{\gamma}_{\ \rho\gamma} + \Gamma^{\rho\ \gamma}_{\ \sigma}\,\Gamma^{\sigma}_{\ \rho\gamma}\right), \tag{6.142}$$

where the first term comes from the connection-dependence buried inside the covariant derivatives of the $\Gamma^{\alpha}_{\ \mu\nu}$ terms in (6.139).

We could calculate the derivative of the connection w.r.t. $g^{\mu\nu}$ and use it as part of the stress tensor, but as it turns out, we don't actually need it (the resulting contribution to the field equations does not itself generate any additional source-like terms). Dropping this term, and computing the trace-reversed form, we find:

$$\bar{T}_{\mu\nu} = T_{\mu\nu} - \frac{1}{2}g_{\mu\nu}\,T = 2\,\Gamma^{\rho}_{\ \mu\nu}\,\Gamma^{\gamma}_{\ \rho\gamma} - 2\,\Gamma^{\rho}_{\ \sigma\mu}\,\Gamma^{\sigma}_{\ \rho\nu}. \tag{6.143}$$

If we want to "couple" our $H_{\mu\nu}$ field theory to itself (which we must do before coupling it to anything else), we need the action to include the term $\mathfrak{H}^{\mu\nu}\,\bar{T}_{\mu\nu}$ (recall the E&M version of the source term: $\sqrt{-g}\,A_\mu\,J^\mu$). Let's return with this information to the flat space-time form – we have the relevant additional terms, and it is easier to introduce them on a Minkowski background (in Cartesian coordinates):

$$\bar{\mathcal{L}} = H^{\mu\nu} \left(\Gamma^{\alpha}{}_{\mu\nu,\alpha} - \Gamma^{\alpha}{}_{\mu\alpha,\nu} \right) + g^{\mu\nu} \left(\Gamma^{\alpha}{}_{\mu\nu} \Gamma^{\gamma}{}_{\alpha\gamma} - \Gamma^{\alpha}{}_{\beta\mu} \Gamma^{\beta}{}_{\alpha\nu} \right)$$
$$+ H^{\mu\nu} \left(\Gamma^{\alpha}{}_{\mu\nu} \Gamma^{\gamma}{}_{\alpha\gamma} - \Gamma^{\alpha}{}_{\beta\mu} \Gamma^{\beta}{}_{\alpha\nu} \right). \tag{6.144}$$

Notice that, by construction, the second and third terms differ only by the factor in front, so we can combine these in a form with overall $(g^{\mu\nu} + H^{\mu\nu})$. In addition, we can add a term that looks like $g^{\mu\nu} \left(\Gamma^{\alpha}{}_{\mu\nu,\alpha} - \Gamma^{\alpha}{}_{\mu\alpha,\nu} \right)$ with impunity, since the derivatives can be flipped onto the underlying flat metric which will vanish. Overall, then, we find that the original theory for $H_{\mu\nu}$, when made self-consistent by coupling to its own stress tensor, becomes:

$$\boxed{\bar{\mathcal{L}} = (g^{\mu\nu} + H^{\mu\nu}) \left(\Gamma^{\alpha}{}_{\mu\nu,\alpha} - \Gamma^{\alpha}{}_{\mu\alpha,\nu} + \Gamma^{\alpha}{}_{\mu\nu} \Gamma^{\gamma}{}_{\alpha\gamma} - \Gamma^{\alpha}{}_{\beta\mu} \Gamma^{\beta}{}_{\alpha\nu} \right).} \tag{6.145}$$

In this final form, we see that the metric and the field $H^{\mu\nu}$ appear in indistinguishable additive combination. The final step is to reform the Lagrangian density, by taking the combination $g^{\mu\nu} + H^{\mu\nu}$ to be, itself, a density. If we simply relabel the metric-plus-field combination $\mathfrak{g}^{\mu\nu}$ a second-rank contravariant density,[14] then we have the final, correct scalar action:

$$\boxed{S = \int d\tau \, \mathfrak{g}^{\mu\nu} \left(\Gamma^{\alpha}{}_{\mu\nu,\alpha} - \Gamma^{\alpha}{}_{\mu\alpha,\nu} + \Gamma^{\alpha}{}_{\mu\nu} \Gamma^{\gamma}{}_{\alpha\gamma} - \Gamma^{\alpha}{}_{\beta\mu} \Gamma^{\beta}{}_{\alpha\nu} \right).} \tag{6.146}$$

The content of this equation, as a first-order action for the second-rank, symmetric tensor $\mathfrak{g}^{\mu\nu}$, can be found by varying with respect to $\Gamma^{\alpha}{}_{\mu\nu}$, from which we will learn that the "field momenta" are related to the field precisely as a Christoffel connection is related to the metric, allowing an interpretation of $\mathfrak{g}^{\mu\nu}$ as a metric. Once this has been established, the variation of the action with respect to $\mathfrak{g}^{\mu\nu}$ will return Einstein's equation: $R_{\mu\nu} - \frac{1}{2} g_{\mu\nu} R = 0$.

Full field equations

The scalar Lagrangian associated with (6.146) can be written as:

$$\bar{\mathcal{L}} = \mathfrak{g}^{\mu\nu} \underbrace{\left(\Gamma^{\alpha}{}_{\mu\nu,\alpha} - \Gamma^{\alpha}{}_{\mu\alpha,\nu} + \Gamma^{\alpha}{}_{\mu\nu} \Gamma^{\gamma}{}_{\alpha\gamma} - \Gamma^{\alpha}{}_{\beta\mu} \Gamma^{\beta}{}_{\alpha\nu} \right)}_{= R_{\mu\nu} \equiv R^{\alpha}{}_{\mu\alpha\nu}}, \tag{6.147}$$

where we emphasize that the term in parentheses is the full Ricci tensor (provided the Γ can be understood as connections). We have soaked the dynamical field $H^{\mu\nu}$ into $\mathfrak{g}^{\mu\nu}$ – the point is that $H^{\mu\nu}$ always appears in combination with $g^{\mu\nu}$ (the background metric), and these together play the role of a metric, so we cannot actually distinguish between the two. From now on, then, $\mathfrak{g}^{\mu\nu}$ is the whole story –

---

[14] A typical second-rank density would be, for example, $\mathfrak{g}^{\mu\nu} = \sqrt{-g} \, g^{\mu\nu}$.

our second-rank field theory has, as its field, the metric of some pseudo-Riemannian space-time. The action that this comes from is, again:

$$S = \int d\tau \sqrt{-g} \, g^{\mu\nu} R_{\mu\nu}(\Gamma),$$

(6.148)

and we have been careful to put in the density $\sqrt{-g}$ to counteract $d\tau$. The integrand is clearly a scalar, so we will obtain a valid covariant theory.

Let's do the variation one last time, with the full $g^{\mu\nu}$ in place – we will take the first-order point of view (in this context, (6.148) is called the "Palatini" action), where as always, the field $g^{\mu\nu}$ and momenta $\Gamma^{\alpha}{}_{\beta\gamma}$ are varied separately. Varying w.r.t. $g^{\mu\nu}$ is easy[15] if we view $R_{\mu\nu}$ as a function of the momenta $\Gamma^{\alpha}{}_{\beta\gamma}$ and its derivatives, then there is no metric dependence and:

$$\delta S = \int d\tau \left(\sqrt{-g} \, R_{\alpha\beta} \, \delta g^{\alpha\beta} - \frac{1}{2} R_{\mu\nu} \, g^{\mu\nu} \frac{\partial g}{\partial g^{\alpha\beta}} \, \delta g^{\alpha\beta} \right)$$

$$= \int d\tau \left(\sqrt{-g} \left(R_{\alpha\beta} - \frac{1}{2} g_{\alpha\beta} \, R \right) \delta g^{\alpha\beta} \right).$$

(6.149)

This tells us that the Einstein tensor $G_{\mu\nu} \equiv R_{\mu\nu} - \frac{1}{2} g_{\mu\nu} R = 0$, which is precisely Einstein's equation in vacuum. We do not yet know that $g_{\mu\nu}$ is a metric – there is currently no relationship between the metric and the momenta, in terms of which $R_{\mu\nu}$ (and hence $G_{\mu\nu}$) is written. As usual, the relation is set by the $\Gamma^{\alpha}{}_{\beta\gamma}$ (momentum) variation. This is a little more involved, but still relatively easy. If we introduce the metric density $\mathfrak{g}^{\mu\nu} \equiv \sqrt{-g} \, g^{\mu\nu}$, then variation gives:

$$\delta S = \int d\tau \left(-\mathfrak{g}^{\mu\nu}{}_{,\alpha} + \mathfrak{g}^{\mu\gamma}{}_{,\gamma} \, \delta^{\nu}_{\alpha} + \Gamma^{\gamma}{}_{\alpha\gamma} \, \mathfrak{g}^{\mu\nu} - 2 \Gamma^{\mu}{}_{\alpha\rho} \, \mathfrak{g}^{\nu\rho} + \mathfrak{g}^{\rho\sigma} \, \Gamma^{\mu}{}_{\rho\sigma} \, \delta^{\nu}_{\alpha} \right) \delta\Gamma^{\alpha}{}_{\mu\nu},$$

(6.150)

and we require that this be zero – there is a slight catch here – the connection (momenta) $\Gamma^{\alpha}{}_{\mu\nu}$ is symmetric in $\mu \leftrightarrow \nu$, so that only the symmetric part of the term in parentheses needs to be zero for the variation to vanish. Our field equation becomes:

$$- \mathfrak{g}^{\mu\nu}{}_{,\alpha} + \frac{1}{2} \left(\mathfrak{g}^{\mu\gamma}{}_{,\gamma} \, \delta^{\nu}_{\alpha} + \mathfrak{g}^{\nu\gamma}{}_{,\gamma} \, \delta^{\mu}_{\alpha} \right) + \Gamma^{\gamma}{}_{\alpha\gamma} \, \mathfrak{g}^{\mu\nu} - \Gamma^{\mu}{}_{\alpha\rho} \, \mathfrak{g}^{\nu\rho} - \Gamma^{\nu}{}_{\alpha\rho} \, \mathfrak{g}^{\mu\rho}$$

$$+ \frac{1}{2} \mathfrak{g}^{\rho\sigma} \left(\Gamma^{\mu}{}_{\rho\sigma} \, \delta^{\nu}_{\alpha} + \Gamma^{\nu}{}_{\rho\sigma} \, \delta^{\mu}_{\alpha} \right) = 0.$$

(6.151)

---

[15] Note that here, $\frac{\partial g}{\partial g_{\mu\nu}} = g \, g^{\mu\nu}$, which directly implies that $\frac{\partial g}{\partial g^{\mu\nu}} = -g \, g_{\mu\nu}$.

The long road is to introduce:[16]

$$\mathfrak{g}^{\mu\nu}{}_{,\alpha} = \sqrt{-g}_{,\alpha}\, g^{\mu\nu} + \sqrt{-g}\, g^{\mu\nu}{}_{,\alpha} = \sqrt{-g}\left(\frac{1}{2}\, g^{\rho\sigma}\, g^{\mu\nu}\, g_{\rho\sigma,\alpha} + g^{\mu\nu}{}_{,\alpha}\right), \quad (6.152)$$

then rewrite all partial derivatives in terms of their covariant derivatives, massage, and voila – we get:

$$\boxed{g_{\mu\nu;\alpha} = 0,} \quad (6.153)$$

which yields the usual relationship between connection and metric as you showed in Problem 3.5b.

This is no surprise, just a check of what we already knew. The logic is: variation w.r.t. $g^{\mu\nu}$ gives the Ricci-flat condition ($R_{\mu\nu} = 0$), a statement about allowed geometry, while variation w.r.t. $\Gamma^{\alpha}{}_{\mu\nu}$ connects the field $g^{\mu\nu}$ to a metric space. In vacuum, it is easy to see that the vanishing of the Einstein tensor demands that $R_{\mu\nu} = 0$, so that here, we need not keep the full varied form: $R_{\mu\nu} - \frac{1}{2}\, g_{\mu\nu}\, R = 0$, but this is useful for coupling.

The "first-order" formalism allowed us to treat this action in a Hamiltonian fashion, just as we did with E&M. In general relativity, we can work from this so-called Palatini action, or, if we prefer, from the equivalent (second-order) "Einstein–Hilbert" action, obtained by viewing the integrand as a function only of the metric and its derivatives:

$$S_{EH} = \int \sqrt{-g}\, R(g, \partial g, \partial^2 g)\, d\tau. \quad (6.154)$$

In the Einstein–Hilbert form, we have already introduced $\Gamma^{\alpha}{}_{\mu\nu}$ as a function of the metric and its derivatives. This second-order form presupposes a geometrical interpretation for the metric. The situation is similar to the "second-order" form of the E&M action: $S \sim \int F_{\mu\nu}(A_{\alpha})\, F^{\mu\nu}(A_{\alpha})\, d\tau$, where we must assume the existence of potentials making up the field strength tensor in order to vary effectively.

6.7.3 Matter coupling

Since we now have an interpretation for our second-rank symmetric tensor field as a metric, it is clear that any field theory that can be written in terms of a Lagrange density will automatically couple to the $g^{\mu\nu}$ field, and will have field equations that involve the metric field. So we should no longer write field theory actions in isolation, since all actions contribute to the determination of the metric, all of them require the implicit additive term: $\int d\tau \sqrt{-g}\, g^{\mu\nu}\, R_{\mu\nu}(\Gamma)$. We know what happens to the field equations for the metric when we introduce additional actions

---

[16] The coordinate derivative of the density g is $g_{,\alpha} = g\, g^{\rho\sigma}\, g_{\rho\sigma,\alpha}$.

(we pick up the stress tensors as sources), but what about the field equations for the external fields themselves? Suppose we have a Lagrange density for a field ϕ (could be scalar or vector): $\mathcal{L}(\phi, g_{\alpha\beta})$ (typically of the form $\mathcal{L} = \sqrt{-g}\,\bar{\mathcal{L}}(\phi, g_{\alpha\beta})$). We suppose that the Lagrange density for ϕ does not depend on the derivatives of the metric, but it could depend on derivatives of ϕ (this has been true for both the Klein–Gordon and Maxwell actions). Then we form the full Lagrange density:

$$\mathcal{L} = \sqrt{-g}\, g^{\mu\nu} R_{\mu\nu}(\Gamma) + \alpha\, \mathcal{L}_\phi(\phi, g_{\alpha\beta}). \tag{6.155}$$

Variation w.r.t. $g^{\mu\nu}$ will return the Einstein tensor from the first term (as in (6.149)), and the second term will contribute $\frac{\partial \mathcal{L}_\phi}{\partial g^{\mu\nu}}$, which is related trivially to the stress tensor – remember that $\frac{1}{2}\sqrt{-g}\,T_{\mu\nu} = \frac{\partial \mathcal{L}}{\partial g^{\mu\nu}}$ in general. The end result of $g^{\mu\nu}$ variation is:

$$\sqrt{-g}\left[\left(R_{\mu\nu} - \frac{1}{2} g_{\mu\nu} R\right) + 2\alpha\, T_{\mu\nu}\right] = 0, \tag{6.156}$$

where α can be used to set the final units. The variation with respect to $\Gamma^\alpha_{\mu\nu}$ will again tell us that the $g^{\mu\nu}$ field has a metric connection. That was all by construction, and the new element is the variation of the full action with respect to ϕ.

Suppose the density for ϕ depends explicitly on ϕ and its first derivatives $\phi_{,\mu}$, and that we can write $\mathcal{L}(\phi, g_{\alpha\beta}) = \sqrt{-g}\,\bar{\mathcal{L}}(\phi, g_{\alpha\beta})$, then the variation of (6.155) is (keep in mind that there is no ϕ in the first term):

$$\delta S = \int d\tau \left(\sqrt{-g}\,\frac{\partial \bar{\mathcal{L}}}{\partial \phi} - \left(\sqrt{-g}\,\frac{\partial \bar{\mathcal{L}}}{\partial \phi_{,\mu}}\right)_{,\mu}\right) \delta\phi, \tag{6.157}$$

where we have used integration by parts to write the second term. Now, using the property of the covariant derivative with densities $(\sqrt{-g}\,f^\alpha)_{,\alpha} = \sqrt{-g}\,f^\alpha_{\;;\alpha}$ from (5.73), we can write the field equation as:

$$\frac{\partial \bar{\mathcal{L}}}{\partial \phi} - \left(\frac{\partial \bar{\mathcal{L}}}{\partial \phi_{,\mu}}\right)_{;\mu} = 0. \tag{6.158}$$

This tells us that the field equation for ϕ has been modified only by the replacement of the partial derivative with the covariant derivative. Think, for example, of the actual Klein–Gordon Lagrange scalar $\bar{\mathcal{L}} = \frac{1}{2}\phi_{,\alpha}\, g^{\alpha\beta}\, \phi_{,\beta} + \frac{1}{2} m^2 \phi^2$. In this concrete case, (6.158) becomes:

$$m^2 \phi - \phi_{;\mu}{}^{\mu} = 0. \tag{6.159}$$

So, we can either always include the Palatini action in any field theory's action, vary everything with respect to *all* fields, including $g^{\alpha\beta}$ and Γ, or we can separate the gravity portion, and consider just the field of interest. To do this correctly, we just need to construct the stress tensor for the field, and include it as a source for the

Einstein tensor, and, when dealing with the field itself, replace partial derivatives with covariant derivatives in the field equations. This means that every field theory modifies the metric, and then is itself modified (via the metric-dependence inside the covariant derivative). We get away with the usual form of Maxwell's E&M, for example, by assuming that we have a locally flat space-time, with no modification.[17] The process of implicitly including $g^{\mu\nu}$ in all field theories (hence coupling all theories to general relativity) is known as "strong equivalence". We can enforce strong equivalence by taking commas to semicolons and inserting metrics where appropriate, sometimes referred to as "minimal substitution".

As a final point – we assumed above that any external Lagrange density was free from dependence on derivatives of the metric. That is certainly true for the scalar case, since $\phi_{,\mu} = \phi_{;\mu}$, and we have no connections to introduce. If we want to couple a vector, like E&M, to gravity, we would take the Lagrangian:

$$\bar{\mathcal{L}} = F^{\mu\nu}\left(A_{\nu,\mu} - A_{\mu,\nu}\right) - \frac{1}{2}\,F^{\mu\nu}\,g_{\alpha\mu}\,g_{\beta\nu}\,F^{\alpha\beta}. \tag{6.160}$$

This Lagrangian also lacks dependence on the connection because of the difference $A_{\nu,\mu} - A_{\mu,\nu}$. When we vary, then, Maxwell's equations will be in terms of covariant derivatives, and the stress tensor of E&M will source the gravitational field $g_{\mu\nu}$. Because of the presence of a metric in any Lagrangian (we must use the metric to form scalars), *everything* couples to gravity – any Lagrangian we write down will provide a source, since $\frac{\partial \mathcal{L}}{\partial g^{\mu\nu}} \neq 0$. This is the mathematical manifestation of the idea that all forms of energy (mass and otherwise) have an effect on the geometry of space-time.

6.8 Source-free solutions

Now that we have seen how matter couples, we will retreat to the vacuum equations: $R_{\mu\nu} = 0$. In electrodynamics, we solve the vacuum problem almost immediately – the Coulomb field is defined everywhere (except at the origin), and satisfies $\nabla^2 V = 0$ for $r \neq 0$. This spherically symmetric field is the simplest possible solution to Laplace's equation (except for $V = 0$). Of course, because we have a delta function source, we can build up solutions for more complicated distributions. That's the role of superposition, an observational fact that is included in Maxwell's equations (they are linear). The general solution for the electrostatic potential, V, given a static distribution of charge ρ is obtained by summing the points solutions:

$$V = \int \frac{\rho(\mathbf{r}')}{4\pi\,\epsilon_0\,\imath}\,d\tau'. \tag{6.161}$$

---

[17] This is, in the end, an approximation motivated by local flatness. Strictly speaking, electric and magnetic fields serve to determine the space-time in which they exist, and they must respond to that space-time.

In general relativity, the situation is quite different – Einstein's equation is nonlinear, so we do not expect superposition to hold (it doesn't). Even the simple spherically symmetric analogue of the Coulomb field of E&M is not immediately apparent. We will have to solve $R_{\mu\nu} = 0$ for a spherically symmetric metric, and once we have the solution, we can do "nothing" with it (in the sense of building other solutions).

The salvation is that there is really only one astrophysically interesting solution to $R_{\mu\nu} = 0$. The metric appropriate to rotating, spherical massive bodies, and the "Coulomb" solution (spherically symmetric static source) is then the slow-motion limit of this. So we will spend some time with the "Schwarzschild" metric, understanding what it has to say above and beyond Newtonian gravity.

Problem 6.3

(a) Show that:

$$\Gamma^{\mu}_{\ \mu\alpha} = \frac{1}{\sqrt{-g}}\, \partial_\alpha \left(\sqrt{-g}\right). \tag{6.162}$$

(b) Recall from Section 5.8.1 that the stress-tensor *density* for a free particle can be written as:

$$T^{\mu\nu} = m \int \dot{x}^\mu\, \dot{x}^\nu\, \delta^4(x^\alpha - x^\alpha(s))\, ds, \tag{6.163}$$

where s is the proper time of the particle. This is nothing but the usual $\rho\, \dot{x}^\mu\, \dot{x}^\nu$ for a degenerate mass distribution (the δ^4 ensures that there is contribution only along the world-line of the particle). Form the stress tensor by appropriate introduction of $\sqrt{-g}$. As a stress tensor, we expect $T^{\mu\nu}_{\ \ ;\mu} = 0$, show that this statement of conservation directly implies the geodesic equation (use the result from part a to simplify) – it is easiest to establish the geodesic equation in the form:

$$\ddot{x}^\rho + \Gamma^{\rho}_{\ \alpha\beta}\, \dot{x}^\alpha\, \dot{x}^\beta = 0. \tag{6.164}$$

Problem 6.4

Consider the single, massless Klein–Gordon field ϕ coupled to E&M via the Lagrangian (working in Cartesian spatial coordinates):

$$\mathcal{L} = F^{\mu\nu}\left(A_{\nu,\mu} - A_{\mu,\nu}\right) - \frac{1}{2} F^{\alpha\beta}\, F_{\alpha\beta} + \frac{1}{2}\phi_{,\mu}\, g^{\mu\nu}\, \phi_{,\nu} + \alpha\, \phi_{,\mu}\, g^{\mu\nu}\, A_\nu. \tag{6.165}$$

Find the field equations by varying with respect to $F^{\mu\nu}$, A_μ, and ϕ separately. Show that the sourced $F^{\alpha\beta}_{\ \ ,\beta}$ equation has the self-consistent property that $F^{\alpha\beta}_{\ \ ,\beta\alpha} = 0$ provided $\partial^\mu A_\mu = 0$ (so we are forced, as with the Proca equation, to "add" $\partial^\mu A_\mu = 0$ to our field equations).

Problem 6.5

Using your field equations from the previous problem, find the constraints you must place on p_μ, q_μ, and P_μ in the plane-wave ansatz:

$$\phi = \Phi_0 \, e^{i \, p_\mu x^\mu} \qquad A_\mu = P_\mu \, e^{i \, q_\gamma x^\gamma} \tag{6.166}$$

so that ϕ and A_μ are valid solutions.

Problem 6.6

Suppose we were presented with an action that had a disguised total derivative – working in flat space-time with Cartesian coordinates, we're given:

$$\mathcal{L} = \frac{1}{2} A_{\mu,\nu} \, g^{\mu\alpha} \, g^{\nu\beta} \, A_{\alpha,\beta} + \left(A_{\mu,\nu} \, g^{\mu\nu} \right)^2 + A_\beta \, A_{\alpha,\nu\mu} \, g^{\alpha\nu} \, g^{\beta\mu}. \tag{6.167}$$

Aside from direct determination of the divergence that leads to the second two terms, establish through variation that this Lagrangian gives the same field equations as:

$$\mathcal{L} = \frac{1}{2} A_{\mu,\nu} \, g^{\mu\alpha} \, g^{\nu\beta} \, A_{\alpha,\beta}. \tag{6.168}$$

When in doubt, then, form the field equations themselves and see what you end up with.

*Problem 6.7

In three dimensions (two spatial, one temporal), there is an analogue of the trivial term (in four dimensions): $\epsilon^{\alpha\beta\gamma\delta} A_{\beta,\alpha} A_{\delta,\gamma}$ appearing in the E&M Lagrangian that is *not* a total divergence, and hence can change the field equations of E&M (such a form is called a "Chern–Simon" term).

(a) Find this term in $D = 3$, and by adding it to the usual E&M Lagrangian, find the field equations. You should work in Cartesian coordinates (in Minkowski space-time) to simplify life.

(b) You have the source-free theory in the above, suppose we put a static (time-independent) point source at the origin of the spatial coordinates ($\rho = q \, \delta^2(\mathbf{x})$) – this gives a source for the right-hand side that looks like: $j^0 = q \, \delta^2(\mathbf{x})$ with $j^x = j^y = 0$. Solve this for the zero component of the vector A^μ (the "potential" in this theory – you may assume it is time-independent).

Problem 6.8

We have motivated the form of the E&M action by constraining the resulting field equations. In light of our approach for the second-rank tensor case, we'll generate the vector action from the field equations themselves. For a vector A_μ, the only second derivative, linear terms we can form are (ignoring the mass term, as usual) $\partial_\mu \partial^\nu A_\nu$ and $\partial_\nu \partial^\nu A_\mu$. So the most general field equation is:

$$\alpha \, \partial_\mu \partial^\nu A_\nu + \beta \, \partial_\nu \partial^\nu A_\mu = 0. \tag{6.169}$$

(a) Enforce the requirement that the field equations are free of scalars – that is, take $A_\mu \longrightarrow A_\mu + \psi_{,\mu}$, and find the relation between α and β that kills any reference to ψ in the field equations.

(b) Define a momentum tensor $F^{\mu\nu}$ such that your field equations read: $\partial_\mu F^{\mu\nu} = 0$.

(c) Start with a Lagrangian that is of the form $\mathcal{L} = F^{\mu\nu} A_{\mu,\nu} - \hat{\mathcal{L}}$ as in (6.119). Construct $\hat{\mathcal{L}}$ so that your momentum definition from part b is enforced as a field equation (obtained by $F^{\mu\nu}$ variation). The final Lagrangian should be precisely the one for E&M (up to an overall constant) in first-order form.

Problem 6.9

In D dimensions, we want to construct the trace-reversed tensor $H_{\mu\nu} = h_{\mu\nu} - \alpha g_{\mu\nu} h$ for $h_{\mu\nu}$. What must α be if we demand that the inversion returns $h_{\mu\nu} = H_{\mu\nu} - \alpha g_{\mu\nu} H$? Check that you recover $h_{\mu\nu} = H_{\mu\nu} - \frac{1}{2} g_{\mu\nu} H$ when $D = 4$.

Problem 6.10

In our work varying Lagrangians with respect to $H_{\mu\nu}$, we have used the symmetry of $H_{\mu\nu}$ implicitly. In this problem, we will establish that this approach was justified.

(a) Consider a piece of a Lagrangian governing a symmetric field $S^{\mu\nu} = S^{\nu\mu}$, in Cartesian coordinates:

$$\mathcal{L} = S^{\mu\nu} B_{\mu\nu}, \tag{6.170}$$

where $B_{\mu\nu}$ is some arbitrary tensor field. Compute $\frac{\partial \mathcal{L}}{\partial S^{\mu\nu}}$ for this action, ignoring the symmetry of $S^{\mu\nu}$. This approach is incorrect because $S^{\mu\nu} = S^{\nu\mu}$, so that $\frac{\partial \mathcal{L}}{\partial S^{12}} = \frac{\partial \mathcal{L}}{\partial S^{21}}$, and this has implications for the part of $B^{\mu\nu}$ that will actually appear in the field equations.

(b) Since $S^{\mu\nu}$ is symmetric, we can write the Lagrangian as:

$$\mathcal{L} = S^{\mu\nu} B_{\mu\nu} = \frac{1}{2} S^{\mu\nu} \left(B_{\mu\nu} + B_{\nu\mu} \right). \tag{6.171}$$

Compute $\frac{\partial \mathcal{L}}{\partial S^{\mu\nu}}$ in this case, again ignoring the symmetry of $S^{\mu\nu}$. Note that this second approach is the correct one, we have used the symmetry of $S^{\mu\nu}$ to take just the symmetric part of $B_{\mu\nu}$, now we can take $\frac{\partial \mathcal{L}}{\partial S^{\mu\nu}}$ while ignoring its symmetry.

(c) Finally, the shortcut we have been using (to obtain, for example, (6.134) from the variation of (6.132)). Compute $\frac{\partial \mathcal{L}}{\partial S^{\mu\nu}}$ using (6.170), then symmetrize the resulting field equation in $\mu \leftrightarrow \nu$, giving a result identical to that of part b.

Problem 6.11

The combination of connections we get from variation of (6.132) with respect to $H^{\mu\nu}$:

$$\Gamma^\alpha{}_{\mu\nu,\alpha} - \Gamma^\alpha{}_{\mu\alpha,\nu} = 0 \tag{6.172}$$

must be symmetric in $\mu \leftrightarrow \nu$ interchange – the first term clearly is, by the symmetry of the momenta. Show that $\Gamma^\alpha{}_{\mu\alpha,\nu}$ is also symmetric using its connection to $h_{\mu\nu}$ from (6.135).

*Problem 6.12

We use the trace-reversed field $H^{\mu\nu}$ to simplify some of the manipulations involved in going from the field equations for $h_{\mu\nu}$ to the appropriate first-order action. So, for example, in (6.132), we vary with respect to $H^{\mu\nu}$ rather than $h^{\mu\nu}$, yet our result is that the field $h^{\mu\nu}$ is the one that is involved in the connection. To make it clear that $H^{\mu\nu}$ is a valid place-holder, try starting from (6.132) written in terms of $H^{\mu\nu} = h^{\mu\nu} - \frac{1}{2} g^{\mu\nu} h$:

$$\bar{\mathcal{L}} = \left(h^{\mu\nu} - \frac{1}{2} g^{\mu\nu} h \right) \left(\Gamma^\alpha_{\mu\nu,\alpha} - \Gamma^\alpha_{\mu\alpha,\nu} \right) + g^{\mu\nu} \left(\Gamma^\alpha_{\mu\nu} \Gamma^\gamma_{\alpha\gamma} - \Gamma^\alpha_{\beta\mu} \Gamma^\beta_{\alpha\nu} \right). \quad (6.173)$$

Find the field equation obtained by varying the action with respect to $h^{\mu\nu}$. Note that you should end up with precisely (6.133) (to bring your answer into this form, you may need to take a trace of the field equation and use the resulting information to simplify).

Problem 6.13

This problem will set up the linearized Riemann and Ricci tensors, so assume that we have a metric, with connection:

$$\Gamma_{\alpha\beta\gamma} = \frac{1}{2} \left(g_{\alpha\beta,\gamma} + g_{\alpha\gamma,\beta} - g_{\beta\gamma,\alpha} \right), \quad (6.174)$$

and that the Riemann and Ricci tensors are defined as in Chapter 4:

$$R^\alpha{}_{\rho\gamma\beta} = \Gamma^\alpha{}_{\gamma\sigma} \Gamma^\sigma{}_{\beta\rho} - \Gamma^\alpha{}_{\beta\sigma} \Gamma^\sigma{}_{\gamma\rho} + \Gamma^\alpha{}_{\beta\rho,\gamma} - \Gamma^\alpha{}_{\gamma\rho,\beta}, \quad (6.175)$$

and $R_{\rho\beta} = R^\gamma{}_{\rho\gamma\beta}$.

(a) Suppose the metric can be written as $g_{\mu\nu} = \eta_{\mu\nu} + \epsilon h_{\mu\nu}$, where $\eta_{\mu\nu}$ is the Minkowski metric (in Cartesian spatial coordinates), and $h_{\mu\nu}$ represents an additional piece, with $\epsilon \ll 1$. If $h_{\mu\nu}$ is diagonal, show that:

$$g^{\mu\nu} = \eta^{\mu\nu} - \epsilon h_{\mu\nu} + O(\epsilon^2), \quad (6.176)$$

where $g^{\mu\nu} = (g_{\mu\nu})^{-1}$ (i.e. the contravariant metric is defined to be the matrix inverse of the covariant form). This equation holds numerically for each element (as a tensor statement, it is meaningless), so that, for example: $g^{00} = -1 - \epsilon h_{00} + O(\epsilon^2)$. The result (6.176) is generally true, but easiest to establish for diagonal $h_{\mu\nu}$.

(b) Show that the connection $\Gamma^\alpha{}_{\beta\gamma} = \epsilon \, [\ldots]$, so that it is of order ϵ.

(c) Isolate the terms in the Riemann tensor that are of order ϵ, write them in terms of $h_{\mu\nu}$ and its derivatives. Finally, construct the "linearized" Ricci tensor from $R_{\rho\beta} = R^\gamma{}_{\rho\gamma\beta}$ (again, keep only terms of order ϵ) – you should end up with (6.137) (note similarities with the normal coordinate form from Section 4.5).

*Problem 6.14

Let's fill in the blanks taking us from (6.151) to (6.153). Start by showing that:

(a) $g_{,\alpha} = g\, g^{\rho\sigma}\, g_{\rho\sigma,\alpha}$ using $\frac{\partial g}{\partial g_{\mu\nu}} = g^{\mu\nu}\, g$, which gives us the derivative of $\mathfrak{g}^{\mu\nu}{}_{,\alpha}$ in (6.152).

(b) By rewriting partial derivatives in terms of covariant ones, taking traces, and generally simplifying, show that the content of (6.151) is indeed $g_{\mu\nu;\alpha} = 0$.

7

Schwarzschild space-time

We begin our understanding of the Einstein equations by developing the spherically symmetric vacuum solution, the Schwarzschild metric. This will involve, from a PDE point of view, constants of integration that we must be able to interpret physically. One way to make the physical associations necessary is to position ourselves sufficiently far away from the source that gravitational effects take the form of familiar effective forces, namely the Newtonian force outside of a spherically symmetric distribution of mass.

Weyl's method of varying an action after an ansatz incorporating target symmetries is introduced; this can be a useful tool for finding solutions with specific symmetries, and we will discuss the method initially in the context of spherically symmetric solutions to vacuum electrodynamics. Our first general relativistic job is to actually generate the solution to Einstein's equation in vacuum (i.e. away from source) given a spherically symmetric central body of mass M. We will use Weyl's method to accomplish this. That exploitation of symmetry will leave us with a single (physically relevant) undetermined constant of integration. Using the far-away (linearized) approximation to Einstein's equation, we will see that this constant can naturally be associated with the mass of the central body.

To develop the linearized theory, we start with:

$$R_{\mu\nu} - \frac{1}{2} g_{\mu\nu} R = 8\pi T_{\mu\nu}, \tag{7.1}$$

and introduce a metric perturbation $h_{\mu\nu}$ (small) that sits on top of a Minkowski background $\eta_{\mu\nu}$, so that: $g_{\mu\nu} = \eta_{\mu\nu} + h_{\mu\nu}$. As a result of keeping only first-order terms in the perturbation $h_{\mu\nu}$, we raise and lower indices using $\eta_{\mu\nu}$ and form the linearized version of the Einstein tensor to insert in the above. From a perturbative point of view, this is all operational, keeping track of the order of the perturbation amounts to an interpretation of $h_{\mu\nu}$ as a second-rank symmetric tensor field on a flat background. But we have seen that this interpretation is natural apart from the perturbative approach. The same equation results from taking a general source-free second-rank tensor field on a flat background, as in Section 6.6.1. In addition, we understand the interpretation (metric perturbation) naturally from this point of view, and we also know roughly where problems will occur – we need to go beyond this linear form when we demand self-consistency.

After pinning down the lone constant of integration obtained from Weyl's method using the linearized correspondence with Newtonian gravity, we will develop the usual physical predictions beyond Newton that Einstein laid out in his original paper [11].

Problem 7.1

This problem begins to develop the GR analogue of the "Coulomb" solution from E&M (general spherically symmetric solution to Maxwell's equations).

(a) Construct the most general spherically symmetric *vector* field **A** (in what direction must it point? What variable can its magnitude depend on? See Problem 5.8). Again, find the solution(s) to $\nabla^2\mathbf{A} = 0$ in this setting.

(b) Now we want to construct the most general spherically symmetric second-rank tensor, g_{ij}. This will end up being the spatial portion of the metric ansatz we will use to develop the so-called Schwarzschild solution. As a second-rank symmetric tensor, your ansatz must have $g_{ij} = g_{ji}$, and as a portion of a spatial metric, it *must* be invertible. Hint: start by requiring that the direction of g_{1j} be x_j, then demand that g_{i1} also have x_i direction – the resulting metric will not be invertible (via Problem 1.14), and you must use δ_{ij} to generate an invertible covariant tensor.

(c) Take your expression for g_{ij} and form the line-element in "spherical coordinates" for your spherically symmetric space.

Problem 7.2

Continuing our work on the spherical metric ansatz.

(a) You determined that, for a spatial tensor interpreted as a metric, spherical symmetry means $ds^2 = A(r)\,dr^2 + B(r)r^2\,(d\theta^2 + \sin^2\theta\,d\phi^2)$. Show that this can, with appropriate redefinition of r, be reduced to $ds^2 = \tilde{A}(\bar{r})\,d\bar{r}^2 + \bar{r}^2\,(d\theta^2 + \sin^2\theta\,d\phi^2)$ for a function $\tilde{A}(\bar{r})$ of the new radial coordinate $\bar{r}$. One can "finish" the job of specifying spherical symmetry in $D = 3 + 1$ by introducing a function $F(r)$ for g_{00}.

(b) Show that in vacuum (where $T^{\mu\nu} = 0$), Einstein's equation reduces to $R_{\mu\nu} = 0$.

(c) Using Mathematica (or by hand), compute and solve $R_{\mu\nu} = 0$ for your ansatz above. This is the Schwarzschild solution of Einstein's equations, appropriate to space-time outside of a spherically symmetric source.

(d) Is your solution flat?

(e) Show explicitly that $g_{\mu\nu;\alpha} = 0$ for your metric.

7.1 The Schwarzschild solution

Unlike Maxwell's equations, which in the static limit can be developed directly from the observation of point charges and simple current distributions, Einstein's equation is easier to develop and understand in a more abstract setting. Historically, the equation itself predates its first exact solution, so it is not surprising that we are only now getting to the details of solving it for the simplest possible source geometry. The Schwarzschild solution is the GR analogue of the Coulomb field – it is the field (metric) *outside* of a spherically symmetric mass distribution.

Spherically symmetric charge distributions are among the first encountered in E&M, and this is motivated in part by their utility. Since Maxwell's equations are linear, solutions can be superimposed (an observable phenomenon), and any static distribution can be made up out of point sources. This is not the case in GR, so while the Schwarzschild solution is a great place to study the physical implications of the theory, it does not directly inform our ideas about more general distributions.

Our derivation of the Schwarzschild solution will employ "Weyl's method", a procedure for inputting symmetry in an action prior to variation. We will work directly from the Einstein–Hilbert form of the action, in which we take:

$$S_{EH} = \int d\tau \sqrt{-g}\, R(g_{\mu\nu}) \tag{7.2}$$

and vary w.r.t. the components of the metric. We are, of course, assuming that the metric is related to the connection in the usual way (that's the information we must input in the Einstein–Hilbert form, so we assume the usual geometric interpretation of the field $g_{\mu\nu}$). In order to solve for the field, we need an ansatz – a spherically symmetric metric with undetermined entries, and we need to be able to form the Ricci scalar. Once we have those two elements, variation will provide the explicit form of Einstein's equation in vacuum $R_{\mu\nu} = 0$ appropriate to our ansatz.

We'll start by applying the Weyl method to the second-order electromagnetic action (in flat space-time, now):

$$S_{EM} = \int d\tau \sqrt{-g}\, F^{\mu\nu}(A_\alpha)\, F_{\mu\nu}(A_\alpha) \tag{7.3}$$

to better understand the assumptions and simplifications that symmetry provides. Notice that we are, appropriately, using the "second-order" form for the electromagnetic action – here we need to introduce the potentials by hand, using $F_{\mu\nu} = A_{\nu,\mu} - A_{\mu,\nu}$, just as we needed to input $\Gamma(g)$ in the Einstein–Hilbert action.

7.1.1 The Weyl method and E&M

The Weyl method requires a symmetry assumption. We will take a vector potential A^μ with the usual interpretation: $A^0 = V$ (in units where $c = 1$), the electric potential, while the spatial components A^i form the magnetic vector potential. The physical setup consists of a point charge sitting at the origin for all time. From this, we will make the following assumptions: $V(x, y, z) = V(r)$, $\mathbf{A} = A(r)\hat{\mathbf{r}}$, and implicit in these two, the solution is time-independent. The spatial assumptions stem from the source symmetry, and are inherited directly from the implicit boundary conditions (continuity) associated with the point source (or spherically symmetric extended body, for that matter). We know that a spherically symmetric scalar field depends only on the distance to the origin, hence $V(r)$, and it is plausible that a spherically symmetric vector field should point radially away from the source everywhere, and depend on distance for its magnitude.

Thinking ahead, it is clear that the magnetic field associated with $\mathbf{A}$ is automatically zero, since $\nabla \times (A(r)\hat{\mathbf{r}}) = 0$. Then we can make a clever gauge choice – here is a spherically symmetric vector potential that gives rise to the same magnetic field, $\mathbf{A} = 0$. So our assumption of spherical symmetry has already whittled down the ingredients – evidently, all we need is $V(r)$.

Turning now to the action itself, we know that the scalar in the integrand is $F^{\mu\nu} F_{\mu\nu} \sim E^2 - B^2$. For the volume factor, if we are using spherical coordinates (in Minkowski space-time): $\sqrt{-g} = r^2 \sin\theta$. Our action is:

$$S_{EM} = \int \left(E^2 - B^2\right) r^2 \sin\theta \, dr d\theta d\phi dt. \tag{7.4}$$

Remember that since we are using the second-order form, we need to insert the definition of $\mathbf{E}$ and $\mathbf{B}$ in terms of the potentials V and $\mathbf{A}$ by hand, the variation does not do this for us (as it does in first-order form). So we use the usual $\mathbf{E} = -\nabla V - \frac{\partial \mathbf{A}}{\partial t}$ and $\mathbf{B} = \nabla \times \mathbf{A}$ with our minimal set. Putting this into S_{EM} and performing the angular integrals, we are left with:

$$S_{EM} = 4\pi \int \left(\frac{dV}{dr}\right)^2 r^2 \, dr dt. \tag{7.5}$$

The variation is now straightforward. Call the integrand $\mathcal{L}$, then the Euler–Lagrange equations are:

$$\frac{d}{dr}\left(\frac{\partial \mathcal{L}}{\partial V'}\right) = 0 \tag{7.6}$$

since there is no V dependence, only $V' \equiv \frac{dV}{dr}$ contributes. The field equation is just:

$$\frac{d}{dr}\left(2r^2\,V'\right) = 0 \longrightarrow V' = \frac{\alpha}{r^2} \tag{7.7}$$

for constant α, and the solution follows immediately:

$$V = -\frac{\alpha}{r}. \tag{7.8}$$

We have dropped the constant solution on the grounds that it will not vanish at spatial infinity.

Well, that's a fancy way to "derive" the Coulomb potential from an action. It's a good example, though, since the exact same logic and simplification occur when we carry out the same program for the Einstein–Hilbert action. In particular, notice that nothing in the process itself tells us how to interpret or set the constant of integration α. In order to find the physical content of α, we need additional information. That will also be true for the Schwarzschild solution obtained using the Weyl method – we will have a solution with an undetermined constant of integration that we must interpret. The interpretation there will come from the linearized form of the solution and its interaction with test particles.

7.1.2 The Weyl method and GR

Spherically symmetric ansatz

To use the Weyl method for the action $S = \int d\tau \sqrt{-g}\,R$, we need to first make a spherically symmetric ansatz for the metric. Think of the spatial portion of the metric first, the g_{ij} components form a three-by-three sub-matrix. In spherical coordinates, we could write the Minkowski metric as in Figure 7.1.

If we want a metric to be unchanged under modifications to the angular portion (our implicit definition of spherical symmetry), then we can, at most, change the angular part of η_{ij} by an overall factor that depends on r – that amounts to changing the definition of r itself, though, via a simple relabeling. We conclude that we should

$$\eta_{\mu\nu} = \begin{pmatrix} -1 & 0 & 0 & 0 \\ 0 & 1 & 0 & 0 \\ 0 & 0 & r^2 & 0 \\ 0 & 0 & 0 & r^2 \sin^2\theta \end{pmatrix}$$

$$\eta_{ij}$$

Figure 7.1   The three-dimensional spatial components of a four-dimensional space-time metric (Minkowski space-time written in spherical coordinates).

be able to write the spatial portion of our metric ansatz as:

$$g_{ij} = \begin{pmatrix} B(r) & 0 & 0 \\ 0 & r^2 & 0 \\ 0 & 0 & r^2 \sin^2\theta \end{pmatrix}. \tag{7.9}$$

For the temporal part, we can introduce another arbitrary function of r only – this makes our full metric ansatz, written in terms of the line element:

$$ds^2 = -A(r)\,dt^2 + B(r)\,dr^2 + r^2\left(d\theta^2 + \sin^2\theta\,d\phi^2\right). \tag{7.10}$$

Our metric ansatz contains two unknown functions. For reasons that will be clear when we calculate the Ricci scalar, it is beneficial to switch our two functions $\{A(r), B(r)\}$ for the combination $\{a(r), b(r)\},$[1] for which (7.10) reads:

$$\boxed{ds^2 = -a(r)\,b(r)^2\,dt^2 + a(r)^{-1}\,dr^2 + r^2\left(d\theta^2 + \sin^2\theta\,d\phi^2\right),} \tag{7.11}$$

and this is our final ansatz.

The Ricci scalar and action

From the metric implied by the line element (7.11), the determinant is just the product of the diagonal elements, so that:

$$\sqrt{-g} = r^2\,b(r)\,\sin\theta. \tag{7.12}$$

Combining this with the Ricci scalar (computed via the `Mathematica` package), we have the full action:

$$S_{EH} = \int d\tau \sqrt{-g}\,R$$

$$= \int \left(2b - 2ab - 4rba' - 4rab' - 3r^2a'b' - r^2ba'' - 2r^2ab''\right)$$

$$\times\,dr d\theta d\phi dt. \tag{7.13}$$

---

[1] This is motivated by our desire for a tractable form for $\sqrt{-g}$.

The angular and temporal integrals can be performed to give some overall constant that doesn't matter. In the end, what we have is a one-dimensional variation problem (with "parameter" r) for an effective Lagrangian that looks like:

$$\mathcal{L} = \left(2b - 2ab - 4rba' - 4rab' - 3r^2a'b' - r^2ba'' - 2r^2ab''\right). \quad (7.14)$$

The Euler–Lagrange equations are, as always:

$$\frac{d^2}{dr^2}\left(\frac{\partial \mathcal{L}}{\partial a''}\right) - \frac{d}{dr}\left(\frac{\partial \mathcal{L}}{\partial a'}\right) + \frac{\partial \mathcal{L}}{\partial a} = 0, \quad (7.15)$$

and we have another copy of these for $a \longrightarrow b$. So we get two field equations, one for $a(r)$, one for $b(r)$. Taking the relevant derivatives, these read:

$$0 = 2rb'$$
$$0 = -2\left(-1 + a + ra'\right) = -2\left(-1 + (ar)'\right). \quad (7.16)$$

These equations are simple, which was the motivation for using our $\{a(r), b(r)\}$ set instead of $\{A(r), B(r)\}$ – the final solutions read:

$$\boxed{b(r) = b_0 \quad a(r) = 1 + \frac{a_0}{r}.} \quad (7.17)$$

The solution, then, has line element given by:

$$ds^2 = -\left(1 + \frac{a_0}{r}\right)b_0^2\,dt^2 + \left(1 + \frac{a_0}{r}\right)^{-1}dr^2 + r^2\left(d\theta^2 + \sin^2\theta\,d\phi^2\right). \quad (7.18)$$

We see that the constant b_0 can be eliminated just by rescaling the temporal coordinate. The form of the Schwarzschild metric, with its single constant of integration, is:

$$g_{\mu\nu} \doteq \begin{pmatrix} -\left(1 + \frac{a_0}{r}\right) & 0 & 0 & 0 \\ 0 & \left(1 + \frac{a_0}{r}\right)^{-1} & 0 & 0 \\ 0 & 0 & r^2 & 0 \\ 0 & 0 & 0 & r^2\sin^2\theta \end{pmatrix}. \quad (7.19)$$

That's it, the metric appropriate to the space-time *outside* a static, spherically symmetric massive body. It applies, in approximate form, to all planets (the earth, for example) and to other slowly rotating spherical objects in the universe. This is an exact solution to Einstein's equation in vacuum: $R_{\mu\nu} = 0$, which we obtained by simplified construction. At this point, it is a good idea to *check* that this metric is indeed a solution to $R_{\mu\nu} = 0$.

The only problem is the undetermined constant a_0. In order to give it some physical meaning, we will linearize the metric, viewing the above as a perturbation to a flat background, and demand that we recover an effective Newtonian potential that interacts in the usual way with test particles. This is the additional information we need to understand before the physical interpretation of a_0 is available.

Problem 7.3

What is the metric for a three-dimensional sphere of radius r (add the angle ψ to θ and ϕ in order to define the coordinates)? Assume that spherical symmetry for $D = 4 + 1$ means that we have separate radial functions scaling dt^2 and dr^2 in the line element, and an unmodified angular portion associated with a sphere in three spatial dimensions. Find the analogue of the Schwarzschild solution in $D = 4 + 1$ from Einstein's equation in vacuum.

Problem 7.4

There is another natural "scalar" we could add to the Einstein action, one that is not R, and not quadratic (or higher) in R. Add this term and find the spherically symmetric solution to this modified theory of gravity (in $D = 3 + 1$). Write your solution so that it reduces to Schwarzschild in the absence of the additional term.

Problem 7.5

We can find axially symmetric vacuum solutions for GR. Think of the line element for Minkowski space-time written in cylindrical coordinates:

$$d\tau^2 = -dt^2 + ds^2 + dz^2 + s^2\,d\phi^2. \tag{7.20}$$

Axial symmetry means (among other things) that we'll generalize this line element using functions only of s and z (no ϕ-dependence, and for now, no temporal-dependence either). We know that any two-dimensional subspace can be written in conformally flat form (see Problem 4.7), so it is reasonable to take an unknown function $e^{2b(s,z)}$ (the exponential is just for computational simplicity) as a factor in front of $ds^2 + dz^2$. We'll introduce an additional function $a(s,z)$ to provide the rest of the axial freedom.

(a) Take as your starting point:

$$d\tau^2 = -e^{2a(s,z)}\,dt^2 + e^{-2a(s,z)}\left[e^{2b(s,z)}\left(ds^2 + dz^2\right) + s^2\,d\phi^2\right]. \tag{7.21}$$

From the metric implied by this line element, construct $R_{\mu\nu}$, the Ricci tensor (use the `Mathematica` package). Vacuum solutions to Einstein's equation have $R_{\mu\nu} = 0$. Show that the content of these *a priori* 10 equations is that any a satisfying $\nabla^2 a = 0$ with:

$$\nabla^2 a = \frac{\partial^2 a}{\partial z^2} + \frac{1}{s}\frac{\partial a}{\partial s} + \frac{\partial^2 a}{\partial s^2} \tag{7.22}$$

(the flat-space Laplacian, in other words) is a solution, and b can be found once a is specified. This (infinite) family of metrics are called "Weyl metrics".

(b) Using the monopole solution $a = \frac{\alpha}{r}$ with $r = \sqrt{s^2 + z^2}$, find $b(s, z)$ and hence, the metric for this choice.

Problem 7.6

Using the Weyl method applied to E&M, it is possible to establish that the spherically symmetric solution *must* be time-independent, in addition to finding its functional form. The same idea holds in general relativity, where the lack of time-dependence for spherically symmetric vacuum solutions is known as "Birkhoff's theorem".

(a) Start with $V = V(r, t)$ and $\mathbf{A} = A(r, t)\hat{\mathbf{r}}$, construct $\mathbf{E} = -\nabla V - \frac{\partial \mathbf{A}}{\partial t}$ and $\mathbf{B} = \nabla \times \mathbf{A}$, then form the action in (7.4).

(b) Vary your action with respect to both $V(r, t)$ and $A(r, t)$, then set $A(r, t) = 0$ (we are enforcing our gauge choice *after variation*) and show that you learn both the functional dependence of $V(r, t)$ on r, and that $\frac{\partial V(r,t)}{\partial t} = 0$, hence the vacuum solution is time-independent.

Notice that by fixing the gauge before varying, as we did in Section 7.1.1, we lost this additional information (having assumed it from the start).

Problem 7.7

The Born–Infeld nonlinear modification of E&M was originally developed to put a finite cut-off on the energy associated with a point charge. The form of the action for this theory is:

$$\mathcal{L}_{BI} = \alpha \sqrt{-g} \left(-1 + \sqrt{1 - \frac{E^2 - B^2}{b^2}} \right), \tag{7.23}$$

where we recover the usual $E^2 - B^2$ Lagrangian when $b \longrightarrow \infty$. We'll work out the spherically symmetric vacuum solution for the field equations that come from the Lagrangian.

(a) Assuming spherical symmetry, so that $\mathbf{E} = V'(r)\hat{\mathbf{r}}$ and setting $\mathbf{B} = 0$, write the Born–Infeld action in terms of $V'(r)$ by itself (use spherical coordinates when calculating $\sqrt{-g}$) – show that for $E^2/b^2 \ll 1$, you recover the appropriate E&M Lagrangian by appropriate choice of α.

The resulting Lagrangian is reminiscent of the action for relativistic particle mechanics – the idea is that just as the particle action reduces to the classical one in the small-speed limit, and enforces a finite maximum speed, the Born–Infeld Lagrangian should yield a theory that looks like Coulomb for E^2/b^2 small, and puts a finite cut-off on E^2.

(b) Use the Euler–Lagrange field equations to find the ODE governing $V(r)$ – replace $V'(r)$ with $E(r)$ and solve for $E(r)$. Check that you recover Coulomb as $r \longrightarrow \infty$ (so that the Born–Infeld solution looks like Coulomb far away from the source). From the usual $\int E^2 \, d\tau$ integration over all space, verify that your solution has finite energy.

7.2 Linearized gravity

We have seen, in disguised form, the equations of linearized gravity in (6.108). Now we will pick a gauge (coordinate choice) for our linearized field equations. As with E&M, this gives us a way to discuss the physically relevant portion of the solutions as opposed to leaving an infinite family. With the gauge choice in place, we will obtain, by comparison with the Newtonian result, an interpretation for the constant a_0 in (7.19).

In addition, we will learn that there are two separate sources for perturbations to Minkowski – the usual Newtonian scalar potential, and a new vector potential associated with moving mass (or more generally, moving energy density). By considering the (linearized) geodesic equations, we can associate these naturally with the gravitational analogues of the scalar and magnetic vector potential from E&M as was foreshadowed by the purely special relativistic considerations of Section 2.6.

7.2.1 Return to linearized field equations

The linearized field equations can be obtained by writing the Einstein tensor $G_{\mu\nu}$ in terms of a metric perturbation $h_{\mu\nu}$, where the full metric is $g_{\mu\nu} = \eta_{\mu\nu} + h_{\mu\nu}$ for $\eta_{\mu\nu}$ Minkowski. Alternatively, we can simply take the field equations we got when we considered the most general action for a second-rank, symmetric field theory. This latter point of view enforces the idea that the linear field equations are really meant to be interpreted as field equations on an explicitly Minkowski background (meaning that we will use $\eta_{\mu\nu}$ to raise and lower indices). In trace-reversed form,[2] the linearized Einstein equation reads:[3]

$$\tilde{G}_{\mu\nu} = \partial^\rho \, \partial_{(\mu} H_{\nu)\rho} - \frac{1}{2} \partial_\rho \partial^\rho \, H_{\mu\nu} - \frac{1}{2} \eta_{\mu\nu} \, \partial^\alpha \partial^\beta \, H_{\alpha\beta} = 8\pi \, T_{\mu\nu}, \qquad (7.24)$$

with $H_{\mu\nu} = h_{\mu\nu} - \frac{1}{2} \eta_{\mu\nu} \, h$, and $h_{\mu\nu}$ the metric perturbation in $g_{\mu\nu} = \eta_{\mu\nu} + h_{\mu\nu}$. Compare (7.24) with (6.109) to see that the content is the same. We are going to use our gauge-invariance, $h_{\mu\nu} \longrightarrow h'_{\mu\nu} = h_{\mu\nu} + f_{(\mu,\nu)}$ (analogous to $A_\mu \longrightarrow A_\mu + f_{,\mu}$ in E&M) to get rid of the divergence terms above. We want $\partial^\rho H'_{\nu\rho} = 0$ (that's right, the gauge condition acts on the metric perturbation, but our target is a

---

[2] There is no particular need for the move to trace-reversed perturbation, it is only a matter of convenience inspired by our work on second-rank tensor field theory.

[3] Recall the "symmetrization" notation: $f_{(\mu,\nu)} = \frac{1}{2} \left(f_{\mu,\nu} + f_{\nu,\mu} \right)$.

divergence-less $H'_{\nu\rho}$):

$$\partial^\rho H'_{\nu\rho} = \partial^\rho \left(h_{\nu\rho} - \frac{1}{2}\eta_{\nu\rho} h + \partial_{(\nu} f_{\rho)} - \frac{1}{2}\eta_{\nu\rho}\partial^\alpha f_\alpha \right)$$

$$= \partial^\rho H_{\nu\rho} + \frac{1}{2}\partial^\rho\partial_\nu f_\rho + \frac{1}{2}\partial^\rho\partial_\rho f_\nu - \frac{1}{2}\partial_\nu\partial^\alpha f_\alpha \qquad (7.25)$$

$$= \partial^\rho H_{\nu\rho} + \frac{1}{2}\partial^\rho\partial_\rho f_\nu.$$

To set this equal to zero, we need to solve Poisson's equation for f_ν: $\partial^\rho\partial_\rho f_\nu = -2\,\partial^\rho H_{\nu\rho}$. Suppose we do that (as with, for example, Lorentz gauge in E&M), then the (really final) linearized Einstein tensor reads:

$$\boxed{\tilde{G}'_{\mu\nu} = -\frac{1}{2}\partial_\rho\partial^\rho H'_{\mu\nu},} \qquad (7.26)$$

where the primes remind us that we have chosen a gauge and transformed already. This is an interesting equation – it says among other things that in source-free regions, the metric perturbation $H'_{\mu\nu}$ can form waves (from now on, I drop the primes indicating the transformation, but remember the gauge condition). We will return to that later on, for now I want to focus on a matter source of some variety, so we have to think about the right-hand side of Einstein's equation in a weak limit.

What should we choose as the form for the matter generating the metric perturbation? Let's consider rigid-body sources – here we mean that the internal stresses, the T_{ij} components of the stress tensor, are zero. That's not strictly speaking possible – even for non-interacting dust at rest, we can boost to a frame in which there are diagonal components aside from T_{00}, but we are taking these to be small (they have $(v/c)^2$ factors associated with them). We have a mass density and a rigid velocity, so generically, the stress tensor takes the form:

$$T_{\mu\nu} \doteq \begin{pmatrix} \rho & -j_1 & -j_2 & -j_3 \\ -j_1 & 0 & 0 & 0 \\ -j_2 & 0 & 0 & 0 \\ -j_3 & 0 & 0 & 0 \end{pmatrix} \qquad (7.27)$$

(remember that it is $T^{\mu\nu}$ that we are used to, the negatives come from lowering using the Minkowski metric) and Einstein's equation in matrix form looks like:

$$\partial^\alpha\partial_\alpha \begin{pmatrix} H_{00} & H_{01} & H_{02} & H_{03} \\ H_{01} & H_{11} & H_{12} & H_{13} \\ H_{02} & H_{12} & H_{22} & H_{23} \\ H_{03} & H_{13} & H_{23} & H_{33} \end{pmatrix} = -16\,\pi \begin{pmatrix} \rho & -j_1 & -j_2 & -j_3 \\ -j_1 & 0 & 0 & 0 \\ -j_2 & 0 & 0 & 0 \\ -j_3 & 0 & 0 & 0 \end{pmatrix}. \qquad (7.28)$$

There is an implicit ϵ on the left ($h_{\mu\nu}$, and hence $H_{\mu\nu}$, is "small" by assumption), meaning that the left-hand side is part of an equation governing a small correction to flat space-time. In order for the above to make sense, there must be a corresponding ϵ on the right, i.e. the *sources* must be small – we agree to assume the source is small, and then the ϵ on both sides cancel. All the spatial–spatial terms on the right vanish by assumption, so the spatial–spatial components of $H_{\mu\nu}$ must satisfy:

$$\partial^\alpha \partial_\alpha H_{ij} = 0 \longrightarrow \nabla^2 H_{ij} = 0, \tag{7.29}$$

the time-derivatives of the above are small if the motion of the *source* is small (compared to the speed of light[4]), and I have discarded them. But consider the boundary conditions, as $r \longrightarrow \infty$, we want the metric perturbation to vanish so that we are left with normal Minkowski space-time very far away from the source. The solution to Laplace's equation can have no minima or maxima on the interior of the domain, so we conclude that the spatial components H_{ij} are identically zero.

Moving on to the less trivial components. The time–time and time–spatial equations read:

$$
\begin{array}{|l|}
\hline
\\
\nabla^2 H_{00} = -16\pi\,\rho \longrightarrow \nabla^2 \phi = 4\pi\,\rho \text{ with } \phi \equiv -\dfrac{h_{00}}{2} \\
\\
\nabla^2 H_{0i} = 16\pi\,j_i \longrightarrow \nabla^2 \mathbf{A} = 4\pi\,\mathbf{j} \text{ with } A_i \equiv \dfrac{h_{0i}}{4} \\
\\
\hline
\end{array}
\tag{7.31}
$$

(keeping in mind that $h_{\mu\nu} = H_{\mu\nu} - \frac{1}{2}\eta_{\mu\nu} H$). These equations are (module signs) just like the relation of the electromagnetic four-potential (in Lorentz gauge) to its sources ρ and $\mathbf{j}$. Here the sources are mass distribution and "mass current" distribution. This says that in GR, a moving mass generates a gravitational field just as a moving charge generates a magnetic field in E&M, a result we predicted by special relativistic arguments in Section 2.6.

We must connect these to motion – what we want is effectively the force equation $\mathbf{F} = m\,\mathbf{a}$ but practically, what we will do is generate it from the geodesic equation of motion for a test particle. So, referring to our classical intuition and our geodesic discussions, we know that the Lagrangian for a test particle in GR is:

$$L = \frac{1}{2} g_{\mu\nu}\,\dot{x}^\mu\,\dot{x}^\nu. \tag{7.32}$$

Varying this, we get equations of motion that look like:

$$\ddot{x}^\alpha + \Gamma^\alpha{}_{\beta\gamma}\,\dot{x}^\beta\,\dot{x}^\gamma = 0, \tag{7.33}$$

---

[4] With units, we have $dx^i = v^i\,dt = \frac{v^i}{c}\,dx^0 = \epsilon^i\,dx^0$ and then:

$$\epsilon\,\frac{\partial H}{\partial x^i} \sim \frac{\partial H}{\partial x^0}. \tag{7.30}$$

and what we'll do is move the Γ term over to the right and interpret it as a force. Remember that our dots here refer to the proper time τ of the particle, but if we assume the particle is moving slowly, the proper time and coordinate time t coincide (to within our ϵ error), so we will just view $\dot{x}$ as $\frac{d}{dt}x$, etc. These velocities are themselves small (a time-derivative is $\epsilon \times$ "spatial derivatives"), so we will also drop terms quadratic in the spatial velocity. That leaves us relatively few terms to consider – we set $\alpha = i$ in order to look at the spatial components.

$$\Gamma^i_{\ \beta\gamma}\, \dot{x}^\beta\, \dot{x}^\gamma = \frac{1}{2}\, \eta^{i\mu}\left(h_{\beta\mu,\gamma} + h_{\gamma\mu,\beta} - h_{\beta\gamma,\mu}\right)\dot{x}^\beta\, \dot{x}^\gamma, \tag{7.34}$$

but because of the diagonal form of the Minkowski metric, $\eta^{i\mu}$ is equivalent to η^{ij}, purely spatial, since $\eta^{i0} = 0$ (it's even better, since η^{ij} is just δ^i_j, but let's leave it in metric form for now). Once again, because temporal derivatives go as epsilon times spatial ones (now applied to the slow motion of the *particle*), we will drop terms quadratic in the velocities from the above, and set time-derivatives of the metric to zero. With these approximations, the sum looks like:

$$\Gamma^i_{\ \beta\gamma}\, \dot{x}^\beta\, \dot{x}^\gamma = \frac{1}{2}\, \dot{x}^0\, \dot{x}^0\, \eta^{ij}\left(-h_{00,j}\right) + \dot{x}^0\, \dot{x}^k\, \eta^{ij}\left(h_{0j,k} - h_{0k,j}\right) \tag{7.35}$$

with $\dot{x}^0 = 1$ in these units. We put this into the equation of motion (7.33) to get:

$$\ddot{x}^i = -\left(-\frac{1}{2}\, \partial^i\, h_{00} + \dot{x}^k\, \eta^{ij}\left(\partial_k\, h_{0j} - \partial_j\, h_{0k}\right)\right), \tag{7.36}$$

or, in the usual (Cartesian) vector notation:

$$\ddot{\mathbf{x}} = \frac{1}{2}\, \nabla h_{00} + \mathbf{v} \times (\nabla \times \mathbf{h}_0). \tag{7.37}$$

Finally, we have from (7.31) the connection to what we think of as potentials, the equation of motion for a test particle in linearized gravity can be written:

$$\boxed{\ddot{\mathbf{x}} = -\nabla \phi + 4\,\mathbf{v} \times (\nabla \times \mathbf{A}),} \tag{7.38}$$

with the potentials determined by the source according to (7.31).

There's no real way for me to ensure that balloons fall from the ceiling at the moment you see the above equation, which is too bad. We've taken a full (second-rank) tensor theory, linearized and massaged it into a vector theory which is precisely E&M with the wrong signs (the sources on the right of (7.31) do not appear with the minus sign we get in the electromagnetic case) and a counting factor of four. Indeed, it answers a question that everyone has in E&M – when you look at how close the mass potential is to the electrostatic one, you imagine that it is possible to write an electro-magneto-static-like theory – but then shouldn't there be an analogue of **B** for moving masses? Classically, this is not the case, but we see here that GR predicts a gravitational interaction with mass "currents", called a gravitomagnetic force.

Our job, way back when, was to give meaning to the integration constant found in the Schwarzschild metric (7.19). That's a vacuum solution of Einstein's equation, it's static (no time-dependence and no off-diagonal terms) and reduces to Minkowski for large values of r. So viewing it from afar, what we see is a weak gravitational field. We know the correspondence between a metric's linearized components and the Newtonian potential associated with a spherical source from (7.31). For a spherically symmetric central body of mass M, the Newtonian gravitational potential (in our units) is $\phi(r) = -\frac{M}{r}$. Equating this form of $\phi(r)$ with $-\frac{1}{2}h_{00}$ (from (7.31)) gives:

$$\frac{1}{2}h_{00} = \frac{M}{r}. \tag{7.39}$$

For the Schwarzschild solution, $h_{00} = -\frac{a_0}{r}$ from (7.19), and we can make the final connection: $a_0 = -2\,M$. The Schwarzschild metric written with this substitution takes its usual form:

$$g_{\mu\nu} \doteq \begin{pmatrix} -\left(1 - \frac{2M}{r}\right) & 0 & 0 & 0 \\ 0 & \frac{1}{1-\frac{2M}{r}} & 0 & 0 \\ 0 & 0 & r^2 & 0 \\ 0 & 0 & 0 & r^2 \sin^2\theta \end{pmatrix}, \tag{7.40}$$

and this is what gives the standard interpretation to the metric, it is the spherically symmetric space-time generated by a point mass M.

Problem 7.8

We used gauge-invariance (arbitrary coordinate choice) to make a trace-reversed metric perturbation with vanishing four-divergence. Show explicitly that the linearized Einstein tensor in (7.24) is insensitive to:

$$h_{\mu\nu} \longrightarrow h_{\mu\nu} + f_{(\mu,\nu)}. \tag{7.41}$$

7.3 Orbital motion in Schwarzschild geometry

We have seen, through the study of the weak field solutions of Einstein's equation that in fact, we can, far away from the source, replace GR with Newtonian gravity. What was more interesting was the new "force of nature" akin to a magnetic force generated by a magnetic vector potential $\mathbf{A}$ that coupled to test particles in the familiar way $\mathbf{a} = 4\mathbf{v} \times \mathbf{B}$ with $\mathbf{B} = \nabla \times \mathbf{A}$. Of course, from a GR point of view, there is no new force of nature here, only what looks like one from the point of view of a flat background – in reality, this "mass current" coupling is nothing more than a perturbative effect of geometry. It is a correction to geodesic motion that comes from linearization of a general metric.

All we have gained, in terms of the full solution (7.19), is an interpretation for the constant a_0 in terms of mass. Now we are ready to leave the linearized discussion, and return to the new physics implied by the full metric, and its associated geodesics. So we will begin with massive orbital motion, followed by light-like "orbital motion" (where we confirm the bending of light by spherically symmetric massive bodies). We can also study the radial geodesics, the GR version of infall for both massive particles and light.

7.3.1 Newtonian recap

Remember the situation in Newtonian gravity from (1.4) – we had a Lagrangian, three-dimensional of course, and a potential. Using the Lagrange approach, we recovered an equation of motion $\rho(\phi)$ (where $\rho = 1/r$, the inverse of the radial coordinate). Alternatively, we looked at the Hamiltonian, identified constants of the motion from Killing vectors, and used those to develop the identical equation.

Either way, we end up with a second-order oscillator ODE (1.78):

$$\rho''(\phi) = -\rho(\phi) + \frac{M}{J_z^2}, \tag{7.42}$$

the solutions to which give us back elliptical orbits. We have established that far away, Schwarzschild geodesics can be viewed in terms of a Newtonian effective force, so we expect the above equation to appear as the leading order equation of motion for the geodesics of massive bodies in Schwarzschild geometry. What sort of additional physics do we expect from a perturbation to this equation? Well, almost any spherically symmetric perturbation induces a "precession" (of both the perihelion and aphelion), a motion of the point of closest (and furthest) approach to the central body – it shifts itself.

7.3.2 Massive test particles in Schwarzschild geometry

Now we begin the process of understanding the Schwarzschild solution by the introduction of test particles. We have the metric form, and we know the geodesic equation for a point particle. If we drop a particle in somewhere, what happens to it? There are two interesting options: it could "fall", or it could "orbit". Let's take the orbiting case first, since this provides the celebrated perihelion precession result that was one of the first successes of Einstein's theory. We have also built up a lot of machinery for understanding orbital solutions in Newtonian gravity, and there are nice parallels there.

We have the geodesic Lagrangian:[5]

$$L = \frac{1}{2} \dot{x}^\alpha \, g_{\alpha\beta} \, \dot{x}^\beta, \tag{7.43}$$

for a particle in a gravitational field described by the metric $g_{\alpha\beta}$ which we will now take to be Schwarzschild. There is no potential, so the Lagrangian is equal in value to the Hamiltonian. One of the first things we can do is describe, for example, all the Killing vectors of the space-time. It is interesting that the spatial Killing vectors of Schwarzschild are precisely those of Newtonian gravity (a consequence of our assumed spherical symmetry), all three components of angular momentum (defined in the usual spherical way) lead to conserved quantities, and we can again set motion in a plane by choosing $\{J_x, J_y\}$ appropriately. The axis of the orbit is again aligned with the $\theta = 0$ line, and we have one component of angular momentum left: J_z. The problem is that since we are now in four dimensions, we need one extra constant of the motion. We have it automatically, but it takes a moment to see – the issue is that while we have our Lagrangian and our Hamiltonian $H = \frac{1}{2} p_\alpha \, g^{\alpha\beta} \, p_\beta$, it is no longer the case that $H = E$, the classical "energy" of the system. H (and L) are still conserved, of course, but their value is somewhat arbitrary, reflecting more about what we mean by parametrization than anything else.

Let me be clear: all we are doing is finding the geodesics of the Schwarzschild geometry, the geodesic equation is really defined in terms of the arc-length:

$$ds^2 = dx^\alpha \, g_{\alpha\beta} \, dx^\beta, \tag{7.44}$$

and by choosing τ (proper time) as our parameter for describing the motion, $x^\alpha(\tau)$, we are just setting the unit of measurement. For "proper time", the geodesic equation is in arc-length parametrization and we write $ds^2 = d\tau^2$ (again, with $c = 1$).[6] As usual, it is time translation-invariance that sets the energy. If we look at the quantity conserved under coordinate time translation, it's just the t momentum (since t is an ignorable coordinate), so that:

$$p_0 = \frac{\partial}{\partial \dot{t}} L = -\left(1 - \frac{2M}{r}\right) \dot{t} \equiv -E. \tag{7.45}$$

Once the motion is set in the $\theta = \frac{\pi}{2}$ plane, we have the constants E, J_z and H. Time-like particles, in proper time parametrization, have $H = L = -\frac{1}{2}$.

---

[5] Note that geodesic Lagrangians do not refer to mass – that is as it should be for gravity. If you want to connect to the classical mechanical Lagrangian, imagine setting $m = 1$.

[6] The Hamiltonian then is half the four-momentum invariant: in Minkowski space-time, the relativistic relation is $p_\alpha \, p^\alpha = -E^2 + \mathbf{p} \cdot \mathbf{p} \, c^2 = -m^2 \, c^4 = -1$ (in our units), and the energy of the particle at rest is $m \, c^2 = 1$.

With four constants of the motion, we can directly parallel our work on orbits in Newtonian fields. The first step is to introduce a new radial coordinate $\rho \equiv \frac{1}{r}$ inducing the transformed metric:

$$
g_{\mu\nu} \doteq \begin{pmatrix} -(1 - 2M\rho) & 0 & 0 & 0 \\ 0 & \frac{1}{\rho^4(1 - 2M\rho)} & 0 & 0 \\ 0 & 0 & \rho^{-2} & 0 \\ 0 & 0 & 0 & \rho^{-2}\sin^2\theta \end{pmatrix}. \tag{7.46}
$$

Next we identify the J_z angular momentum – we know the answer, but it can be trivially obtained (again) by calculating the p_ϕ momentum:

$$
p_\phi = \frac{\partial L}{\partial\dot\phi} = \dot\phi\,\rho^{-2}\sin^2\theta \equiv J_z, \tag{7.47}
$$

and setting $\theta = \frac{1}{2}\pi$, $\dot\theta = 0$ corresponding to choices for J_x and J_y. With two additional constants E and $H = -\frac{1}{2}$, we can reduce the entire problem to an equation involving a single derivative of ρ:

$$
-\frac{1}{2} = L = \frac{1}{2}\left(J_z^2\,\rho^2 - \frac{E^2\,\rho^4 - \dot\rho^2}{\rho^4(1 - 2M\rho)}\right)
$$

$$
\downarrow \tag{7.48}
$$

$$
\dot\rho^2 = \rho^4\left(E^2 - (1 - 2M\rho)(1 + J_z^2\,\rho^2)\right).
$$

To get the equation (1.78), for bound orbits, we had to transform to ϕ parametrization: $\rho(t) \longrightarrow \rho(\phi)$, under which $\dot\rho = \frac{d\rho}{d\phi}\frac{d\phi}{dt} = \rho'\,J_z\,\rho^2$, then we can write our current expression for $\dot\rho^2$ in terms of ϕ derivatives:

$$
\boxed{(\rho')^2 = \left(\frac{(E^2 - 1)}{J_z^2} + \frac{2M\rho}{J_z^2} - \rho^2 + 2M\rho^3\right).} \tag{7.49}
$$

So far, our approach has been similar to the Hamiltonian discussion of our earlier work on Newtonian gravity – the problem is, rather than something like the effective Newtonian potential which is quadratic in r (and ρ, for that matter), the above is cubic. We can proceed by factoring the cubic and identifying the turning points of the radial equation, setting one equal to the perihelion, the other to the aphelion, and ignoring a spurious extra root. But our goal is to predict the perturbation in the orbits of planets due to GR effects, and the perturbation parameter will be M. This leads to a degenerate perturbation (as M goes to zero, the above cubic loses a root, so there is a discontinuity in the space of solutions even for small values of M). It turns out that it is easier to treat this situation by taking a ϕ-derivative on both sides; this gives us an equation comparable to (7.42). We lose constants of integration, and will have to put them back as appropriate, but this is a reasonable price to pay to keep the perturbation analysis simple. Differentiating gives:

$$\rho'' = -\rho + \frac{M}{J_z^2} + 3\,M\,\rho^2.$$ (7.50)

The term on the left and first two terms on the right are familiar from (7.42), but now there is a new term: $3\,M\,\rho^2$. Let me pause here before going further, (7.50) is exact, you can solve it formally with (the aptly named) elliptic integrals, or numerically. Both options are important for realistic GR calculations. What we are about to do is linearize, and hence, specialize our approach.

Take a perturbative solution $\rho = \tilde{A} + B\,\tilde{\rho}(\phi)$, and expand in powers of B – that is, we are assuming the orbit we want to describe is close to a circle $B = 0$. Then we have, defining $A \equiv \frac{M}{J_z^2}$, inserting our ρ into (7.50), and collecting in powers of B:

$$B^0 : \tilde{A} - 3\,\tilde{A}^2\,M - A = 0 \Rightarrow \tilde{A} = \frac{1 - \sqrt{1 - 12\,A\,M}}{6\,M} \Rightarrow \tilde{A} \approx A + O(A^2 M)$$

$$B^1 : \tilde{\rho}'' = -\tilde{\rho}\,(1 - 6\,\tilde{A}\,M) \Rightarrow \tilde{\rho} = \cos\left(\phi\sqrt{1 - 6\,\tilde{A}\,M}\right)$$

$$\Rightarrow \tilde{\rho} \approx \cos(\phi(1 - 3\,A\,M)),$$ (7.51)

where the approximations come in assuming that the effect of the additional term AM is small, as indeed it must be – for most planetary orbits, we see only very small corrections to Keplerian ellipses. Our approximate solution reads:

$$\rho(\phi) \approx A + B\,\cos(\phi(1 - 3\,A\,M)),$$ (7.52)

so the inverse is:

$$r(\phi) \equiv \frac{1}{\rho(\phi)} = \frac{1}{A + B\,\cos(\phi\,(1 - 3\,A\,M))}.$$ (7.53)

Precession

Before finishing the job, let us look at the sorts of trajectories we can get out of this type of equation – again, we identify A and B as defining the radial turning points, but now we have an additional parameter δ:

$$r(\phi) = \frac{1}{A + B\,\cos(\phi\,(1 - \delta))}.$$ (7.54)

A is related to the lateral extent of the orbit, and B tells us the "circularity", while δ represents a phase shift of sorts – if we start at $\phi = 0$, $r = 1/(A + B)$, and go 2π in ϕ, we will not end up back at $r = 1/(A + B)$, but rather at $r = 1/(A + B\,\cos(2\pi\,\delta))$. In order to get back to $r = 1/(A + B)$, the starting

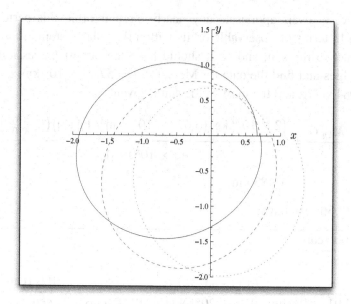

Figure 7.2   First three "orbits" of a precessing ellipse $r = \frac{1}{A + B \, \cos(\phi \, (1 - \delta))}$ with $A = 1$, $B = 0.5$, and $\delta = 0.1$.

point, we must go $\bar\phi \equiv 2\pi/(1 - \delta)$ radians. The observational effect is shown in Figure 7.2.

Now for the perturbation coming from GR, we have from (7.52): $\delta = 3\,A\,M$. For δ small, $\bar\phi \approx 2\pi\,(1 + \delta)$, so the advance (since $\bar\phi > 2\pi$) per orbit is $\Delta\phi \equiv \bar\phi - 2\pi = 2\pi\,\delta = 2\pi\,(3\,A\,M)$. A refers to an orbital feature – typically, planetary orbits are described in terms of their eccentricity e and semi-major axis a related to our A and B variables by $e = B/A$, and:

$$\frac{1}{2}\left(\frac{1}{A\,(1 + e)} + \frac{1}{A\,(1 - e)}\right) = a \Rightarrow A = \frac{1}{a\,(1 - e^2)}. \qquad (7.55)$$

So we have $\Delta\phi = 6\pi\,M/(a\,(1 - e^2))$ per orbit. Because it is a small effect, we accumulate this over 100 years – the observable quantity is:

$$\frac{\Delta\phi}{T} \times \frac{100 \text{ years}}{\text{century}} = \frac{6\pi\,M}{T\,a\,(1 - e^2)} \times \frac{100 \text{ years}}{\text{century}}, \qquad (7.56)$$

where a is the semi-major axis of the orbit, e its eccentricity, T its period in years, and M must be measured in units of length. Fortunately, people do measure M in units of length,[7] and we can either look it up (we'll take the sun as our central body) or generate it using appropriate constants.

---

[7] A direct result of setting $\frac{G}{c^2} = 1$.

The above analysis applies to any nearly circular orbit, this perturbation is ubiquitous. In terms of observable orbits within the solar system, Mercury is the closest planet to the sun, and so it should have the largest precession. We can go to the tables and find the data for Mercury: $a = 57.91 \times 10^6$ km, $e = 0.2056$, $T = 0.24084$ years, and the mass of the sun is given by:

$$M_{km} \equiv \frac{M_{kg}\, G}{c^2} = \frac{\left(2 \times 10^{30}\ \text{kg}\right)\left(6.672 \times 10^{-11}\ \text{m}^3/(\text{kg s}^2)\right)\left(\frac{1\ \text{km}}{1000\ \text{m}}\right)}{\left(3 \times 10^8\ \text{m/s}\right)^2} \qquad (7.57)$$

$$= 1.483\ \text{km},$$

so back in (7.56), we have:

$$\frac{\Delta\phi}{T} \times \frac{100\ \text{years}}{\text{century}}$$

$$= \frac{6\pi\,(1.483\ \text{km})}{\left(57.91 \times 10^6\ \text{km}\right)\left(1 - (0.2056)^2\right)(0.24084\ \text{years})} \times \frac{100\ \text{years}}{\text{century}} \qquad (7.58)$$

$$= 2.093 \times 10^{-4} \left(\frac{360°}{2\pi}\right)\left(\frac{60'}{1°}\right)\left(\frac{60''}{1'}\right)\frac{1}{\text{century}}$$

$$\approx 43''/\text{century},$$

which agrees well with the "observational" result (this is a tricky thing to observe) of $42.56'' \pm 0.94''$.

As an *a posteriori* check of the validity of our perturbation calculation, let's go back and calculate A in its natural units (1/length), from (7.55), and for Mercury's orbital parameters. We have:

$$A = \frac{1}{a\left(1 - e^2\right)} = \frac{1}{\left(57.91 \times 10^6\ \text{km}\right)\left(1 - 0.2056^2\right)} = 1.8 \times 10^{-8}\frac{1}{\text{km}}, \qquad (7.59)$$

and the claim, in for example (7.51), is that $A^2\, M$ is small – we've kept terms of order $A\, M = (1.8 \times 10^{-8}) \times 1.483 \approx 2.67 \times 10^{-8}$, but then $A^2\, M \approx 4.8 \times 10^{-16}$ 1/km. Our other approximation was that B should also be small – $B = A\, e = 1.8 \times 10^{-8}$ 1/km $\times\, 0.2056 \approx 3.7 \times 10^{-9}$ 1/km. We've kept terms of order B, and again, the corrections to this are order $B^2 \approx 10^{-16}$ 1/km^2. Evidently, our expansion is justified.

This is the famous first test of general relativity, one which Einstein worked out in his paper introducing the subject (reproduced in [11]). It is a beautiful result, and especially so because almost any perturbation to an elliptical orbit causes precession to first order – that he got just the right form to account for the precession is shocking. It is also important, historically, that because everything perturbing an ellipse makes it precess, the actual observed precession of Mercury is $\approx 5000''$ per century, but for years previous to Einstein's paper, people had been whittling down the excess by doing precise calculations on the perturbative effects of the other planets in the solar system, comets, etc. One imagines that had the GR calculation been done a century earlier, it would not have seemed so great.

7.3.3 Exact solutions

With our linearized approximation, it is tempting to close the book on massive bodies "falling" (along geodesics) in the Schwarzschild geometry. There are a number of issues left to consider, not the least of which is: for the full equations of motion, what happens if we are not in a "linearized" regime? This is a question that can be answered by proceeding with the linearization to higher and higher-order terms (that process is part of the "post-Newtonian" approximation).

At some point, the procedure becomes tedious (depends on your stamina), and we might imagine moving to a purely computational solution. This is easier done than said – the ODEs we get are exceedingly well-behaved, both from a function-theoretic point of view and as a point of numerical analysis.

What is the difference between an exact solution and a perturbative one? How much more exotic can the "full" (numerically exact, anyway) solution get? In Schwarzschild geometry, we can *look* at the orbits, measure their precession, etc. But the general observation is that not much new (aside from precession) is going on.

In more complicated situations, as we shall see, there are far more exotic trajectories available. In particular, again taking the subset of GR solutions that are of astrophysical interest (namely the metric associated with spinning massive spheres, the Kerr metric) one can get "zoom-whirl" trajectories in which the orbiting body approaches the central body, wraps around many times in a tight circle, and then moves away in a high-eccentricity ellipse – this is an extreme case of precession. We will explore some of the computational issues in Chapter 9. In addition to precession effects, motion is no longer constrained to occur in a plane, leading to additional orbital parameters.

Problem 7.9

The perturbative approach we used for the ODE in precession of perihelion can also be used for polynomials. This makes some of the degeneracy issues that one encounters in such settings more explicit.

(a) We have $x^2 + \epsilon x - 1 = 0$ for small $\epsilon \ll 1$. By taking $x = x_0 + \epsilon x_1$, inputting into the polynomial, and solving the equations you get at the ϵ^0 and ϵ^1 level, find the first-order approximation to the roots. Check your answer using the quadratic formula and expanding the exact roots for small ϵ.

(b) Try the same procedure for $\epsilon\, x^2 + b\, x + c = 0$ with $b > 0$ (find the approximate solution through order ϵ) – this time, your perturbative expansion will only find *one* root. Why? Can you modify your starting ansatz from $x = x_0 + \epsilon\, x_1$ to capture the second root? Check your answer by taking the expansion of the exact roots from the quadratic.

7.4 Bending of light in Schwarzschild geometry

Now we look at a new class of geodesic available in the Schwarzschild geometry – light. In classical, Newtonian gravity, the motion of light is not something we can sensibly discuss – but as a geodesic on a manifold, light is not that strange – we ask, for the physical particulars of the photon (null tangent vectors), what a "straight line" looks like. There is some confusion about the physical interpretation of the "proper time" – that is naturally built in to the discussion and had better exist. After all, the proper time of light is an ill-defined concept.[8] It is worth noting the null-vector ($ds^2 = 0$) nature of these photon trajectories, they come up again when we talk about coordinate transformations and one can draw an interesting parallel between the coordinate time and the "proper time".

Our setup is the same as for massive test particle orbits – we have a spherically symmetric central body of mass M that is generating the Schwarzschild metric. Outside of this central body, we have a "test particle" of mass $m = 0$, and we want to find its geodesic trajectory. Refer to Figure 7.3.

Light is "just another" test particle in GR, on a semi-equal footing with (time-like) massive particles. It does make some sense to separate the treatments, but as you will see, there is almost no difference in the calculation.

How do we describe light? Well, its rest mass is zero, i.e. the invariant four-momentum magnitude (squared) is (in Minkowski space-time) $-E^2 + \mathbf{p} \cdot \mathbf{p} = 0$. That tells us in our Lagrange setting, that $L = H = 0$ rather than the $-\frac{1}{2}$ we had in the previous section. On the metric side, $H = 0$ is just another option for us – for a metric with a time component (typified by the -1 in η_{00}) we cannot say that a "length" of zero implies that the vector is zero. Material particles must have "time-like" or $d\tau^2 = -1$ character (for arc-length parametrization with τ the proper time), then "light-like" means $d\tau^2 = 0$[9] and "space-like" has $d\tau^2 = 1$, the normalization of spatial coordinate axes. Regardless, here we will deal with light-like "particles":

---

[8] Proper time, remember, refers to the time read on clocks in the rest frame of a particle – there is no Lorentz boost that can put us in the rest frame of light.

[9] This is clear for anything that travels at the speed of light. In a time dt, $dx = c\, dt$, and we have $-c^2\, dt^2 + dx^2 = -c^2\, dt^2 + c^2\, dt^2 = 0$.

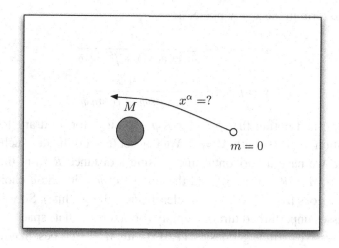

Figure 7.3 Setup for light-like geodesics in Schwarzschild space-time.

light. Notice that since $d\tau^2 = 0$, our notion of proper time for light is no longer sensible, we can still parametrize the curve, though, and without confusion, let me call this parameter τ.

The setup is identical to Section 7.3.2, so we can cut to the chase – start with the Schwarzschild metric written in the ρ-coordinate, we again set the motion in the $\theta = \frac{\pi}{2}$ plane, identify J_z as the canonical momentum associated with ϕ (and so constant), and $-E$, the canonical momentum for t, also constant. The first difference occurs at the "$L = -\frac{1}{2}$" of massive particle analysis. For light, with $L = 0$, we have:

$$0 = L = \frac{1}{2}\left(J_z^2 \rho^2 - \frac{E^2 \rho^4 - \dot{\rho}^2}{\rho^4 (1 - 2M\rho)} \right)$$

$$\downarrow \qquad\qquad (7.60)$$

$$\dot{\rho}^2 = \rho^4 \left(\frac{E^2}{J_z^2} - \rho^2 + 2M\rho^3 \right).$$

Switching again to ϕ-derivatives via $\dot{\rho} = \rho' J_z \rho^2$ gives us the analogue of $\rho'' = -\rho + \frac{M}{J_z^2} + 3M\rho^2$. Here the constant term is missing and we just have:

$$\boxed{\rho'' = -\rho + 3M\rho^2.} \qquad (7.61)$$

Now let's think a bit about the meaning of the $M = 0$ solution, the one we would get from, for example, special relativity. It says that the trajectory of light, written in parametric form in the $x - y$ plane, is:

$$x = r \cos \phi = \frac{\cos \phi}{\alpha \cos \phi + \beta \sin \phi}$$

$$y = r \sin \phi = \frac{\sin \phi}{\alpha \cos \phi + \beta \sin \phi},$$

(7.62)

using $r = 1/\rho$ and noting that $\rho = \alpha \cos \phi + \beta \sin \phi$ for arbitrary $\{\alpha, \beta\}$ is the general solution to (7.61) with $M = 0$. We can set the coordinates such that in the $M = 0$ limit, we have a horizontal line passing a distance R from the origin by putting $\alpha = 0$, $\beta = R^{-1}$ (never mind the funny $\cot \phi$, with these choices $y = R$ always, and x goes from $-\infty \longrightarrow \infty$, clearly describing a line). So we would say that light passes unperturbed through empty (or otherwise flat) space.

As with massive particle geodesics, (7.61) (with $M \neq 0$) is best handled through perturbation – let us take a solution of the form:

$$\rho = \frac{\sin \phi}{R} + \epsilon \, \tilde{\rho},$$

(7.63)

then we see that:

$$-\frac{1}{R} \sin \phi + \epsilon \, \tilde{\rho}'' = -\left(\frac{1}{R} \sin \phi + \epsilon \, \tilde{\rho} \right) + 3 M \left(\frac{1}{R} \sin \phi + \epsilon \, \tilde{\rho} \right)^2$$

(7.64)

and if we associate our ϵ expansion parameter with the mass M of the gravity-producing object (this is the point here: for $M = 0$, we get a line), then to first order in $\epsilon = M$, we can read off the equation for $\tilde{\rho}$:

$$\tilde{\rho}'' = -\tilde{\rho} + \frac{3}{R^2} \sin^2 \phi.$$

(7.65)

Aside: generic solution of inhomogenous, linear ODEs

I just want to take a moment to talk about the simple relationship between source-free solutions of first-order ODEs and their driven solutions – take:

$$f'(x) = A \, f(x)$$

(7.66)

where we are (here and only here) viewing $f(x)$ as a vector and A as a matrix (x is just the single parameter in the problem, time for example). The solution is:

$$f(x) = e^{Ax} \, f(0),$$

(7.67)

where we mean matrix exponentiation (defined in terms of the powers of matrices just as with normal exponentials).

Suppose we add a source $G(x)$, some function of x, then we have:

$$f'(x) = A \, f(x) + G(x),$$

(7.68)

but that looks a lot like a change of equilibrium for an oscillator, as if we added a constant to a spring equation of motion, for example. Motivated by this observation, let's see what happens if we add a function that depends on x (call it $g(x)$) to the ansatz:

$$\bar{f}(x) \equiv e^{Ax}(f(0) + g(x)) \longrightarrow \bar{f}'(x) = A\,\bar{f} + e^{Ax}\,g'(x). \tag{7.69}$$

That's perfect, compare with the above, and we see that $\bar{f}(x)$ solves (7.68) if we set:

$$e^{Ax}\,g'(x) = G(x) \longrightarrow g'(x) = e^{-Ax}\,G(x) \longrightarrow g(x) = \int_0^x e^{-A\bar{x}}\,G(\bar{x})\,d\bar{x}, \tag{7.70}$$

where we have made some assumptions effectively about the invertibility of A. Going back, the final form for $\bar{f}(x)$ is:

$$\bar{f}(x) = e^{Ax}\,f(0) + e^{Ax}\int_0^x e^{-A\bar{x}}\,G(\bar{x})\,d\bar{x}. \tag{7.71}$$

Solving inhomogenous ODEs amounts to exponentiating a matrix and doing some integration. Let's apply this idea to (7.65), the homogenous part is given by:

$$\frac{d}{d\phi}\begin{pmatrix} \tilde{\rho} \\ \tilde{\rho}' \end{pmatrix} = \begin{pmatrix} 0 & 1 \\ -1 & 0 \end{pmatrix}\begin{pmatrix} \tilde{\rho} \\ \tilde{\rho}' \end{pmatrix}, \tag{7.72}$$

and we can immediately tell what the matrix A in the above has as its exponential – we know the solution is $\rho(\phi) = \rho(0)\cos\phi + \rho'(0)\sin\phi$, so:

$$e^{A\phi} = \begin{pmatrix} \cos\phi & \sin\phi \\ -\sin\phi & \cos\phi \end{pmatrix} \longrightarrow e^{-A\phi} = \begin{pmatrix} \cos\phi & -\sin\phi \\ \sin\phi & \cos\phi \end{pmatrix}. \tag{7.73}$$

With this "fundamental" solution, let's calculate the source integral:

$$e^{A\phi}\int_0^\phi e^{-A\bar{\phi}}\begin{pmatrix} 0 \\ \frac{3\sin^2\bar{\phi}}{R^2} \end{pmatrix} d\bar{\phi} = \frac{1}{R^2}\,e^{A\phi}\int_0^\phi \begin{pmatrix} -3\sin^3\bar{\phi} \\ 3\cos\bar{\phi}\,\sin^2\bar{\phi} \end{pmatrix} d\bar{\phi}$$

$$= \left(\frac{1}{R^2}\right)\begin{pmatrix} 4\sin^4(\frac{1}{2}\phi) \\ 2\sin\phi - \sin(2\phi) \end{pmatrix}. \tag{7.74}$$

All right, you could have done it faster, but this is generic. Anyway, the point of all this:

$$\tilde{\rho}(\phi) = \alpha\,\cos\phi + \beta\,\sin\phi + \frac{4\sin^4(\frac{1}{2}\phi)}{R^2}$$

$$= \alpha\,\cos\phi + \beta\,\sin\phi + \frac{3 - 4\cos\phi + \cos(2\phi)}{2R^2} \tag{7.75}$$

$$= \left(\alpha - \frac{2}{R^2}\right)\cos\phi + \beta\,\sin\phi + \frac{1 + \cos^2\phi}{R^2}.$$

The first two terms, with their $\cos\phi$ and $\sin\phi$ dependence, can be dropped, if we like – these just augment the homogenous solution (the first term of (7.63)) and are already included at the ϵ^0 level. So set $\beta = 0$ and $\alpha = \frac{2}{R^2}$ (our choice).

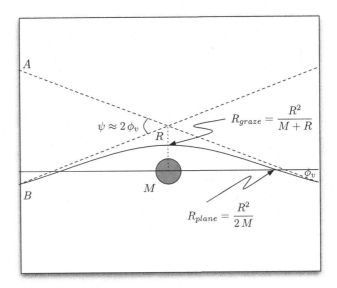

Figure 7.4   The path of light, deflected about a massive object. Light that comes from B arrives at our viewing platform at infinity at an angle ϕ_v with respect to horizontal. If we interpret the light as traveling in a straight line, then we would say that the light is coming from location A. Then with respect to the actual location, the light is "deflected" by the angle ψ.

The punch line of the aside is that:

$$\rho(\phi) = \frac{\sin\phi}{R} + M\frac{(1+\cos^2\phi)}{R^2} \qquad (7.76)$$

solves (7.61) to first order in $\epsilon \sim M$. What do these trajectories look like? They are bent about the $\phi = \frac{1}{2}\pi$ axis – reaching a closest approach to the massive body at $R_{graze} = \frac{R^2}{M+R}$ (setting $\phi = \frac{1}{2}\pi$) and crossing the body's plane at $\phi = 0 \longrightarrow R_{plane} = \frac{R^2}{2M}$. Now part of our assumption of small M is that the bending is slight. We can see this at the axis crossing, $R_{plane} \sim M^{-1}$, so if we are sitting far away (at spatial infinity effectively), the angle made w.r.t. the axis is small, as shown in Figure 7.4 – this is our viewing angle ϕ_v.

At spatial infinity, $\rho = 0$, and we assume this is where our viewing platform is set up. Then for $\phi = -\phi_v$ (ϕ in this picture starts at zero on the axis, so we are at a negative value for the angle) and ϕ_v small, we can Taylor expand (7.76):

$$0 = \rho(-\phi_v) \approx \frac{2M}{R^2} - \frac{\phi_v}{R} \longrightarrow \phi_v = \frac{2M}{R} \qquad (7.77)$$

and by symmetry, here, we have the deflection angle $\psi \approx 2\phi_v$, so:

$$\boxed{\psi = \frac{4M}{R}.} \tag{7.78}$$

Using $R = R_{sun} = 6.96 \times 10^5$ km and $M = 1.483$ km, the mass of the sun in kilometers:

$$\psi \approx 8.523 \times 10^{-6} = \left(4.8 \times 10^{-4}\right)^\circ \times \frac{\left(60^2\right)''}{1^\circ} \tag{7.79}$$

$$\approx 1.76''.$$

In other words, if we assume we are looking along the straight dotted line of our viewing angle in Figure 7.4, we would say that the light we see comes from point A, while due to bending it came from point B, a deflection of ψ. In the case of grazing deflection (that's why we set $R = R_{sun}$), this is still an incredibly small effect.

Problem 7.10

Do there exist circular orbits for light? If so, at what radii could they occur?

Problem 7.11

For the matrix:

$$A = \begin{pmatrix} 0 & 1 \\ 3 & 2 \end{pmatrix}, \tag{7.80}$$

find e^A using, at most, a calculator. You can check your answer using `Mathematica`'s `MatrixExp` function.

7.5 Radial infall in Schwarzschild geometry

Focusing on orbital and bending solutions for material particles and light (respectively) follows a natural Newtonian progression. We can also look at pure radial infall – it is interesting to note that, while physically distinct from Newtonian infall, the equations of motion for this case are . . . similar to the Newtonian ones. We will discuss radial geodesics for both massive particles and light. Of course, for light, there is no classical analogue, so we are in a difficult spot when it comes to the interpretation of radial infall – in order to give some physical meaning to the solutions, we will need to perform a coordinate transformation for the Schwarzschild metric, which we do in the next section.

So far, we have looked at "orbital" solutions for test particles and grazing solutions for light. Radial infall is actually an easier case, corresponding to simple quadrature in Newtonian gravity, so we are going backwards in difficulty. These solutions can be thought of as zero angular momentum degenerate solutions to the orbital equations of motion.

7.5.1 Massive particles

For a time-like trajectory, with the Schwarzschild metric, the Lagrangian is:

$$L = \frac{1}{2} \dot{x}^\mu g_{\mu\nu} \dot{x}^\nu = \frac{1}{2} \left(-\dot{t}^2 \left(1 - \frac{2M}{r} \right) + \frac{\dot{r}^2}{1 - \frac{2M}{r}} + r^2 \dot{\theta}^2 + r^2 \sin^2 \theta \, \dot{\phi}^2 \right).$$

(7.81)

For zero angular momentum, we have $J_x = J_y = J_z = 0$, which can be used to set $\dot{\theta} = \dot{\phi} = 0$ and $\theta = \frac{1}{2}\pi$. The temporal canonical momentum is, as always, constant, so we set $E = -\frac{\partial L}{\partial \dot{t}}$. For material particles in arc-length parametrization, $L = H = -\frac{1}{2}$, and the value of L is the only difference between the light and matter radial solutions. Putting it all together, we have an equation of motion for r:

$$\frac{d}{d\tau} \frac{\partial L}{\partial \dot{r}} - \frac{\partial L}{\partial r} = 0 \longrightarrow \ddot{r} = -\frac{M}{r^2}.$$

(7.82)

This is surprisingly similar (which is to say "identical") to the equation for New-tonian infall – the interpretation is slightly different, we have r the Schwarzschild radial coordinate rather than a flat spherical one, and the derivatives are w.r.t. τ, the proper time of the particle rather than t, but the forms are the same. The solution is obtained by multiplying through by $\dot{r}$ and setting the boundary conditions at spatial infinity (where we start from rest, by assumption) to find an expression for $\dot{r}$:

$$\dot{r}\ddot{r} = -\frac{M\dot{r}}{r^2} \longrightarrow \frac{1}{2} \frac{d}{d\tau} (\dot{r}^2) = \frac{d}{d\tau} \left(\frac{M}{r} \right),$$

(7.83)

and we can directly integrate (losing a constant of integration) the above to write $r(\tau)$ or vice versa as:

$$\boxed{\tau(r) = \frac{\pm 2}{3\sqrt{2M}} \left(r^{3/2} - r_0^{3/2} \right),}$$

(7.84)

where we "begin" at r_0. Because of the square root we took, there is a choice of sign – if we start at r_0 and fall toward $r \longrightarrow 0$, then the correct sign is negative (so that time is positive, since $r_0 > r$). That's just the normal classical result: we start off at r_0 and go toward the center. What an observer at rest at spatial infinity sees, however, is quite a bit different. We must switch from $\tau(r)$ to $t(r)$, the coordinate

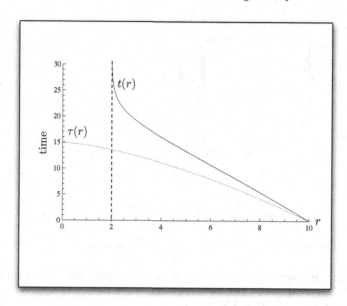

Figure 7.5   Proper time $\tau(r)$ and coordinate time $t(r)$ as a function of distance. The mass here is $M = 1$, and the asymptote $r = 2M$ is never reached for $t(r)$.

(observer at infinity) time, to find out how the situation has changed:

$$\frac{d\tau}{dr} = \frac{dt}{dr}\frac{d\tau}{dt} = -\sqrt{\frac{r}{2M}}$$

$$\frac{dt}{dr} = -\sqrt{\frac{r}{2M}}\left(\frac{Er}{2M-r}\right) \tag{7.85}$$

(we calculate $\frac{d\tau}{dt}$ from the definition of t-momentum, which is constant). This is easily integrated (by switching to $y \equiv \frac{2M}{r}$), and the end result for coordinate time is:

$$t(r) = \frac{E}{3}\sqrt{\frac{2}{M}}\left(\sqrt{r}\,(6M+r) - \sqrt{r_0}\,(6M+r_0)\right)$$

$$+ 2ME\left[\log\left(\frac{1-\sqrt{\frac{2M}{r}}}{1+\sqrt{\frac{2M}{r}}}\right) - \log\left(\frac{1-\sqrt{\frac{2M}{r_0}}}{1+\sqrt{\frac{2M}{r_0}}}\right)\right]. \tag{7.86}$$

Now we can compare the two – our traveling version $\tau(r)$ which looks Newtonian, and the observed version $t(r)$. The result for $r_0 = 10\,M$ is shown in Figure 7.5. Apparently, while we happily travel toward the center of the source, it looks externally as if we are constantly slowing down, stopping, as $t \longrightarrow \infty$ at $r = 2M$.

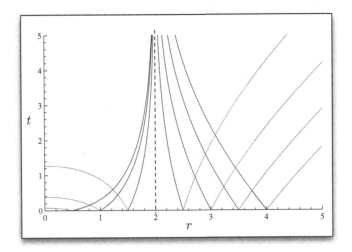

Figure 7.6 Ingoing and outgoing radial infall for light (black lines are plotted using the upper sign in (7.88), light gray lines have the lower sign) – notice the boundary at $r = 2M$ ($M = 1$ here).

7.5.2 Light

Exactly the same type of analysis works for the case of light, $L = H = 0$, and we have to solve the radial equations again. If we take the usual, $J_x = J_y = J_z = 0$ solution with $L = 0$ and $\dot{t} = \frac{E\,r}{r - 2M}$, then the equation of motion for r is just $\dot{r}^2 = E^2$. We switch to a $t(r)$ parametrization as with massive particles:

$$\frac{dt}{d\tau} = \frac{dt}{dr}\frac{dr}{d\tau} = \frac{E\,r}{r - 2M}$$
$$\frac{dt}{dr} = \pm\frac{r}{r - 2M}.$$

(7.87)

Integrating this expression, we have:

$$\boxed{t(r) = \mp(r + 2M\log(r - 2M)) \pm (r_0 + 2M\log(r_0 - 2M)).}$$

(7.88)

We can see in Figure 7.6 the trajectories for a variety of r_0 – we have taken both the "$-$" solutions (these go toward the $r = 2M$ point) and "$+$" solutions (going away) on either side of the discontinuous $r = 2M$ surface.

From the form of the Schwarzschild metric, it is clear that something happens at $r = 2M$, the "Schwarzschild radius". As observers at infinity, we never see anything fall below this surface, while as travelers, we go right through it. Light itself is trapped inside, unable to escape, so a person caught at $r < 2M$ would not have contact with the outside universe. Such a surface is called an "event horizon". In fact, our intuition about light, to the extent that we have it, is incomplete in this picture – it is possible to show that light can cross the event horizon, but cannot leave. The issue with the picture in Figure 7.6 is effectively the lack of an analogous proper time which would allow us to see the trajectory of the light in its own frame.

7.6 Light-like infall and coordinates

Here, we will take a closer look at the issue of time for photon trajectories. Remember that the line element for light is given by $ds^2 = 0$. This means that the trajectory has tangent vector $\dot{x}^\alpha$ that is null ($\dot{x}^\alpha \, \dot{x}_\alpha = 0$), and leads to some difficulties of interpretation (and interpretation only – we still know how to generate a well-defined trajectory, of course) for the parameter τ that shows up in the description of the equations of motion.

What I will suggest is that the coordinate system we have chosen does not represent the situation entirely accurately. We found, with real materials, a discrepancy between an observer "falling" toward the horizon and an observer watching the fall. From the faller's point of view, everything is fine (so to speak) until the final singularity at the center of Schwarzschild coordinates is encountered. From the point of view of the observer at infinity, the faller slows and never crosses the horizon at all. This seeming paradox was neatly resolved by shifting attention from the proper time to the coordinate time.

With light, there is only one physically meaningful "time" – coordinate. Yet, we imagine the same basic situation should hold: light should cross unperturbed into the inner region (i.e. cross the horizon), while an observer should see it getting slower and slower (redder and redder) without ever going away. One of these directions was provided by coordinate time, but we need to get a handle on the other. Without a natural proper time, the analysis is a little more involved.

7.6.1 Transformation

So far, we have been examining the Schwarzschild solution in a single coordinate system. That's equivalent to studying classical mechanics in Cartesian coordinates only. The advantage, of course, is that these coordinates are well-adapted to the spherical symmetry of the space-time (meaning that the angular Killing tensors retain their Euclidean form). They do not reveal all aspects of the geodesics, though. For example, the metric written in Schwarzschild coordinates has line element:

$$ds^2 = -\left(1 - \frac{2M}{r}\right) dt^2 + \left(1 - \frac{2M}{r}\right)^{-1} dr^2 + r^2 \left(d\theta^2 + \sin^2\theta \, d\phi^2\right),$$

$$(7.89)$$

and for $r > 2\,M$, the dt component (g_{00}) is always negative, allowing us to make the association with a time coordinate, if only by comparison to Minkowski. For $r < 2\,M$, it is the radial g_{rr} that is negative, so evidently inside this radius, the metric looks more like Minkowski with r playing the role of time. What's worse, the metric *at* $r = 2\,M$ isn't even well-defined.

What's going on here? We know that $r = 2\,M$ has little effect on the motion of massive particles. There is strange behavior there from our viewing station at infinity, but a particle itself passes right through $r = 2\,M$ without even realizing GR exists. In fact, this coordinate singularity (as it turns out) is very much akin to the singularity at the poles of the two-dimensional spatial metric for a sphere:

$$ds^2 = r^2 \left(d\theta^2 + \sin^2\theta \, d\phi^2\right), \tag{7.90}$$

where the length of vectors at $\theta = 0$, and π is effectively undetermined. That's easy enough to handle there, we just re-coordinatize so the pole is somewhere else and happily continue to measure. What is the equivalent procedure for Schwarzschild coordinates?

The problem is really with the two-dimensional $\{t, r\}$ subspace, so our goal is to find new coordinates $\{u, v\}$ that are not singular at the point corresponding to $r = 2\,M$. Clearly, this will involve a mixing of the original t and r-coordinates, and we take the simplest possible ansatz (although others are available, and preferable depending on the setting). This will lead us to a precisely linear (in M) representation of the metric – i.e. in the coordinates we are about to develop, the full Schwarzschild metric takes the form $g_{\mu\nu} = \eta_{\mu\nu} + \frac{M}{r} h_{\mu\nu}$ with $h_{\mu\nu}$ of order 1. This is a simple example of "analytic extension", the process by which we attempt to cover as much of space-time as possible with a single coordinate system.

Consider the two-dimensional portion of the Schwarzschild metric (using the pre-superscript to identify this two-dimensional subspace):

$$^2ds^2 \equiv -\left(1 - \frac{2\,M}{r}\right) dt^2 + \left(1 - \frac{2\,M}{r}\right)^{-1} dr^2. \tag{7.91}$$

Define new coordinates via $u = t + f(r)$, $v = r$. The motivation for this choice is simple – because $^2ds^2$ is independent of time, only the element dt transforms, and we don't have to worry about complicated implicit definitions of v in terms of r. We are allowing t to change by any function of r in an additively separable manner. We demand that in the $\{u, v\}$ coordinates, the line element take the form:

$$^2ds^2 = -\left(1 - \frac{2\,M}{r}\right) du^2 + \left(1 + \alpha\,\frac{M}{r}\right) dv^2 + \beta\,\frac{M}{r}\, du\, dv, \tag{7.92}$$

again a minimalist approach. We are requiring that the new "time" coordinate u play the same role as t in $^2ds^2$, which is already linear in $\frac{M}{r}$ and asking that the remaining portion of the metric also separate into Minkowski plus a linear correction. Then inputting the transformation $u = t + f(r)$, $v = r$ in (7.91), we have:

$$^2ds^2 = -\left(1 - \frac{2M}{r}\right) du^2 + \left(\left(-1 + \frac{2M}{r}\right) f'^2 + \left(1 - \frac{2M}{r}\right)^{-1}\right) dv^2$$

$$+.\, 2\left(1 - \frac{2M}{r}\right) f'\, du\, dv, \tag{7.93}$$

and comparing with the target (7.92), we have an equation for the cross term $2\left(1 - \frac{2M}{r}\right) f' = \beta \frac{M}{r}$ that can be solved:

$$f' = \frac{\beta M}{2(r - 2M)} \longrightarrow f(r) = f_0 + \frac{1}{2}\beta M \log(r - 2M), \tag{7.94}$$

introducing an integration constant f_0. Turning to the dv^2 term, we want:

$$\left(\left(-1 + \frac{2M}{r}\right) f'^2 + \left(1 - \frac{2M}{r}\right)^{-1}\right) = \frac{\left(r - \frac{1}{2}\beta M\right)\left(r + \frac{1}{2}\beta M\right)}{r(r - 2M)}, \tag{7.95}$$

and we can get the desired form by setting $\beta = 4$, at which point we see that $\alpha = 2$. Using $f(r)$ from (7.94), and $r = v$, the new two-dimensional subspace has line element:

$$^2ds^2 = -\left(1 - \frac{2M}{r}\right) du^2 + \left(1 + \frac{2M}{r}\right) dr^2 + \frac{4M}{r} du\, dr, \tag{7.96}$$

and the new "time" coordinate u is related to t and r via:

$$u = t + (f_0 + 2M \log(r - 2M)). \tag{7.97}$$

The full metric is given by the line element:

$$\boxed{\begin{aligned} ds^2 = &-\left(1 - \frac{2M}{r}\right) du^2 + \left(1 + \frac{2M}{r}\right) dr^2 \\ &+ \frac{4M}{r} du\, dr + r^2\left(d\theta^2 + \sin^2\theta\, d\phi^2\right), \end{aligned}} \tag{7.98}$$

which is free of singularities except at $r = 0$. This coordinate choice defines the "Eddington" coordinates. We can once again calculate the radial null geodesics using this new coordinate system. As before, we set the angular momenta to zero (to get the radial portion), and the Lagrangian to zero (to get null geodesics, appropriate to a description of light). The equations for $\dot{u}$ and $\dot{r}$ become the pair:

$$\dot{u} = E \quad \dot{r} = -E \tag{7.99}$$

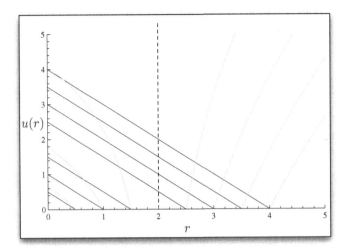

Figure 7.7   Radial infall of light for the Eddington form of the Schwarzschild metric – black lines come from taking the solution in (7.101), light gray lines come from (7.102).

and

$$\dot{u} = \frac{E\,(r + 2\,M)}{r - 2\,M} \qquad \dot{r} = E. \tag{7.100}$$

Switching to the $u(r)$ parametrization, we get the two solutions:

$$u(r) = r_0 - r, \tag{7.101}$$

for (7.99), and:

$$u(r) = r - r_0 + 4\,M\,\log\!\left(\frac{r - 2\,M}{r_0 - 2\,M}\right), \tag{7.102}$$

for (7.100), as shown in Figure 7.7. From the geodesics here, it is clear that light-like particles can cross the horizon at $r = 2\,M$, unlike the equivalent picture in Schwarzschild coordinates which indicates that light never reaches the (event) horizon. In neither case can the light emerge from the interior, cutting off information for points with $r < 2\,M$ from the outside world.

   The procedure we have carried out in this section makes the physical picture for radial infall of both light and massive particles clear – from the particle's point of view, the fall is "just like" Newtonian gravity (proper time) yet an observer would see the particle slowing down and never reaching the $r = 2M$ horizon (coordinate time). The case for light is slightly more involved because there is no notion of proper time for a photon – to discuss the photon's point of view, we must transform the metric, effectively changing what we mean by the temporal coordinate. Then we see that once again, the photon goes right through the horizon (coordinate time for the metric in (7.98)) and for an observer at infinity, the photon never reaches the horizon (coordinate time for Schwarzschild metric).

In addition to making the connection between the two points of view (observers traveling with different types of particles, and observers at infinity), the Eddington form of the metric is manifestly free of discontinuity, and (most surprisingly, perhaps) is linear in the source mass M. We demanded that the Eddington coordinates arrange this linear form, but the fact that we can actually construct a metric that is of the perturbative form: $g_{\mu\nu} = \eta_{\mu\nu} + h_{\mu\nu}$ that solves the full Einstein equations, not just the linearized ones, is interesting.

7.7 Final physics of Schwarzschild

We have studied a few properties of the Schwarzschild metric and its geodesics. For completeness, we'll cover some final topics that are related to the physical interpretation of this space-time. First, a physical configuration that is embedded in the metric itself – GR allows for the existence of black holes. Then we'll look at some of the issues of observation in general relativity, specifically in the context of the Schwarzschild solution. By constructing local, Minkowski basis vectors, and allowing them to move along geodesics, we are able to turn local observations into global predictions (redshift of light is our example). Finally, we can use the notion of a local "laboratory" to measure the effective tidal force of GR.

7.7.1 Black holes

We have been focused on the very special location, $r = 2M$. It is generic in the sense that any spherically symmetric space-time has a distance $2M$ from its center. In particular, the earth has one at ≈ 1 cm, the sun has one at ≈ 3 km, everything has a distance $r = 2M$ from its center, so why don't we observe any of the crazy effects (light and particles can enter but not exit, for example)? Because the Schwarzschild radius is *inside* the object, where Einstein's equation is not source-free – we're using the wrong solution to Einstein's equation.[10] There is a very special class of objects for which the Schwarzschild radius is *outside* the matter distribution, these are called black holes, the GR equivalent of point particles. We saw that light (a.k.a. information) cannot escape the "boundary" (the event horizon) at $r = 2M$, hence the name black hole.

Currently, black holes are known to exist, although this is a relatively recent development – they have been a theoretical prediction since the Schwarzschild

---

[10] In E&M, you cannot use an exterior solution for a uniformly charged sphere as the solution inside the sphere.

solution was introduced. Some of the properties of black holes are not as exotic as they sound – for example, we can ask, in Newtonian gravity, what the escape speed of light is. The escape speed is defined to be the minimum speed of a particle such that the kinetic energy of the particle overcomes the gravitational energy (ending, at rest, at spatial infinity). In our units, the defining equation is:

$$\frac{1}{2} v^2 = \frac{M}{r},$$

(7.103)

using this we can find the radius corresponding to a particular escape velocity. For light, $v = c = 1$ (again, in our units), so that the radius associated with the entrapment of light is $r = 2 M$. We would expect, even for Newtonian gravity, that for a mass M, light is trapped inside a spherical surface of Schwarzschild radius. This is not as deep a parallel as we might like. The flaw is in the starting point, the definition of escape speed should really read:

$$\frac{1}{2} m v^2 = \frac{m M}{r}.$$

(7.104)

For massive particles, the particle mass m cancels. For light, with $m = 0$, the cancellation is less compelling.

7.7.2 Gravitational redshift in Schwarzschild geometry

As a warmup for the next section, let's consider the redshift of light. We will emit light from our lab at rest at r (with $r > 2 M$), to be observed by a lab (at rest) very far away at infinity. What does this mean physically? Well, first consider our four-velocity. For a massive body that is not moving spatially, there must still be temporal evolution – so $\dot{r} = \dot{\theta} = \dot{\phi} = 0$, and we have $\dot{x}^\alpha \dot{x}_\alpha = -1$ (for time-like particles), so:

$$-\dot{t}^2 \left(1 - \frac{2 M}{r}\right) = -1 \longrightarrow \dot{t} = \frac{1}{\sqrt{1 - \frac{2 M}{r}}},$$

(7.105)

where r is the Schwarzschild radial position of the lab (we can set $\theta = \frac{1}{2} \pi, \phi = 0$). Our full four-velocity is:

$$\dot{x}^\alpha_r \doteq \begin{pmatrix} \left(1 - \frac{2 M}{r}\right)^{-\frac{1}{2}} \\ 0 \\ 0 \\ 0 \end{pmatrix}.$$

(7.106)

We are sitting at some Schwarzschild distance r, and this velocity represents the (normalized) tangent to our motion "curve".

Now we emit a photon of energy $\hbar\omega_r$ in our frame – what does that mean? Well, the photon's four-momentum has its energy as the zero component. In our laboratory, we have a natural set of spatial basis vectors,[11] and our own four-velocity is tangent to our geodesic in the temporal direction, so provides the unit "time" basis vector. Our measurement of the photon energy proceeds by projecting its momentum onto our local temporal basis vector, that is:

$$\hbar\omega_r = \dot{x}_r^\alpha\, p_\alpha = \left(1 - \frac{2M}{r}\right)^{-\frac{1}{2}} p_0. \qquad (7.107)$$

This is, after all, what we mean by the energy of a particle.

Now, the photon travels off to infinity along a linght-like radial geodesic – remember that as the Schwarzschild radial coordinate $r \longrightarrow \infty$, the metric becomes Minkowski, so at infinity, a stationary observer has four-velocity:

$$\dot{x}_\infty^\alpha \doteq \begin{pmatrix} 1 \\ 0 \\ 0 \\ 0 \end{pmatrix}, \qquad (7.108)$$

and this expression can also be generated directly from (7.106) for $r \longrightarrow \infty$. The observer at infinity measures the photon energy w.r.t. the temporal basis vector out there (i.e. $\dot{x}_\infty^\alpha$):

$$\hbar\omega_\infty = \dot{x}_\infty^\alpha\, p_\alpha = p_0. \qquad (7.109)$$

From our discussion of the geodesics of light (or anything else, for that matter) we know, by virtue of the cyclic nature of t in the Hamiltonian governing geodesics, that p_0 is conserved (it's what we traditionally call $-E$), so:

$$\boxed{\hbar\omega_\infty = \left(1 - \frac{2M}{r}\right)^{\frac{1}{2}} \hbar\omega_r,} \qquad (7.110)$$

and the observed frequency ω_∞ is thus less than the observed emitted frequency ω_r by a familiar factor.

7.7.3 Real material infall

Previously, we considered the motion of a test particle falling radially inward toward a central body generating the Schwarzschild geometry (a study of massive radial

---

[11] Remember that components of tensors can be projected onto basis vectors, giving us our geometrical "vector" objects as in (3.15) in Section 3.1. For four-dimensional space-time, we must include one additional basis vector for the coordinate time, and it is the projection of a tensor onto this basis vector that returns the measured energy.

geodesics). This is an artificial situation, for more realistic extended bodies (like "people"), the question of what happens as one travels towards the Schwarzschild radius, or the singularity at $r = 0$ is more complicated since there are many massive geodesics that make up an extended body. We will compute the tension on an extended body (us) as it falls radially in toward a spherically symmetric massive body. This calculation provides a nice vehicle for further consideration of local "laboratory" frames, and is standard (see [9], [31], for example).

What we really want to know is the deviation of two different parts of the body as it falls freely – at some point, our feet are accelerated faster than our heads, and there is an overall stretching. Sounds like a job for the equation of geodesic deviation from Section 4.7.3, which is:

$$\frac{D^2}{D\tau^2} \eta^\alpha = -R^\alpha{}_{\rho\gamma\beta} \dot{x}^\beta \dot{x}^\rho \eta^\gamma, \tag{7.111}$$

where $\dot{x}^\beta$ is the geodesic four-velocity, so that $\dot{x}^\gamma \dot{x}^\beta{}_{;\gamma} = 0$, since $x^\beta(\tau)$ is a geodesic. The separation η^γ is a four-separation, we'll take care of that in a moment. The idealization we have in mind: there are single point particles representing your head and feet, and we'll calculate the relative acceleration between them.

Our two points are traveling along (particle) radial geodesics, so from our discussion of massive (point) particle geodesics in Section 7.5.1, we know the geodesic form immediately. All we need are the velocities – the following expression comes from assuming we start at infinity with zero velocity and head in (how do you know?) toward a central body located at $r = 0$:

$$\dot{x}^\mu \doteq \begin{pmatrix} \frac{1}{1-\frac{2M}{r}} \\ -\sqrt{\frac{2M}{r}} \\ 0 \\ 0 \end{pmatrix}. \tag{7.112}$$

Using this and the Riemann tensor appropriate to the Schwarzschild metric, we can calculate the right-hand side of (7.111):

$$\frac{D^2}{D\tau^2} \eta^\gamma \doteq \begin{pmatrix} -\frac{4M^2}{(r-2M)r^3} & -\frac{2M\sqrt{2M}}{(r-2M)^2 r^{3/2}} & 0 & 0 \\ \frac{2M\sqrt{2M}}{r^{3/2}} & \frac{2M}{(r-2M)r^2} & 0 & 0 \\ 0 & 0 & -\frac{M}{r^3} & 0 \\ 0 & 0 & 0 & -\frac{M}{r^3} \end{pmatrix} \begin{pmatrix} \Delta t \\ \Delta r \\ \Delta\theta \\ \Delta\phi \end{pmatrix}, \tag{7.113}$$

where the deviation vector η^γ is expressed as the vector on the right above.

The difficulty in solving (7.113) comes from its non-diagonal form. Our first goal is to diagonalize the matrix via a change of basis. This change amounts to the introduction of appropriate basis vectors and the projection of η^γ onto them. We have discussed bases in the context of general transformations before, back in Section 3.1.3, the only new element, as in the previous section, is the notion that physically measurable quantities must be projections onto an orthonormal set of basis vectors. Orthonormality in this setting means Minkowski orthonormality.

In order to construct a set of four independent, normalized basis vectors, we pick, somewhere along the geodesic, four orthogonal vectors and parallel transport them around with us (they will thus *remain* orthogonal). In this case, since we want to do a local calculation of forces, we need to construct four vectors $\{\hat{\mathbf{e}}_0^\alpha, \hat{\mathbf{e}}_r^\alpha, \hat{\mathbf{e}}_\theta^\alpha, \hat{\mathbf{e}}_\phi^\alpha\}$ with the following properties:

$$
\begin{aligned}
-1 &= \hat{\mathbf{e}}_0^\alpha \, \hat{\mathbf{e}}_0^\beta \, g_{\alpha\beta} \\[4pt]
\delta_{ij} &= \hat{\mathbf{e}}_i^\alpha \, \hat{\mathbf{e}}_j^\beta \, g_{\alpha\beta} \\[4pt]
0 &= \hat{\mathbf{e}}_0^\alpha \, \hat{\mathbf{e}}_j^\beta \, g_{\alpha\beta} \\[4pt]
0 &= \frac{D}{D\tau} \hat{\mathbf{e}}_0^\alpha = \frac{D}{D\tau} \hat{\mathbf{e}}_i^\alpha .
\end{aligned}
\tag{7.114}
$$

The first two lines just tell us that the basis vectors, at each point, are orthogonal and normalized (in the Minkowski sense), and the third line says that the spatial and temporal bases are independent. The fourth line is the parallel transport statement for all four. We make measurements in the lab via projection, and our lab basis vectors satisfy the four-dimensional requirement $\hat{\mathbf{e}}_i^\alpha \, g_{\alpha\beta} \, \hat{\mathbf{e}}_j^\beta = \eta_{ij}$, where η_{ij} is the usual four-dimensional Minkowski metric. So our lab is equipped with three spatial axes and a temporal axis.

There is still the question of choice – in theory the above do not determine four unique vectors. But Schwarzschild, with its structure of $M \times S^2$ (M refers to the two-dimensional $t - r$ space, and S^2 is the 2-sphere, just a fancy way of pointing out that the deviation from spherical Minkowski is in the r and t directions), has angular unit vectors built-in, i.e. take:

$$
\hat{\mathbf{e}}_\theta \doteq \begin{pmatrix} 0 \\ 0 \\ \frac{1}{r} \\ 0 \end{pmatrix} \qquad \hat{\mathbf{e}}_\phi \doteq \begin{pmatrix} 0 \\ 0 \\ 0 \\ \frac{1}{r \sin\theta} \end{pmatrix},
\tag{7.115}
$$

the usual $\hat{\theta}$ and $\hat{\phi}$ for spherical coordinates. We still need a vector for time and the radial coordinate – but we have $\dot{x}^\mu \dot{x}_\mu = -1$ with $\dot{x}^\mu$ parallel transported along x^μ by construction of the radial geodesic. For that reason alone, we might set $\hat{e}_0^\alpha = \dot{x}^\alpha$. More importantly, this choice puts us in our rest frame – we measure our velocity to be zero (instantaneously, the basis vectors change as we move of course). Mathematically, this has the effect of setting the time-separation $\eta^0 = \Delta t$ to zero so we can measure the relative spatial acceleration between two points simultaneously in the "lab" of either one. With three of the four basis vectors fixed, the fourth is actually unique – using the definition of basis vectors from above, and specializing to the $t - r$ plane, the radial basis vector has the form:

$$
\hat{e}_r \doteq \begin{pmatrix} \frac{\sqrt{2Mr}}{2M-r} \\ 1 \\ 0 \\ 0 \end{pmatrix}. \tag{7.116}
$$

That's all very well and good, we have constructed a suitable basis, but now we have to project the deviation equation onto that basis. This is not so bad, using the transport property of the basis vectors:[12]

$$
\boxed{ g_{\alpha\gamma}\, \hat{e}_i^\alpha\, \frac{D}{D\tau}\, \eta^\gamma = \frac{D}{D\tau}\left(g_{\alpha\gamma}\, \hat{e}_i^\alpha \eta^\gamma \right) } \tag{7.117}
$$

for each of the $\hat{e}_i$ ($i = 0, 1, 2, 3$). So we can *define* the projected deviation vector η'^β by its components:

$$
\begin{aligned}
\Delta t' &= \hat{e}_0^\alpha\, g_{\alpha\beta}\, \eta^\beta \\
\Delta r' &= \hat{e}_r^\alpha\, g_{\alpha\beta}\, \eta^\beta \\
\Delta \theta' &= \hat{e}_\theta^\alpha\, g_{\alpha\beta}\, \eta^\beta \\
\Delta \phi' &= \hat{e}_\phi^\alpha\, g_{\alpha\beta}\, \eta^\beta.
\end{aligned} \tag{7.118}
$$

Now multiply both sides of (7.111) by $\hat{e}_i^\sigma\, g_{\alpha\sigma}$, and you get four equations, one for each value of i. Because of the parallel transport property of the basis vectors in (7.117), we know that $\hat{e}_i^\sigma\, g_{\alpha\sigma} \frac{D^2}{D\tau^2} \eta^\alpha = \frac{D^2}{D\tau^2}\left(\hat{e}_i^\sigma\, g_{\alpha\sigma} \eta^\alpha \right)$. Then using (7.118) to write the right-hand side in terms of the primed separations, we get the following

---

[12] Basically, what we have done is diagonalize the right-hand side of the deviation equation with eigenvectors that are parallel transported (read "constant" w.r.t. the $\frac{D}{D\tau}$ operation), then we can do the usual – redefine η on both sides of the equation so as to diagonalize the system. In this setting, the eigenvalues of the matrix on the right are 0, $\frac{2M}{r^3}$ and $-\frac{M}{r^3}$, precisely what we get through the basis transformation.

four equations:

$$\frac{D^2}{D\tau^2}(\Delta t') = 0$$

$$\frac{D^2}{D\tau^2}(\Delta r') = \frac{2M}{r^3}\Delta r'$$

$$\frac{D^2}{D\tau^2}(\Delta\theta') = -\frac{M}{r^3}\Delta\theta'$$

$$\frac{D^2}{D\tau^2}(\Delta\phi') = -\frac{M}{r^3}\Delta\phi',$$

(7.119)

where the first equation comes from choosing $\hat{\mathbf{e}}_0^\alpha = \dot{x}^\alpha$, and tells us that we will be measuring events simultaneously in this frame. We are primarily concerned with the radial equation, the radial acceleration of the two particles (head and feet, say) is what will cause the tension felt by our body.

To go further, we need to refine our idealization – point particles at the head and feet is too coarse to get a material tension, let's use the standard assumption:[13] the human body is approximated as a rectangular box with height h, and width w, we take as our total mass m, then $\rho = \frac{m}{h\,w^2}$ is our density. The relative acceleration of two nearby parcels of mass, viewed as a force, is counteracted by our bodies, so that:

$$d\mathbf{F}' = dm\,\frac{2M}{r^3}\Delta r'\,\hat{\mathbf{e}}_r,$$

(7.120)

with the rest frame set at the center of mass, so $\Delta r'$ is just the distance to the center of mass of the body. What is the net force across the square cross-section at the center due to the mass below? This is what we need to calculate to get the pressure (force per unit area). Using $dm = \rho\,dV$, we have:

$$\mathbf{F}' = \hat{\mathbf{e}}_r\int_{-\frac{h}{2}}^0 \frac{2M}{r^3}r'\,(\rho\,w^2\,dr') = \hat{\mathbf{e}}_r\int_{-\frac{h}{2}}^0 \frac{2M\,r'}{r^3}\frac{m}{h}\,dr' = -\frac{M\,m\,h}{4\,r^3}\hat{\mathbf{e}}_r,$$

(7.121)

that's the net force on the center of mass, to get pressure (tension, really, because negative), we just divide by w^2:

$$P = -\frac{|\mathbf{F}'|}{w^2} = -\frac{M\,m\,h}{4\,w^2\,r^3}.$$

(7.122)

Time to put in the numbers – suppose we assume the black hole has solar mass $M \approx 1.5\,\text{km}$, that we have a width of 50 cm, and a height of 2 m. For our mass, take $75\,\text{kg} \approx 5.6 \times 10^{-29}\,\text{km}$, then we just have to convert pressure into appropriate

---

[13] This problem appears in a variety of places, the assumptions used here are from [31].

units (we currently have $G = c = 1$).

$$1 \text{ Atm} = 101,325 \text{ Pa} = (101,325)\frac{\text{kg}}{\text{m s}^2}$$

$$= 101,325 \times \frac{G}{c^4}\frac{\text{kg}}{\text{m s}^2} = 8.344 \times 10^{-40}\frac{1}{\text{m}^2} \times \left(\frac{1000 \text{ m}}{1 \text{ km}}\right)^2 \qquad (7.123)$$

$$= 8.344 \times 10^{-34}\frac{1}{\text{km}^2},$$

which has the same units as our pressure formula. The human body can withstand ≈ 100 Atm of pressure, so we solve for the radius:

$$\frac{(1.5 \text{ km})(5.6 \times 10^{-29} \text{ km})(0.002 \text{ km})}{4(0.0005 \text{ km})^2 r^3} = 100\,(8.344 \times 10^{-34})\frac{1}{\text{km}^2} \qquad (7.124)$$

$$r \approx 130 \text{ km}.$$

Maybe not the most enlightening number ever (and you will be incensed when you find out the Newtonian result), the point is, you can't get very close. The physics is the general relativistic form of the (Newtonian) gravitational tidal force – for a gravitational force that goes like r^{-2}, as you fall inwards, your feet are closer to the central body than your head, and so your feet are accelerated more than your head, leading to an overall stretch. Similarly, your left and right sides are not falling parallel to one another, they are both falling toward the center of the central body, and so must approach, leading to a squeeze.

What is more important than the result is the procedure – we used the equation of geodesic deviation to calculate the deviation between two points on nearby geodesics. In order to connect this with our ideas about three-dimensional space, we required that the separation vector be purely spatial (choice of basis), so that we measured distances between geodesics at equal (coordinate) time. In this subspace, we also used an orthonormal basis rather than the un-normalized "coordinate" basis, basically ensuring that our notion of integration was preserved from spherical coordinates. Then we interpreted the deviation caused by the geometry of space-time as an acceleration that would cause *us* to feel a force. This is typical of local measurements: go to the rest frame of the observer to find out what is happening "in the lab".

Why haven't we had to set basis vectors until now? Most of the tests we have discussed involved observing the behavior of light and particles from afar – at a viewing platform located far away from the sources, where space-time is effectively flat (asymptotic limit of the Schwarzschild solution is one way to see it, or else our work on linearized gravity where the linearization is on a Minkowski background).

Even when we did make local measurements, they were scalars, and of course, no basis can change a scalar.

7.8 Cosmological constant

We'll end our discussion of the spherically symmetric vacuum solution by looking at the analogous solution to the minimal modification to general relativity we can make. As you saw in Problem 7.4, it is possible to modify general relativity to include a "cosmological constant", a term Λ that enters the Einstein–Hilbert action via:

$$S_{CC} = \int d\tau \sqrt{-g}\left(R(g_{\mu\nu}) + 2\Lambda\right),\tag{7.125}$$

and this action leads to field equations in the absence of matter that look like:

$$R_{\mu\nu} - \frac{1}{2}g_{\mu\nu}R - \Lambda g_{\mu\nu} = 0.\tag{7.126}$$

There are two views we can take of this equation. First, we can view it as a modification of general relativity, with a tunable parameter. The second view is to move the Λ term over to the right-hand side:

$$R_{\mu\nu} - \frac{1}{2}g_{\mu\nu}R = \Lambda g_{\mu\nu},\tag{7.127}$$

where we now "interpret" the right-hand side as a source for general relativity. That is, we take the cosmological constant to describe an intrinsic, universal distribution of something, and add this source in to any other stress tensors of interest. By analogy with the stress tensor for a perfect fluid (see Section 4.7 and Problems 2.12 and 4.13) with nonzero pressure (see [30] for a discussion of the fluid stress tensor):

$$T^{\mu\nu} = (\rho + p)\, u^\mu u^\nu - p\, g^{\mu\nu},\tag{7.128}$$

we identify the source as a "fluid" with negative pressure (if $\Lambda > 0$), $p = -\rho$ and ρ set by Λ – a "vacuum" energy density associated with no external mass distribution.

We can find the spherically symmetric solutions for Einstein's equation with cosmological constant from (7.125) using the Weyl method, as you did in Problem 7.4. The end result there was that the line element is:

$$ds^2 = -\left(1 - \frac{2M}{r} + \frac{\Lambda r^2}{3}\right)dt^2 + \left(\frac{1}{1 - \frac{2M}{r} + \frac{\Lambda r^2}{3}}\right)dr^2 + r^2\left(d\theta^2 + \sin^2\theta\, d\phi^2\right).$$

$$\tag{7.129}$$

This metric, unlike Schwarzschild, is interesting even in the absence of a massive source – setting $M = 0$ yields a nontrivial space-time.

We cannot do justice here to the interesting physical possibilities of the cosmological constant (nor the associated problems), and refer the interested reader to [4, 20] for further introductory material.

Problem 7.12

Find the basis vectors (orthonormal form, $\hat{\mathbf{e}}_t$, $\hat{\mathbf{e}}_r$, $\hat{\mathbf{e}}_\theta$ and $\hat{\mathbf{e}}_\phi$) for a laboratory falling radially in toward a central body via parallel transport along the radial geodesic with boundary condition:

$$\hat{\mathbf{e}}_t \doteq \begin{pmatrix} 1 \\ 0 \\ 0 \\ 0 \end{pmatrix} \quad \hat{\mathbf{e}}_r \doteq \begin{pmatrix} 0 \\ 1 \\ 0 \\ 0 \end{pmatrix} \quad \hat{\mathbf{e}}_\theta \doteq \begin{pmatrix} 0 \\ 0 \\ \frac{1}{r} \\ 0 \end{pmatrix} \quad \hat{\mathbf{e}}_\phi \doteq \begin{pmatrix} 0 \\ 0 \\ 0 \\ \frac{1}{r \sin \theta} \end{pmatrix}$$

at spatial infinity ($r \longrightarrow \infty$ where the laboratory begins at rest). You are constructing the local basis vectors at all points in Schwarzschild coordinates assuming we begin with the "intuitive" laboratory frame at infinity where the Schwarzschild metric reduces to Minkowski. Observe that the $\hat{\mathbf{e}}_t$ basis vector is indeed $\dot{x}^\alpha$ for radial infall.

What are the "lengths" of these four vectors along the trajectory?

Problem 7.13

The Schwarzschild singularity is an "apparent" (or "coordinate") singularity, $r = 2M$ poses no physical problems, even though the metric appears to be undefined at this point (similar to the north pole of a sphere). One way to see this is to construct invariants of the space – what is the simplest, nonzero scalar (discounting the dimension of the space-time) you can form that involves r, and shows that there is no intrinsic difficulty at the "Schwarzschild radius"?

Problem 7.14

For Newtonian gravity, we have $\nabla^2 \psi = 4\pi\rho$ as the fundamental equation governing the generation of fields from sources. Then the equations of motion follow from the Lagrangian $L = \frac{1}{2} m v^2 + m \psi$. A general class of solution to Laplace's equation (governing fields away from sources) follows from:

$$\psi = \frac{Q_0}{r} + \frac{Q_i\, x_i}{r^3} + \frac{Q_{ij}\, x_i\, x_j}{r^5} + \frac{Q_{ijk}\, x_i\, x_j\, x_k}{r^7} + \ldots. \tag{7.130}$$

Here, we are in flat three-dimensional space with $x_i \doteq (x \quad y \quad z)^T$ and $r \equiv \sqrt{x^2 + y^2 + z^2}$. For the first three terms, find the restriction on $Q_{\ldots}$ that gives a solution with $\nabla^2 \psi = 0$. Using this, generate the most general ψ from the first three terms that has no ϕ-dependence (rewrite $\mathbf{x}$ in spherical coordinates). What is this potential in the equatorial plane, $\theta = \frac{\pi}{2}$?

Problem 7.15

Find the electric potential V and magnetic vector potential $\mathbf{A}$ outside a uniformly charged rotating solid sphere (with charge density ρ, constant angular velocity $\boldsymbol{\omega} = \omega \hat{\mathbf{z}}$ and radius R) – see Example 5.11 in [15].

(a) Write your solution in terms of $\boldsymbol{\ell} = \frac{I\omega}{M}$, the angular momentum per unit mass of the sphere (here I is the moment of inertia for a solid sphere) and Q, the total charge on the sphere. Put the solution in Gaussian units (most appropriate for us).

(b) Now think of a spinning ball of mass M with uniform density ρ. Make the replacement to your "electrostatic" potentials developed in Section 2.6 (particularly (2.117)), write down (in "spherical" coordinates) the metric perturbation associated with this configuration using (7.31).

*Problem 7.16

The Weyl method can be used to develop the electric field and metric for the space-time outside of a charged massive body (that is static, and spherically symmetric). The metric is called the Reissner–Nordstrom metric. This example provides a simple case of a field theory like E&M coupled to gravity.

(a) Starting from the assumption of spherical symmetry, we have, for the metric:

$$ds^2 = -a(r) b(r)^2 dt^2 + a(r)^{-1} dr^2 + r^2 \left(d\theta^2 + \sin^2 \theta \, d\phi^2\right) \qquad (7.131)$$

and for the electrostatic potential: $V(r)$ – we'll work (from the start) with $\mathbf{A} = 0$. Write down the Lagrangian $\mathcal{L}$ associated with the full action:

$$\mathcal{L} = \mathcal{L}_{EH} + \alpha \, \mathcal{L}_{EM} \qquad (7.132)$$

for a coupling constant α – you can use second-order form for the electromagnetic portion $\mathcal{L}_{EM} = \sqrt{-g} \, (E^2 - B^2)$. Refer to (7.14) for the Einstein–Hilbert Lagrangian obtained from this starting point.

(b) Vary the action with respect to $a(r)$, $b(r)$, and $V(r)$ (use the Euler–Lagrange equations directly with $\mathcal{L}$). From the field equations, solve for $a(r)$, $b(r)$, and $V(r)$. The potential should be precisely Coulomb, so we know how to interpret the constant of integration that should have shown up. Verify that you recover the Schwarzschild metric in the absence of charge.

Problem 7.17

Show that the variation of the term $2\sqrt{-g}\,\Lambda$ in the action (7.125) returns $-\Lambda\, g_{\mu\nu}$ in the field equation: $R_{\mu\nu} - \frac{1}{2} g_{\mu\nu} - \Lambda\, g_{\mu\nu} = 0$ (for no matter sources).

Problem 7.18

Show that the metric from (7.129) with $M = 0$ is not flat. What is the Ricci scalar for this space-time? Notice that the scalar curvature is negative if $\Lambda > 0$.

8

Gravitational radiation

While we have not exhausted the vacuum solutions, we are about half-way done with the physically interesting ones (for point sources). So it is time for a brief interlude – radiative solutions. Our study will split into two general categories: waves in vacuum and their sources. In the linearized theory, as was mentioned in our perturbative expansion of Einstein's equation, we have a generic wave equation with and without sources – the solution to this in the tensor setting is well-defined: superimposed (because we are in the linearized limit) waves of definite frequency and two polarizations. We will be studying both the vacuum and source-driven waves predicted by GR. Gravitational waves are the basis for detection in a number of experiments, both current and planned, so it is important to understand the nature of this radiation, and the patterns generated by typical sources. Basically, though, the tools are identical to E&M with one extra polarization.

We know that the metric, as a field, is most directly comparable to the four-potential in E&M. The potential, however, is not the physically relevant field, as is evidenced by the Lorentz force law which makes reference to the fields **E** and **B**. The same is true in general relativity – it is not the metric itself that informs physical observation – after all, at the very least, the metric is governed by Einstein's equation, which is coordinate-independent. In order to make a prediction, one must first choose a coordinate system, and then see what form the metric takes. In GR, the coordinate choice is itself the gauge choice, and we will start by fixing all the gauge freedom for the metric perturbation in linearized GR, then use this completely gauge-fixed metric perturbation to calculate effective forces. Since we can do the same thing in E&M, namely, take a four-potential A_μ and fix all of the gauge freedom, leaving us with a four-potential with only physically relevant content, we'll start with E&M, then parallel the gauge fixing for linearized GR.

8.1 Gauge choice in E&M

Remember how the choice of gauge goes in standard electrodynamics, you start with the field equations for **E** and **B** (source-free):

$$\nabla \cdot \mathbf{E} = 0 \qquad \nabla \times \mathbf{E} = -\frac{\partial \mathbf{B}}{\partial t}$$

$$\nabla \times \mathbf{B} = \frac{\partial \mathbf{E}}{\partial t} \qquad \nabla \cdot \mathbf{B} = 0 \tag{8.1}$$

and introduce the potentials in the usual way $-\nabla \cdot \mathbf{B} = 0 \longrightarrow \mathbf{B} = \nabla \times \mathbf{A}$. Then we observe that the curl of **E**, while not (by itself) zero, can be augmented:

$$\nabla \times \mathbf{E} = -\frac{\partial}{\partial t}(\nabla \times \mathbf{A}) \longrightarrow \nabla \times \left(\mathbf{E} + \frac{\partial \mathbf{A}}{\partial t}\right) = 0, \tag{8.2}$$

so that the combination $\mathbf{E} + \frac{\partial \mathbf{A}}{\partial t}$ is curl-free. So we introduce a potential V here, $\mathbf{E} + \frac{\partial \mathbf{A}}{\partial t} = -\nabla V$, and we have automatically satisfied the above.

Now there are only the "source" equations (the left column of (8.1)):

$$\nabla \cdot \mathbf{E} = \nabla \cdot \left(-\nabla V - \frac{\partial \mathbf{A}}{\partial t}\right) = 0 \longrightarrow \nabla^2 V = -\frac{\partial}{\partial t}(\nabla \cdot \mathbf{A}) \tag{8.3}$$

and the vector one:

$$\left(\nabla^2 \mathbf{A} - \frac{\partial^2 \mathbf{A}}{\partial t^2}\right) = \nabla\left(\nabla \cdot \mathbf{A} + \frac{\partial V}{\partial t}\right). \tag{8.4}$$

The usual observation is: (1) adding the gradient of a scalar to **A** doesn't change $\nabla \times \mathbf{B}$ and (2) adding the time-derivative of that scalar to V doesn't change **E**. So we introduce:

$$\mathbf{A}' = \mathbf{A} + \nabla \phi \qquad V' = V - \frac{\partial \phi}{\partial t}, \tag{8.5}$$

which guarantees that $\mathbf{B}' = \mathbf{B}$ and $\mathbf{E}' = \mathbf{E}$.

The **E**, **B** content of the new potential V' and $\mathbf{A}'$ is identical to V and **A**, but now we have freedom to *choose* properties for the potentials. In particular, we can impose Coulomb gauge, $\nabla \cdot \mathbf{A} = 0$, that simplifies the right-hand side of (8.3), giving us Laplace's equation. But then we're stuck with a tricky vector-potential equation. Of course, viewing these as components of a four-potential makes the situation somewhat clearer – in particular, both the scalar potential and the components of the vector potential satisfy wave equations, throwing out the time portion as we do in Coulomb gauge is not in the spirit of relativity.

So we choose Lorentz gauge, which has $\partial_\mu A^\mu = 0$ with A^μ the usual four-vector. Then, as with our GR wave equation, we have:

$$\Box^2 A^\mu = 0. \tag{8.6}$$

The solutions to the vacuum wave equation are interpreted as light in E&M. They are given this interpretation from the observation that the fundamental solutions are single-frequency fields that can propagate through vacuum with constant speed *c*. Let's review these special solutions, just to remind ourselves of the way in which frequency and polarization show up there, and how we can use the four-potential itself to make the physical properties of the wave (direction of travel, frequency, and speed) explicit.

We start with the wave equation for the four-potential A^μ – already in Lorentz gauge (or we would not *have* a wave equation):

$$\left(-\partial_t^2 + \nabla^2\right) A^\mu = 0. \tag{8.7}$$

The idea is to introduce a Fourier transform solution – i.e. make the assumption $A^\mu(x) = P^\mu\, e^{i\,(\mathbf{k}\cdot\mathbf{x} - \omega t)}$. For this assumed form, the wave equation (8.6) reads:

$$\left(\omega^2 - k^2\right) A^\mu = 0, \tag{8.8}$$

which tells us that the wave four-vector defined as $k_\mu \doteq (\omega, k_x, k_y, k_z)$ has $k^\mu k_\mu = 0$. In addition, we must satisfy the Lorentz gauge condition:

$$\partial_\mu A^\mu = 0 \longrightarrow P^0\,\omega + \mathbf{k}\cdot\mathbf{P} = 0 \tag{8.9}$$

with $P^\mu \doteq (P^0, \mathbf{P})^T$. In other words, we have $k_\mu P^\mu = 0$. This condition gives us a set of algebraic relations.

For concreteness, think of a wave propagating in the z-direction with frequency ω. Demanding that $k_\mu k^\mu = 0$ gives $k_z = \omega$, and then the orthogonality with P^μ tells us that $P^0 + P^z = 0$. This leaves us with two free spatial components, P^x and P^y (think of the P^z-dependence of the electric and magnetic fields) in the polarization vector, just as we expect for light (i.e. the polarization vector is orthogonal to the propagation vector).

There is unfixed gauge freedom left in the potential A^μ, even once it is in divergence-less form. Think of a transformation $A'_\mu = A_\mu + \psi_{,\mu}$ for scalar ψ with $\partial^\mu A_\mu = 0$ already. We know this will leave us in Lorentz gauge provided:

$$\partial^\mu A'_\mu = \partial^\mu A_\mu + \partial^\mu \partial_\mu \psi = 0 \longrightarrow \Box^2 \psi = 0, \tag{8.10}$$

so we can make an *additional* gauge transformation, even when we are in Lorentz gauge. Take $\psi = Q\, e^{i\,k_y\, x^y}$, so that the D'Alembertian is zero – then:

$$A'_\mu = A_\mu + i\, k_\mu\, Q e^{i\,k_y\, x^y}, \tag{8.11}$$

and we can use this to set, for example, $A'_0 = 0$ by appropriate choice of Q.

A four-potential cannot, in general, carry physical information directly. There is gauge freedom, so any physical content ascribed to A^μ can be modified by introducing an arbitrary gradient (for example). Once *all* of the gauge freedom is used up, as we have done above, and provided those gauges lend themselves to physical interpretation, we can identify useful physics from the potential. When we have moved to Lorentz gauge (manifestly relativistic) and set the potential $A_0 = 0$, what we have is a vector plane wave:

$$\mathbf{A} = \mathbf{P} \, e^{i(\mathbf{k} \cdot \mathbf{x} - \omega t)} \text{ with } \mathbf{P} \cdot \mathbf{k} = 0 \qquad (8.12)$$

that is, correctly, transverse to the direction of propagation, and has identifiable frequency. So the vector potential here is as adapted as possible to the physical content of $\mathbf{E}$ and $\mathbf{B}$, and mirrors the relations we find when we generate those fields from it.

8.2 Gauge choice in linearized GR

When we discussed the Newtonian correspondence to GR in Section 7.2, we took a linearized approximation to the full Einstein equations, chose a gauge and found that the metric perturbations, $h_{\mu\nu}$, satisfied the equations:

$$\boxed{\begin{aligned} & 8\pi \, T_{\mu\nu} = -\frac{1}{2} \partial_\rho \partial^\rho \, H_{\mu\nu} \\ & H_{\mu\nu} \equiv h_{\mu\nu} - \frac{1}{2} \eta_{\mu\nu} \, h \\ & \partial^\rho H_{\nu\rho} = 0, \end{aligned}} \qquad (8.13)$$

with the last of the above satisfied by our gauge choice for the coordinates. The linearized equation, in the absence of sources, is just:

$$\partial^\rho \partial_\rho \, H_{\mu\nu} = 0, \qquad (8.14)$$

a four-dimensional wave equation for $H_{\mu\nu}$ written on a Minkowski background (raise and lower with $\eta_{\mu\nu}$). Since $H_{\mu\nu}$ is symmetric, this is just 10 times a scalar wave equation (in $D = 4$). We take the usual plane wave ansatz:

$$H_{\mu\nu} = P_{\mu\nu} \, e^{i \, k_\alpha x^\alpha}, \qquad (8.15)$$

interpret $P_{\mu\nu}$ as the polarization *tensor*, and k_μ as the wave vector. What does this mean in terms of (8.13)? Put it in, we have:

$$\partial^\rho \partial_\rho \, H_{\mu\nu} = -\left(k^\rho k_\rho\right) H_{\mu\nu} = 0 \qquad (8.16)$$

and assuming the metric perturbation itself is nonzero, this requires that $k_\mu k^\mu = 0$. As a Fourier transform, we recognize the frequency k_0 and wave-vector k_i. The condition $k_\mu k^\mu = 0$ tells us that we have a null vector, i.e. the solution is light-like, just as in E&M. In this flat-space setting, we can make the identification of a field with characteristic speed c.

The gauge condition (the third equation in (8.13)) puts constraints on $P_{\mu\nu}$:

$$\partial^\rho H_{\nu\rho} = \partial^\rho \left(P_{\nu\rho} \, e^{i \, k_\mu \, x^\mu} \right)$$

$$= i \, P_{\nu\rho} \, (i \, k^\rho) \, e^{i \, k_\mu \, x^\mu} = 0 \longrightarrow k^\rho \, H_{\nu\rho} = 0,$$

(8.17)

and this is a Poynting vector-like statement: the propagation of the wave, k^ρ is orthogonal to the perturbation $H_{\nu\rho}$.[1] In addition to this interpretation, we have constrained four of the 10 components of $P_{\nu\rho}$ by requiring $P_{\nu\rho} \, k^\nu = 0$.

We're not quite done. Remember the transformation we made to impose the gauge condition[2] – we started with the coordinate transformation $x^\mu \longrightarrow x'^\mu = x^\mu + f^\mu$, inducing the transformation to the field $h_{\mu\nu} \longrightarrow h'_{\mu\nu} = h_{\mu\nu} + f_{(\mu,\nu)}$, and we set:

$$\partial_\rho \partial^\rho f_\nu = -2 \partial^\rho H_{\nu\rho}$$

(8.18)

to achieve $\partial^\rho H_{\nu\rho} = 0$. But to f_ν, we can add any vector w_ν satisfying $\partial^\rho \partial_\rho w_\nu = 0$ without violating the condition that $\partial^\rho H_{\nu\rho} = 0$. Evidently, there is still some gauge freedom left over (precisely scalar insensitivity in the field equations).

So let's finish fixing the gauge. We will choose w_ν to make a further coordinate transformation inducing yet another transformation in $h_{\mu\nu}$:

$$h''_{\nu\rho} = h'_{\nu\rho} - w_{(\nu,\rho)},$$

(8.19)

where the primes refer to the partially gauge-fixed coordinates (our original choice of f_ν used to set $\partial^\rho H^\delta_{\mu\rho} = 0$), and the double primes are our new, further, choice. Since w_ν itself satisfies the wave equation, we may as well take it to be of the form:

$$w_\nu = Q_\nu \, e^{i \, k_\mu \, x^\mu}$$

(8.20)

with the same wave vector k_μ we are using for the metric. Then the (new) metric perturbation reads:

$$h''_{\nu\rho} = h'_{\nu\rho} - i \, e^{i \, k_\mu \, x^\mu} \left(Q_\nu \, k_\rho + Q_\rho \, k_\nu \right)$$

$$= H_{\nu\rho} - \frac{1}{2} \eta_{\nu\rho} \, H^\alpha_\alpha - i \, e^{i \, k_\mu \, x^\mu} \left(Q_\nu \, k_\rho + Q_\rho \, k_\nu \right)$$

(8.21)

$$= \left(P_{\nu\rho} - \frac{1}{2} \eta_{\nu\rho} \, P^\alpha_\alpha - i \left(Q_\nu \, k_\rho + Q_\rho \, k_\nu \right) \right) e^{i \, k_\mu \, x^\mu}.$$

---

[1] The word "orthogonal" has spatial connotation. In a space-time with indefinite signature, we cannot draw spatial vectors at 90° angles and associate that directly with $a^\mu \, b_\mu = 0$ for two four-vectors. In this case, the four-vector orthogonality reduces to spatial orthogonality, as we shall see, but the term is a little premature at this stage.

[2] It is interesting that the gauge choice we have in general relativity stems from the insensitivity of the field equations governing $h_{\mu\nu}$ to a vector contribution $f_{\mu,\nu} + f_{\nu,\mu}$. Such a term is naturally associated, as an additive contribution to the metric, with a coordinate choice as discussed in Section 5.6.1. So in general relativity, unlike in E&M, the gauge choice has a clear meaning: we are choosing a coordinate system.

We have four unknowns in the above, Q_ν. We started with 10 unknown components in the polarization tensor $P_{\nu\rho}$, we fixed four with the original gauge choice, and now we will fix four more, leaving two independent components, implying that in the linearized theory of general relativity, there are two distinct polarizations. To finish specifying the gauge choice, we need four equations to fix Q_ν, which four to take? The most convenient constraint is to eliminate the time-space portion of the metric, that will leave us with just the spatial section. That is, we want $h''_{0\rho} = 0$. Written out in components, the conditions are:

$$2 Q_0 k_0 = -i \left(P_{00} + \frac{1}{2} P_\alpha^\alpha \right)$$

$$Q_0 k_j + Q_j k_0 = -i P_{0j} \quad j = 1, 2, 3$$

(8.22)

or in matrix form:

$$\underbrace{\begin{pmatrix} 2 k_0 & 0 & 0 & 0 \\ k_1 & k_0 & 0 & 0 \\ k_2 & 0 & k_0 & 0 \\ k_3 & 0 & 0 & k_0 \end{pmatrix}}_{\equiv K} \begin{pmatrix} Q_0 \\ Q_1 \\ Q_2 \\ Q_3 \end{pmatrix} = -i \begin{pmatrix} P_{00} + \frac{1}{2} P_\alpha^\alpha \\ P_{01} \\ P_{02} \\ P_{03} \end{pmatrix}.$$

(8.23)

The matrix on the left is nonsingular, so we can invert it, it's not even that bad:

$$K^{-1} = \begin{pmatrix} \frac{1}{2 k_0} & 0 & 0 & 0 \\ -\frac{k_1}{2 k_0^2} & \frac{1}{k_0} & 0 & 0 \\ -\frac{k_2}{2 k_0^2} & 0 & \frac{1}{k_0} & 0 \\ -\frac{k_3}{2 k_0^2} & 0 & 0 & \frac{1}{k_0} \end{pmatrix},$$

(8.24)

and the solution for the coefficients Q_ν is:

$$Q_0 = -\frac{i}{2 k_0} \left(P_{00} + \frac{1}{2} P_\alpha^\alpha \right)$$

$$Q_j = \frac{i}{2 k_0^2} \left(P_{00} + \frac{1}{2} P_\alpha^\alpha \right) k_j - \frac{i}{k_0} P_{0j}.$$

(8.25)

We are done, we have formed a metric perturbation that is "transverse", here meaning that there are no components except for spatial ones so that $k^\rho h_{\nu\rho} = 0$ refers to spatial orthogonality, as in E&M. There is an additional property of this gauge, consider the trace of the $h''_{\mu\nu}$ from its definition (8.19):

$$h''^\mu_\mu = P^\mu_\mu - 2 P^\mu_\mu - 2 i Q_\mu k^\mu$$

$$= -P^\mu_\mu - 2 i \left[\frac{i}{2} \left(P_{00} + \frac{1}{2} P^\mu_\mu \right) + \frac{i}{2 k_0^2} \left(P_{00} + \frac{1}{2} P^\mu_\mu \right) k_j k^j - \frac{i}{k_0} P_{0j} k^j \right],$$

(8.26)

and note from Einstein's equation (linearized), we still have $k^\mu k_\mu = 0 \longrightarrow k_j k^j = k_0^2$ and from the original gauge choice, $P_{\mu\nu} k^\nu = 0 \longrightarrow P_{0j} k^j = P_{00} k^0$, so that we can simplify the second line above:

$$h''^\mu_\mu = -P^\mu_\mu - 2i\left[i\left(P_{00} + \frac{1}{2}P^\alpha_\alpha\right) - i\,P_{00}\right]$$

(8.27)

$$= 0.$$

We see that the metric perturbation is also "traceless". This "transverse traceless" gauge (finally) is completely fixed, with two independent degrees of freedom left. Because the perturbation is traceless, we have $H_{\mu\nu} = h_{\mu\nu}$ without ambiguity, and we are ready to do some physics.

8.3 Source-free linearized gravity

We know already that we can set $h_{\mu\nu}$ to have zero components for $h_{0\nu} = h_{\nu 0}$, and this suggests that we just take the spatial part for our "real" perturbation. We will introduce Cartesian coordinates in Minkowski space-time – a global inertial frame. That is, after all, what we are doing in the linearization process. Let's be explicit, suppose we want to describe a wave with frequency ω traveling in the z-direction according to our laboratory. To sum up the content of all of our gauge fixing, our wave solution, with all side-constraints, satisfies:

$$\begin{aligned} h_{\mu\nu} &= P_{\mu\nu}\, e^{i\,k_\mu x^\mu} \\ 0 &= P_{\mu\nu}\, k^\mu \\ 0 &= k_\mu k^\mu \\ 0 &= P_{0\nu} = P^\alpha_\alpha. \end{aligned}$$

(8.28)

Setting $k_\mu \doteq (\omega, 0, 0, k_z)$ (the wave vector appropriate to z propagation with frequency ω), we have from $k_\mu k^\mu = 0$, $k_z = \pm\omega$. The upshot of the rest of the constraints is a final polarization tensor that looks like:

$$P_{\mu\nu} \doteq \begin{pmatrix} 0 & 0 & 0 & 0 \\ 0 & -P_{yy} & P_{xy} & 0 \\ 0 & P_{xy} & P_{yy} & 0 \\ 0 & 0 & 0 & 0 \end{pmatrix}.$$

(8.29)

Well that's great, but what *is* it? We are in a good position for interpretation, equivalent to the E&M case: a flat background with a field on top of it. We

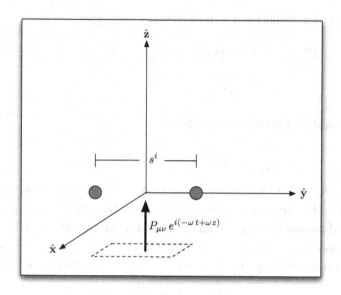

Figure 8.1   A gravitational wave propagating in the $\hat{z}$-direction.

have the polarization tensor, we know the wave vector. The question, as always, is: what happens to particles? We need to go back to our test particle case and see how things move. Our basic physical picture is shown in Figure 8.1, and our goal is to find the properties of particle motion under the influence of this wave.

Referring to Section 4.7.3 and in particular, the equation for geodesic deviation (4.109), we consider two particles separated by s^γ (to avoid using η^γ as the separation vector). These particles each travel along geodesics, and we will assume that the (spatial) velocity of the particles is small compared to the speed of light, so that we can write $\dot{x}^\beta \doteq (1, 0, 0, 0)^T$. Then the relative "acceleration" between the particles becomes:

$$\frac{D^2}{D\tau^2} s^\alpha = -R^\alpha{}_{0\gamma 0}\, \dot{x}^0\, \dot{x}^0\, s^\gamma,\qquad(8.30)$$

leaving us with the task of computing the Riemann tensor to first order in $h_{\mu\nu}$.

Let's take the right-hand side first – we need the linearized Riemann tensor, obtained by taking $g_{\mu\nu} = \eta_{\mu\nu} + \epsilon\, h_{\mu\nu}$, forming the Riemann tensor and dropping all terms quadratic (and higher) in the perturbation parameter ϵ. This procedure amounts to keeping just the derivative of the connection terms in the definition of Riemann (and dropping terms that go like the connection squared):

$$\tilde{R}^\alpha{}_{\beta\gamma\delta} = \frac{1}{2}\,\epsilon\, \eta^{\alpha\rho}\left(h_{\beta\gamma,\delta\rho} + h_{\delta\rho,\beta\gamma} - h_{\beta\delta,\gamma\rho} - h_{\gamma\rho,\beta\delta}\right).\qquad(8.31)$$

The relevant components (in our transverse traceless gauge) are:

$$\tilde{R}^{\alpha}{}_{0\gamma 0} = \frac{1}{2}\epsilon\,\eta^{\alpha\rho}\left(h_{0\gamma,0\rho} + h_{0\rho,0\gamma} - h_{00,\gamma\rho} - h_{\gamma\rho,00}\right)$$

$$= -\frac{1}{2}\epsilon\,\eta^{\alpha\rho}\,h_{\gamma\rho,00},$$

(8.32)

which isn't too bad. On the left, we have:

$$\frac{D^2}{D\tau^2}s^{\gamma} = \dot{x}^{\alpha}\,\dot{x}^{\beta}\,s^{\gamma}{}_{;\alpha\beta} + s^{\gamma}{}_{;\alpha}\,(\dot{x}^{\alpha}{}_{;\beta}\,\dot{x}^{\beta}),$$

(8.33)

the second term is zero because x^{α} is a geodesic by definition. Again, since we are working in the linearized theory, there is no difference to this order between covariant and normal derivatives, and in addition, we can convert $\tau \longrightarrow t$ coordinate time, making the first term on the right in (8.33) $\ddot{s}^{\gamma}$. Putting the left and right together and remembering that there are only spatial components to the metric:

$$\boxed{\ddot{s}^{i} = \frac{1}{2}\epsilon\,\ddot{h}^{i}{}_{j}\,s^{j}.}$$

(8.34)

One last linearization issue – let $s^{i} = \bar{s}^{i} + \epsilon\,\tilde{s}^{i}$, so that we can match orders in the expansion of the deviation and the Riemann tensor. Then we have two equations, one for order ϵ^{0}, one for ϵ^{1} (and then the rest are $O(\epsilon^{2})$):

$$\epsilon^{0}:\ddot{\bar{s}}^{i} = 0$$

$$\epsilon^{1}:\ddot{\tilde{s}}^{i} = \frac{1}{2}\bar{s}^{j}\,\ddot{h}^{i}{}_{j}.$$

(8.35)

The zeroth-order equation tells us that $\bar{s}^{i} = \alpha^{i}\,t + \beta^{i}$, but let's agree to an initial condition: the masses are at rest at time $t = 0$ (before the wave arrives). Then we set $\alpha^{i} = 0$, and the second equation can be integrated (the remaining integration constant(s) β^{i} are just the initial separation):

$$s^{i}(t) = \beta^{i} + \frac{1}{2}\epsilon\,\beta^{j}\,h^{i}{}_{j}.$$

(8.36)

Let's put in our plane-wave solution, with wave vector pointing in the z-direction. Using the form of the polarization tensor from (8.29), the separation vector is given by:

$$\boxed{\begin{aligned} \mathbf{s} &= \left(s^{x}(0) + e^{i\,\omega\,(s^{z}(0)-t)}(s^{y}(0)\,P_{xy} - s^{x}(0)\,P_{yy})\right)\hat{\mathbf{x}} \\ &+ \left(s^{y}(0) + e^{i\,\omega\,(s^{z}(0)-t)}\,(s^{x}(0)\,P_{xy} + s^{y}(0)\,P_{yy})\right)\hat{\mathbf{y}}. \end{aligned}}$$

(8.37)

To see the motion clearly, we take two pairs of test masses. For both pairs, we set $s^{z}(0) = 0$, the wave has no effect on that direction. In addition, it is useful to

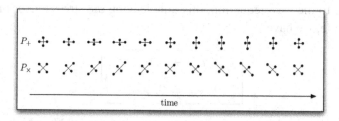

Figure 8.2 The "plus" polarization (top, $P_{xy} = 0$) shows the separation distance as a function of time from (8.37) for initial masses lying on the $\hat{x}$ and $\hat{y}$-axes. The "cross" polarization (bottom, $P_{yy} = 0$) is shown for masses initially separated along the 45° lines relative to the axes.

separate the motion into the two polarizations, obtained by setting $P_{xy} = 0$ and $P_{yy} = 0$, respectively. In the top row of Figure 8.2, we show the displacements for two pairs of test particles: one pair initially positioned on the x-axis, one pair on the y. On the bottom of Figure 8.2, we start with pairs of masses rotated 45° with respect to the axes. By looking at the two polarizations separately, we can see the motivation for the names: $P_{yy} = P_+$, the "plus" polarization, and $P_{xy} = P_\times$, the "cross" polarization.

So, if you want to detect gravitational waves, you need to measure a separation, and indeed, this is fundamentally how current detectors work – there are two test masses (mirrors) separated by a few kilometers, in the shape of an L, and you have effectively a giant interferometer, lasers are shot at the mirrors, you cancel them perfectly at the output and wait for a lack of cancellation, indicating that the mirrors have moved. Sounds easy enough. The problem, as we shall see, is twofold: (1) gravitational waves are incredibly weak, both because of the coupling and the quadrupole nature of the radiation, and (2) the frequency range of interest (for earth-based observations) is on the order of 10 Hz – trucks, for example, are an issue.

8.4 Sources of radiation

We are ready to move on to the source side of linearized waves – what produces these perturbations? I hope that you are appreciating the similarity between gravitational waves and the familiar E&M waves, just an extra index, but it does lead to some interesting physics. On the source side, we will see some of this new physics – in particular, just as monopole sources do not radiate in E&M, monopole and dipole sources do not radiate gravitationally. Structurally, the lack of dipole radiation is a manifestation of the fact that the fundamental field is second rank, together with the gauge conditions. Physically, it can be associated with momentum conservation (just as the lack of monopole radiation "comes from" energy conservation).

Returning to the wave equation, this time leaving the stress tensor intact on the right-hand side, we have:

$$\partial_\rho \partial^\rho h_{\mu\nu} = -16\pi\, T_{\mu\nu}$$
$$\partial^\rho h_{\nu\rho} = 0.$$

(8.38)

The top equation is similar to the source equation for E&M, which would read $\partial_\rho \partial^\rho A^\mu = -4\pi J^\mu$, and can be solved in integral form with a Green's function. The second condition is the analogue of the Lorentz gauge condition, and will, in the end, enforce stress-tensor conservation.

Aside: Green's functions

Here we will develop the Green's function for the D'Alembertian, the solution to $\partial^\rho \partial_\rho \phi = -\delta(x^\mu)$ on a flat background. From this, we can build up solutions to the wave equation that are driven by sources, clearly the next step given the right-hand side of (8.38).

To a certain extent, classical physics, aside from interpretation, is the study of ODEs and PDEs. Most physical laws are expressed as derivative operators on some object of interest – trajectory coordinates, fields, potentials, etc. We have seen this in our discussion of general relativity over and over – Einstein's equation is an example, the linearized Einstein equation another, the equations of motion themselves are an example. Particularly for fields, though, the main tool is the field equations derived from an action – these will be, unless there are simplifying assumptions, partial derivative operators in all four dimensions.

With each linear PDE operator (and boundary conditions) comes an associated "Green's function" – the question is, how do we generically solve a PDE like:

$$\partial^\alpha \partial_\alpha \phi = -\rho(x^\mu)$$

(8.39)

if someone hands us a source function ρ? The somewhat counterintuitive answer is, solve the problem:

$$\partial^\alpha \partial_\alpha G = -\delta(x^\mu - x'^\mu),$$

(8.40)

and then the solution to (8.39) is:

$$\phi(x^\mu) = \int d\tau'\, \rho(x'^\mu)\, G(x^\mu, x'^\mu).$$

(8.41)

That's a bunch of symbols, let's motivate with the familiar three-dimensional Poisson problem from electrostatics. We have:

$$\nabla^2 \phi = -\rho,$$

(8.42)

which means, according to our prescription, we are interested in:

$$\nabla^2 G(\mathbf{x}, \mathbf{x}') = -\delta(\mathbf{x} - \mathbf{x}'), \tag{8.43}$$

and you should think of G as a potential for a point source located at $\mathbf{x}'$ as measured at $\mathbf{x}$ (that's what the "source" $\rho = \delta(\mathbf{x} - \mathbf{x}')$ tells us). In forming the integral $\phi = \int d\tau' \, \rho(\mathbf{x}') \, G(\mathbf{x}, \mathbf{x}')$, we are taking the real distribution of charge (say) and viewing it as a set of points, each point contributes G to the potential and so the total potential is just the sum (integral) of all the points with the appropriate point potential – a statement of superposition. To prove that this ϕ actually solves Poisson's equation, we use "Green's theorem":

$$\int_{\Omega} \left(\phi \, \nabla^2 G - G \, \nabla^2 \phi\right) d\tau' = \int_{\partial\Omega} (\phi \, \nabla G - G \, \nabla \phi) \cdot d\mathbf{a}', \tag{8.44}$$

where the primes tell us the integration is over $\mathbf{x}'$. The integration occurs over whatever volume Ω (and, on the right, its surface, $\partial\Omega$) we like, although it should include the points $\mathbf{x}'$ located inside the distribution if we want to get an interesting answer. Under the usual assumptions about the fields (that they go to zero at spatial infinity), we take the surface at infinity (assuming ϕ and G go to zero out there). This just kills the right-hand side of the above, and then we can evaluate the left because it already involves the Laplacian:

$$\int d\tau' \left(-\phi \, \delta(\mathbf{x} - \mathbf{x}') + G \, \rho\right) = 0 \longrightarrow -\phi(\mathbf{x}) + \int d\tau' \, G \rho = 0 \tag{8.45}$$

recovering (8.41), and we are done.

There are a couple of important facts about Green's functions that one should keep in mind: (1) they depend on a given differential operator *and* boundary conditions – the whole of the specified problem, (2) they depend on the dimension of space (or space-time), (3) they rely on superposition – that is, they only apply to *linear* (or linearized) PDEs. A more useful (and relatively generic) property of Green's functions is their symmetry, $G(\mathbf{x}, \mathbf{x}') = G(\mathbf{x}', \mathbf{x})$. Indeed, for our problems, which have been specified only with the implicit boundary condition at $|\mathbf{x}| \longrightarrow \infty$ (G vanishes there), one can go even further: $G(\mathbf{x}, \mathbf{x}') = G(|\mathbf{x} - \mathbf{x}'|)$.

The Green's function approach, while it looks like a cure-all, requires us to compute the appropriate G for our problem. Because of the δ-function on the right-hand side, this is not trivial, and the calculation of these functions is sometimes tricky. For example, one way to approach the Laplacian solution is to specialize to spherical coordinates, and set the source coordinates $\mathbf{x}'$ to the origin – then the Green's function depends only on r, and away from the origin, we have:

$$\nabla^2 G(r) = 0 \longrightarrow G'' + \frac{2\,G'}{r} = 0 \longrightarrow G(r) = \frac{\alpha}{r} + \beta. \tag{8.46}$$

The PDE $\nabla^2 G(r) = -\delta^3(\mathbf{x})$ *at the origin* has an infinite right-hand side – to get rid of this, we can integrate both sides over an infinitesimal ball (with radius ϵ) enclosing $r = 0$:

$$\int d\tau \, \nabla^2 G(r) = \int d\tau \, \delta(\mathbf{x}) = -1, \tag{8.47}$$

where the right-hand side represents a fundamental property of the δ-function. For the left-hand side, we have:

$$\int d\tau \, \nabla^2 G = \int d\tau \, \nabla \cdot (\nabla G) = \int \nabla G \cdot \hat{\mathbf{n}} \, da = -4\pi\alpha, \tag{8.48}$$

so we set the normalization to $\alpha = \frac{1}{4\pi}$ – the above says nothing about β, but in E&M at least, we know that β just sets the zero of the potential at some point. Our final solution is:

$$G(r) = \frac{1}{4\pi r}. \tag{8.49}$$

If we have the source position at $\mathbf{x}'$ rather than the origin, then the usual shift occurs – take $\boldsymbol{\imath} \equiv \mathbf{x} - \mathbf{x}'$ with magnitude $\imath$. The Green's function solution undergoes the obvious modification:

$$\nabla^2 G = -\delta(\mathbf{x} - \mathbf{x}') \longrightarrow G(|\mathbf{x} - \mathbf{x}'|) = \frac{1}{4\pi\imath}. \tag{8.50}$$

The Green's function approach is useful, because in the linearized Einstein equations with source, we have:

$$\partial^\alpha \partial_\alpha h_{\mu\nu} = -16\pi \, T_{\mu\nu}. \tag{8.51}$$

So we ask the "point-source" question, and develop the Green's function $Q_{\mu\nu}(x^\mu)$ solving:

$$\partial^\alpha \partial_\alpha Q_{\mu\nu} = -\delta(x^\mu) \tag{8.52}$$

(I'm using $Q_{\mu\nu}$ rather than $G_{\mu\nu}$ since the latter refers to the Einstein tensor). Once we have $Q_{\mu\nu}(x^\mu)$ for a source at the origin, we will again shift the source to a different spatial location $\mathbf{x}'$, and now time is also involved, so we will need to introduce t'. As it stands, we are thinking of a source located at the origin that flashes on and off instantaneously at $t = 0$, and the delta function is a product of four one-dimensional ones: $\delta(x^\mu) = \delta(x)\delta(y)\delta(z)\delta(t)$. Clearly, any time-varying distribution can be modeled with these fundamental objects, shifted in time and space appropriately.

Notice that when we expand out the flat-space D'Alembertian in (8.52), we get $-\partial_t^2 + \nabla^2$, so this is almost a Laplacian, and we can make it even more like a Laplacian by a temporal Fourier transform – multiply (8.52) by $e^{i\omega t}$ and integrate over t:

$$-\int_{-\infty}^{\infty} dt \, (\partial_t^2 \, Q_{\mu\nu}(t, \mathbf{x})) \, e^{i\omega t} + \nabla^2 \int_{-\infty}^{\infty} dt \, Q_{\mu\nu}(t, \mathbf{x}) \, e^{i\omega t}$$

$$= -\int_{-\infty}^{\infty} \delta(t) \, \delta(\mathbf{x}) \, e^{i\omega t} \, dt \qquad (8.53)$$

$$-\int_{-\infty}^{\infty} dt \, (-\omega^2) \, Q_{\mu\nu}(t, \mathbf{x}) \, e^{i\omega t} + \nabla^2 \int_{-\infty}^{\infty} dt \, Q_{\mu\nu}(t, \mathbf{x}) \, e^{i\omega t} = -\delta(\mathbf{x}),$$

where we have integrated by parts twice on the left and used $\delta(t)$ on the right to get the second line. Now to set the Fourier transform normalization, define the Fourier transform of $Q_{\mu\nu}$ with respect to time:

$$\tilde{Q}_{\mu\nu}(\omega, \mathbf{x}) = \frac{1}{\sqrt{2\pi}} \int_{-\infty}^{\infty} dt \, Q_{\mu\nu}(t, \mathbf{x}) \, e^{i\omega t}, \qquad (8.54)$$

so that:

$$Q_{\mu\nu}(t, \mathbf{x}) = \frac{1}{\sqrt{2\pi}} \int_{-\infty}^{\infty} d\omega \, \tilde{Q}_{\mu\nu}(\omega, \mathbf{x}) \, e^{-i\omega t}. \qquad (8.55)$$

In terms of these, our wave equation becomes:

$$(\omega^2 + \nabla^2) \, \tilde{Q}_{\mu\nu}(\omega, \mathbf{x}) = -\frac{1}{\sqrt{2\pi}} \delta(\mathbf{x}). \qquad (8.56)$$

The differential operator on the left-hand side of this equation is called the "Helmholtz" operator (its Green's function is known, but let's continue in our derivation). We know, because this PDE comes to us with the implicit condition that $h_{\mu\nu} \longrightarrow 0$ at infinity, that the Green's function depends only on the magnitude of $\mathbf{x}$, denoted r (when we introduce the shifted source-point, we will replace this distance with the relative distance $\imath = |\mathbf{x} - \mathbf{x}'|$ between the source and field point). Then the spherical Laplacian may be used as usual for $\nabla^2 \tilde{Q}_{\mu\nu}(\omega, r)$:

$$\left(\omega^2 + \frac{d^2}{dr^2}\right) \left[r \, \tilde{Q}_{\mu\nu}(\omega, r)\right] = -\frac{r}{\sqrt{2\pi}} \delta(\mathbf{x}). \qquad (8.57)$$

This equation is only valid away from $r = 0$, since we've multiplied through by r to get it in this form – but then we can solve the homogenous (zero right-hand side) problem immediately, it's just a harmonic oscillator:

$$r \, \tilde{Q}_{\mu\nu}(\omega, r) = \alpha \, e^{i\omega r} + \beta \, e^{-i\omega r}. \qquad (8.58)$$

The solution exhibits oscillatory behavior in r. We typically only need one of the two solutions (they will become retarded and advanced potentials in the end), so take:

$$\tilde{Q}_{\mu\nu}(\omega, r) = \alpha \, \frac{e^{\pm i\,\omega\,r}}{r}. \tag{8.59}$$

We need to normalize to a delta function, that will set α as usual. The quickest way to do this is to look at the limit of $\tilde{Q}_{\mu\nu}(\omega, r)$ as $r \longrightarrow 0$ and compare with the Coulomb result (8.49) (we could also work directly from Gauss's law, and normalize by integrating over a sphere as in (8.48), that's equivalent here):

$$\lim_{r \to 0} \tilde{Q}_{\mu\nu}(\omega, r) = \frac{\alpha}{r} \pm \alpha\,(i\,\omega) + O(r), \tag{8.60}$$

so that the leading-order behavior of this function near the origin is identical to the three-dimensional Green's function for the Laplacian (8.49), modulo constants. Then we should set $\alpha = \frac{1}{\sqrt{2\pi}\,4\pi}$ (the extra factor of $\sqrt{2\pi}$ just comes from the modified normalization on the right-hand side of (8.57), detritus from the Fourier transform) to match the $\omega = 0$ Coulomb case, and now we can write the final form for the transformed Green's function:

$$\boxed{\tilde{Q}_{\mu\nu}(\omega, r) = \frac{1}{4\pi\,\sqrt{2\pi}\,r}\, e^{\pm i\,\omega\,r}.} \tag{8.61}$$

To return to t, we Fourier transform back using (8.55):

$$\begin{aligned}
Q_{\mu\nu}(t, r) &= \frac{1}{\sqrt{2\pi}} \int_{-\infty}^{\infty} d\omega\, e^{-i\,\omega\,t} \left(\frac{1}{4\pi\,\sqrt{2\pi}\,r}\, e^{\pm i\,\omega\,r} \right) \\
&= \frac{1}{4\pi\,r}\,\frac{1}{(2\pi)} \int_{-\infty}^{\infty} d\omega\, e^{i\,\omega\,(t\pm r)}.
\end{aligned} \tag{8.62}$$

Using the familiar Fourier transform of the delta function:

$$\frac{1}{2\pi} \int_{-\infty}^{\infty} d\omega\, e^{i\,\omega\,(t-p)} = \delta(t - p), \tag{8.63}$$

we can complete the transformation:

$$Q_{\mu\nu}(t, r) = \frac{\delta(t - r)}{4\pi\,r}. \tag{8.64}$$

Again, this Green's function can be modified to refer to a point source located at $\mathbf{x}'$ at time t' simply by introducing the relative separations $t \longrightarrow t - t', r \longrightarrow \imath \equiv |\mathbf{x} - \mathbf{x}'|$:

$$\boxed{Q_{\mu\nu}(t, \imath) = \frac{1}{4\pi\,\imath}\, \delta\!\left(t - t' \mp \imath\right) = \frac{1}{4\pi\,\imath}\, \delta\!\left(t' - t \pm \imath\right).} \tag{8.65}$$

As usual, both the retarded and advanced contributions are included in the above, selected by appropriate choice of $+$ (retarded) or $-$ (advanced).

8.4.1 Source approximations and manipulation

We have not made any assumptions about the sources themselves yet – and because of the "spatial dependence on time" (through the delta function, as we shall see), we need to be careful to separate out all the spatial dependence before making "far-away" types of approximation. The easiest way to disentangle is to Fourier transform both sides of the integral form of the solution to (8.51) using our Green's function (8.65):

$$h_{\mu\nu} = 4 \int \frac{T_{\mu\nu}(t', \mathbf{x}')}{\imath} \delta(t' - (t \pm \imath)) \, d\tau' \, dt', \tag{8.66}$$

where $d\tau' = dx' \, dy' \, dz'$ (just the spatial portion of the integration).

Let's focus on the right-hand side first – multiply by $e^{i\omega t}$ and integrate over t, using the δ to perform the dt' integral, we have:

$$\int h_{\mu\nu} \, e^{i\omega t} \, dt = 4 \int d\tau' \, dt \, e^{i\omega(t-\imath)} \frac{e^{i\omega\imath}}{\imath} T_{\mu\nu}(t - \imath, \mathbf{x}'), \tag{8.67}$$

where we multiply and divide by the spatial term so that we can define the retarded time $t_- \equiv t - \imath$ (specializing to these causal solutions). A change of variables gives:

$$\int h_{\mu\nu} \, e^{i\omega t} \, dt = 4 \int d\tau' \, dt_- \frac{e^{i\omega\imath}}{\imath} e^{i\omega t_-} T_{\mu\nu}(t_-, \mathbf{x}')$$

$$= 4\sqrt{2\pi} \int d\tau' \frac{e^{i\omega\imath}}{\imath} \tilde{T}_{\mu\nu}(\omega, \mathbf{x}'), \tag{8.68}$$

where $\tilde{T}_{\mu\nu}(\omega, \mathbf{x}')$ is the Fourier transform of the stress tensor.

We make the usual approximation, that $\imath = |\mathbf{x} - \mathbf{x}'|$ is dominated by the observation distance $|\mathbf{x}|$, i.e. we are far away and the origin is centered inside the mass distribution. Then approximately, we have:

$$4\sqrt{2\pi} \int d\tau' \frac{e^{i\omega\imath}}{\imath} \tilde{T}_{\mu\nu}(\omega, \mathbf{x}') \approx 4\sqrt{2\pi} \frac{e^{i\omega|\mathbf{x}|}}{|\mathbf{x}|} \int d\tau' \, \tilde{T}_{\mu\nu}(\omega, \mathbf{x}'). \tag{8.69}$$

We have been working on the Fourier transform of the right-hand side of (8.66), the Fourier transform of the left-hand side is just $\tilde{h}_{\mu\nu}$. The approximation to the Fourier transform of both sides of (8.66) gives (from (8.68)):

$$\boxed{\tilde{h}^{\mu\nu}(\omega, \mathbf{x}) \approx 4 \frac{e^{i\omega|\mathbf{x}|}}{|\mathbf{x}|} \int d\tau' \, \tilde{T}^{\mu\nu}(\omega, \mathbf{x}').} \tag{8.70}$$

The integral manipulations here are important, using the Fourier transform to separate temporal and spatial dependence demands that we appropriately transform the gauge condition. We will first work out the procedure for E&M, and we will then do the same thing for gravitational radiation (see [24] for the E&M version, the gravitational radiation discussion is informed by [4]).

Sources in E&M

The analysis we are about to carry out for the gravitational field perturbation, and its connection to source variability, is similar to the argument made in E&M. Because some of the manipulations to come may be unfamiliar in their current form, let's work out the electromagnetic result (basically the dipole radiation field). In terms of the four-potential A^μ and J^μ (with $A^0 = V/c$, $J^0 = c\,\rho$ and $A^i \doteq \mathbf{A}$, $J^i \doteq \mathbf{J}$ as usual), we have:

$$\partial^\alpha \partial_\alpha A^\mu = -\mu_0 J^\mu, \tag{8.71}$$

and, working in Lorentz gauge:

$$\partial_\alpha A^\alpha = 0. \tag{8.72}$$

By comparing (8.71) with (8.38), we see that the integral form of the Fourier transform of A^μ is, from the analogue of (8.70):

$$\tilde{A}^\mu \approx \frac{\mu_0\, e^{i\,\omega\,|\mathbf{x}|/c}}{4\,\pi\,|\mathbf{x}|} \int d\tau'\, \tilde{J}^\mu(\omega, \mathbf{x}'), \tag{8.73}$$

where I am including all appropriate constants for final comparison (and as an example of how they must be reintroduced).

The Lorentz gauge condition, written out, is:

$$\partial_0 A^0 + \partial_j A^j = 0 \longrightarrow \frac{1}{c}\frac{\partial A^0}{\partial t} + \partial_j A^j = 0, \tag{8.74}$$

where the j index only runs over the spatial coordinates. Now, we need the Fourier transform of this condition, so multiply by $e^{i\,\omega\,t}$, and integrate t from negative infinity to infinity:

$$\frac{1}{\sqrt{2\pi}}\frac{1}{c}\int_{-\infty}^{\infty} \frac{\partial A^0}{\partial t} e^{i\,\omega\,t}\, dt + \partial_j \tilde{A}^j = 0. \tag{8.75}$$

The second term just picks up a tilde, since the spatial coordinates are insensitive to temporal Fourier transformation. The first term can be integrated by parts, assuming A^0 vanishes at temporal infinity, and this leaves us with:

$$-\frac{i\,\omega}{c}\tilde{A}^0 + \partial_j \tilde{A}^j = 0. \tag{8.76}$$

The conservation of charge, $\partial_\rho J^\rho = 0$, gives the same condition on the Fourier side:

$$-\frac{i\,\omega}{c}\tilde{J}^0 + \partial_j \tilde{J}^j = 0. \tag{8.77}$$

Our first observation, from (8.76), is that if we know $\tilde{A}^j$, then we know $\tilde{A}^0$, so we can focus on the spatial components of the four-potential. Then, from (8.77), we see that spatial derivatives of the vector $\tilde{\mathbf{J}}$ are directly proportional to $\tilde{J}^0$, motivating us to try to replace vector components of $\tilde{J}^\mu$ under the integral on the right of (8.73) with $\tilde{J}^0$. To accomplish this goal, note the product rule:

$$\tilde{J}^j(\omega, \mathbf{x}') = \partial_i'\left(\tilde{J}^i\, x'^j\right) - x'^j\, \partial_i' \tilde{J}^i, \tag{8.78}$$

where ∂_i' refers to the derivative with respect to x'^i. Now, inserting this expression for $\tilde{J}^j$ in (8.73) (focusing on the spatial components), the first term will vanish, since it is a total derivative (which turns into a boundary term, evaluated at, say, infinity, where the source vanishes by assumption), and we have:

$$\tilde{A}^j = \frac{e^{i\omega|\mathbf{x}|/c}}{4\pi|\mathbf{x}|} \int d\tau'\left(-x'^j\, \partial_i' \tilde{J}^i\right). \tag{8.79}$$

Notice that our clever use of the product rule has produced the desired divergence of $\tilde{J}^i$, so we can rewrite in terms of $\tilde{J}^0$ using (8.77):

$$\tilde{A}^j = -\frac{i\omega}{c}\, \frac{\mu_0\, e^{i\omega|\mathbf{x}|/c}}{4\pi|\mathbf{x}|} \int d\tau'\left(-x'^j\, \tilde{J}^0\right). \tag{8.80}$$

Finally, we return to A^j, the spatial components of A^μ by Fourier transforming (8.80) – multiply both sides by $e^{-i\omega t}$ and integrate over ω:

$$A^j = \frac{\mu_0}{4\pi c|\mathbf{x}|} \int d\tau'x'^j \left[\frac{1}{\sqrt{2\pi}} \int e^{-i\omega(t-|\mathbf{x}|/c)}\,(-i\omega)\, \tilde{J}^0(\omega, \mathbf{x})\,d\omega\right]$$

$$= \frac{\mu_0}{4\pi c|\mathbf{x}|} \int d\tau'x'^j \frac{\partial}{\partial t}\left[\frac{1}{\sqrt{2\pi}} \int e^{-i\omega(t-|\mathbf{x}|/c)}\, \tilde{J}^0(\omega, \mathbf{x}')\,d\omega\right]. \tag{8.81}$$

The term in square brackets is just the Fourier transform of $J^0(t - |\mathbf{x}|/c, \mathbf{x}')$, so we can write:

$$A^j = \frac{\mu_0}{4\pi|\mathbf{x}|}\, \frac{\partial}{\partial t} \int d\tau'x'^j\, \rho(t_-, \mathbf{x}'), \tag{8.82}$$

with $t_- \equiv t - |\mathbf{x}|/c$, and we recognize $|\mathbf{x}|/c$ as the time it takes light to reach $\mathbf{x}$ from the origin.

When written in terms of the dipole moment of the distribution:

$$p(t)^j = \int d\tau \rho(t, \mathbf{x})\, x^j, \tag{8.83}$$

we have the final result:

$$\boxed{\mathbf{A}(t, \mathbf{x}) = \frac{\mu_0\, \dot{\mathbf{p}}(t_0)}{4\pi|\mathbf{x}|}.} \tag{8.84}$$

To recover the zero component (i.e. the scalar potential, V), we use (8.76), as you will verify in Problem 8.4.

Sources in linearized gravity

From the electrodynamics discussion above, we can return to (8.70) with a view toward the spatial components only – this restricted focus comes (again) from the gauge condition $\partial^\mu h_{\mu\nu} = 0$, and the equivalent statement for the stress tensor (linearized conservation). Specifically, in terms of Fourier components, the gauge condition and conservation read (analogous to (8.76) and (8.77) for E&M):

$$\partial_\mu h^{\mu\nu} = 0 \longrightarrow (-i\,\omega)\,\tilde{h}^{0\nu} + \partial_j\,\tilde{h}^{j\nu} = 0$$
$$\partial_\mu T^{\mu\nu} = 0 \longrightarrow (-i\,\omega)\,\tilde{T}^{0\nu} + \partial_j\,\tilde{T}^{j\nu} = 0. \tag{8.85}$$

Once again, the top equation allows us to solve for the space–space components in terms of the others. Referring only to spatial indices, we can rewrite the integrand on the right in (8.70) as a total divergence and a remaining piece:

$$\int d\tau'\,\tilde{T}^{jk} = \int d\tau'\,\left(\partial_s'\,(\tilde{T}^{sk}\,x'^j) - x'^j\,\partial_s'\,T^{sk}\right), \tag{8.86}$$

where the total divergence is turned into a surface integral, and as long as the distribution is not infinite, can be set to zero (extend the volume of integration to infinity, T^{jk} will be zero outside of some finite region, then the source at infinity is zero). Replacing the spatial derivatives with $\tilde{T}^{0j}$ (using conservation for Fourier modes as in (8.85)) and noting that h^{jk} is symmetric in $j \leftrightarrow k$ interchange, we can rewrite and repeat the process:

$$\int d\tau'\,\tilde{T}^{jk} = -\frac{i\,\omega}{2} \int d\tau'\,\left(x'^k\,\tilde{T}^{0j} + x'^j\,\tilde{T}^{0k}\right)$$
$$= -\frac{i\,\omega}{2} \int d\tau'\,\left(\partial_s'\,(x'^k\,x'^j\,\tilde{T}^{0s}) - x'^k\,x'^j\,\partial_s'\,\tilde{T}^{0s}\right) \tag{8.87}$$
$$= \frac{(i\,\omega)^2}{2} \int d\tau'\,x'^k\,x'^j\,\tilde{T}^{00}.$$

The above is just what we would call (the Fourier transform of) an integral related to the quadrupole moment of the source in E&M (with T^{00} identified as ρ). Again, we see the importance of a second index – in E&M, all of this machinery is applied to the wave equation for the potential $\partial^\alpha \partial_\alpha A^\mu = -\mu_0 J^\mu$. The move from two spatial indices to one via a total derivative and the conservation law for $T^{\mu\nu}$ is the mathematical statement of charge conservation (energy density conservation here). That is identical to E&M, and there, forces us to look at dipole radiation as the first radiating term. With the move from one spatial index to none, as above, we are effectively using momentum conservation in GR, so there is no dipole radiation, in addition to no monopole, and we are left with the quadrupole as the first contributing term to the far-field solutions.

We have, on the Fourier side, the relation:

$$\tilde{h}^{jk}(\omega, \mathbf{x}) = 4 \frac{e^{i\omega|\mathbf{x}|}}{|\mathbf{x}|} \frac{(i\omega)^2}{2} \int d\tau' \, x'^j \, x'^k \, \tilde{T}^{00}(\omega, \mathbf{x}'). \tag{8.88}$$

Multiplying by $e^{-i\omega t}$ and integrating, we can return to time:

$$h^{jk}(t, \mathbf{x}) = \frac{1}{\sqrt{2\pi}} \frac{2}{|\mathbf{x}|} \frac{d}{dt^2} \int d\omega \, e^{-i\omega(t-|\mathbf{x}|)} \int d\tau' \, x'^j \, x'^k \, \tilde{T}^{00}(\omega, \mathbf{x}')$$

$$= \frac{2}{|\mathbf{x}|} \frac{d^2}{dt^2} \left(\int d\tau \, x'^j \, x'^k \, T^{00}(t, \mathbf{x}') \right) \Bigg|_{t=t_- \equiv t - |\mathbf{x}|}. \tag{8.89}$$

Defining the energy density quadrupole to be three times the quantity in parentheses (a normalization), our solution is:

$$\boxed{h^{jk}(t, \mathbf{x}) = \frac{2}{3\,|\mathbf{x}|} \ddot{q}^{jk}|_{t=t_-} \\[2mm] q^{jk} \equiv 3 \int d\tau' \, T^{00}(t, \mathbf{x}') \, x'^j \, x'^k.} \tag{8.90}$$

8.4.2 *Example – circular orbits*

Let's take a simple example to get a picture of the connection between metric perturbations and sources. Our model system is a small body in a circular orbit about a more massive one, as shown in Figure 8.3.

We will take the orbiting body to have a Newtonian circular orbit. In terms of the mass M and orbital radius R, we can find the frequency of the orbit as usual:

$$\frac{m\,v^2}{R} = \frac{M\,m}{R^2} \longrightarrow \boxed{\omega = \frac{v}{R} = \sqrt{\frac{M}{R^3}}}. \tag{8.91}$$

Then in Cartesian coordinates, we can write the trajectory of the orbiting body as:

$$x' = R\,\cos(\omega t)$$
$$y' = R\,\sin(\omega t) \tag{8.92}$$
$$z' = 0.$$

The energy density is just that of a point mass m evaluated along its trajectory:

$$T^{00} = m\,\delta^3(\mathbf{r}') = m\,\delta(z')\,\delta(x' - R\,\cos(\omega t))\,\delta(y' - R\,\sin(\omega t)), \tag{8.93}$$

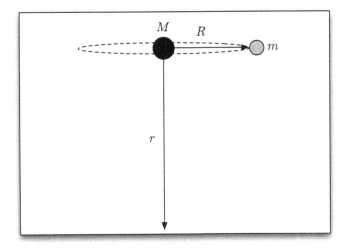

Figure 8.3   A circular orbit for a massive body M and a smaller body m observed on-axis from a distance r away.

and we can calculate the quadrupole moment directly:

$$q^{jk} = 3 \int d\tau'\, T^{00}(x')\, x'^j\, x'^k$$

$$\doteq 3m \int d\tau'\, \delta(z')\, \delta(x' - R\cos(\omega t))\, \delta(y' - R\sin(\omega t)) \begin{pmatrix} x'^2 & x'\,y' & x'\,z' \\ x'\,y' & y'^2 & y'\,z' \\ x'\,z' & y'\,z' & z'^2 \end{pmatrix}$$

$$= 3m \begin{pmatrix} R^2 \cos^2(\omega t) & R^2 \sin(\omega t)\cos(\omega t) & 0 \\ R^2 \sin(\omega t)\cos(\omega t) & R^2 \sin^2(\omega t) & 0 \\ 0 & 0 & 0 \end{pmatrix}.$$

$$\text{(8.94)}$$

Using (8.90), the spatial portion of the metric perturbation is:

$$h^{jk}(t, \mathbf{x}) \doteq \frac{4\, R^2\, m\, \omega^2}{r} \begin{pmatrix} \cos(2\,\omega\, t_-) & \sin(2\,\omega\, t_-) & 0 \\ \sin(2\,\omega\, t_-) & -\cos(2\,\omega\, t_-) & 0 \\ 0 & 0 & 0 \end{pmatrix}. \tag{8.95}$$

For circular orbits, we can replace ω in the magnitude using (8.91), the factor in front of the matrix is $\frac{4\,m\,M}{r\,R}$.

We have constructed the wave to be in the z-direction, and the form of the spatial portion tells us that the perturbation is transverse (already) and traceless for an observer on the z-axis. One of the hallmarks of gravitational waves, owing to the two time-derivatives of the quadrupole, is that oscillations occur with twice the frequency of the source (for circular orbits).

From our discussion of geodesic deviation of test masses under the influence of a metric perturbation in Section 8.3, and in particular (8.36), we know that the separation as a function of time will go roughly like h_{ij} itself times the initial separation ($s(t) \approx \beta\, h(t)$). Given the above form for h_{ij} derived from a circular orbit, we can get an idea of the sensitivity requirements for a gravitational wave detector.

Suppose we are viewing a solar-like orbit, with a massive body (the sun) and a much smaller body (the earth) in an approximately circular orbit. If we view from a platform located at, say, $r = 10\,R$ (10 times the distance to the sun), then we need to measure a separation given by:

$$\Delta s(t) \sim \beta\, h, \tag{8.96}$$

where β is the initial separation of the test masses, and h is:

$$h \sim \frac{4\,m\,M}{r\,R} = \frac{2\,m\,M}{5\,R^2}. \tag{8.97}$$

If we put in $M \approx 1.5$ km for the sun, and the earth has $m \approx 4.4 \times 10^{-6}$ km, with distance between the two $r = 1$ AU $\approx 1.5 \times 10^8$ km, then:

$$h \approx \frac{2 \times (1.5\ \text{km}) \times \left(4.4 \times 10^{-6}\ \text{km}\right)}{5 \times \left(1.5 \times 10^8\ \text{km}\right)^2} \approx 1.2 \times 10^{-22}. \tag{8.98}$$

We take a reasonable initial separation for the test masses, say $\beta \sim 1$ km, then our instrument needs to be sensitive to:

$$\Delta s \sim h\,\beta \approx \left(1.2 \times 10^{-22}\right) \times (1\ \text{km}) \approx 1.2 \times 10^{-22}\ \text{km} \approx 1.2 \times 10^{-9}\text{Å}. \tag{8.99}$$

And one can compare this with the radius of the hydrogen atom, approximately 0.5 Å.

There are better sources, of course, the earth–sun system is a very weak radiator, but while there are more massive, faster systems around, they also tend to be further away. The upshot is, we are pretty much stuck measuring (very) small separations. Amazingly, the ground-based detectors (LIGO, for example – see [22]) are getting close to the sensitivity limit in some frequency ranges.

One last point about gravitational waves. It is not possible to define local energy in the gravitational field – this is just the usual statement about equivalence, how do you separate the background space-time from the portion that would carry energy? To put it another way: in general, we know that space-time is locally flat, that's the starting point of all our work. So in the local frame, where the metric and connection vanish in a small neighborhood, we would say there is no energy associated with the metric. One can define total energy at certain points (like infinity) but not local energy.

That's surprising when we think of linearized gravitational radiation, which has a flat background and waves propagating on it just like E&M – why can't we mimic the definition of field energy there? We can, but this is not a self-consistent procedure in GR, the linearization effectively breaks down, so it's not correct (this was the point of Section 6.7). However, if one insists, and proceeds carefully, it is possible to talk about the energy radiated in gravitational waves from a source like the circular one we have been considering.

8.4.3 Energy carried away in the field

Think of the field $h_{\mu\nu}$ on a flat background (so that in this section, the spatial metric is δ_{ij}), we can compute the energy flux associated with such a field unambiguously.[3] If we knew the stress tensor associated with $h_{\mu\nu}$, $t^{\mu\nu}$, we could use the conservation law for it to find the time rate of change of energy in any volume Ω, that would be:

$$\frac{dE}{dt} = -\oint_{\partial\Omega} t^{0i}\, da_i, \qquad (8.100)$$

as usual (here t^{0i} is playing the role of a gravitational Poynting vector). The $-$ sign ensures that energy flowing out is counted as energy *loss*. If we take, as our volume, a sphere at spatial infinity, then we could compute the energy lost to the gravitational radiation associated with, for example, the circular orbit above. The first problem is attempting to define the stress tensor of the field. The usual prescription from Section 5.6 will work, if we go back to the linearized action, but we can quickly get the relevant form directly from Einstein's equation – remember:

$$G_{\mu\nu} = 8\pi \left(T_{\mu\nu} + t_{\mu\nu}\right), \qquad (8.101)$$

where $T_{\mu\nu}$ is the stress tensor associated with matter or other fields, and $t_{\mu\nu}$ is the stress tensor associated with the field $h_{\mu\nu}$. Now our $h_{\mu\nu}$ perturbation solves the linearized portion of this equation – that means there are "corrective" terms that are quadratic in $h_{\mu\nu}$. We can exploit those terms to define the stress tensor for $h_{\mu\nu}$ – let $G_{\mu\nu} = \tilde{G}_{\mu\nu} + \tilde{\tilde{G}}_{\mu\nu} + \ldots$, where the first term is linear in $h_{\mu\nu}$, the second term quadratic, and so on. Truncating the Einstein tensor expansion on the left of (8.101) (a hallmark of our approximation), we have:

$$\tilde{G}_{\mu\nu} + \tilde{\tilde{G}}_{\mu\nu} = 8\pi\, T_{\mu\nu} + 8\pi\, t_{\mu\nu}, \qquad (8.102)$$

and the linearized equations that we have solved to find $h_{\mu\nu}$ are $\tilde{G}_{\mu\nu} = 8\pi\, T_{\mu\nu}$, so that:

$$\boxed{t_{\mu\nu} = \frac{1}{8\pi}\, \tilde{\tilde{G}}_{\mu\nu}.} \qquad (8.103)$$

---

[3] This is a standard argument, if a little slippery – full treatments can be found in [20, 31, 36].

The task of computing the quadratic portions of the Einstein tensor is involved, but once the dust has settled (see [20] for a good discussion and motivation of the following form), and the transverse-traceless gauge has been enforced, we end up with:

$$t_{\mu\nu} = \frac{1}{32\,\pi} \langle h^{TT\alpha\beta}{}_{,\mu} h^{TT}_{\alpha\beta,\nu} \rangle, \qquad (8.104)$$

where we have introduced the averaging procedure denoted $\langle \circ \rangle$ (over, typically, a full cycle of the radiation) as in E&M. In order to find the energy lost by the system from (8.100), we need the three t_{0k} components of this tensor, so we are interested in:

$$t_{0i} = \frac{1}{32\,\pi} \langle \dot{h}^{TT\alpha\beta} h^{TT}_{\alpha\beta,i} \rangle, \qquad (8.105)$$

where the dot represents the time-derivative.

If we are taking the surface $\partial\Omega$ in (8.100) to be a sphere at spatial infinity, then we also have $da_i = \hat{x}_i\, r^2 \sin\theta\, d\theta d\phi$, with $\hat{x}_i$ the ith component of the radial unit vector, and we can form the integrand:

$$t^{0i}\, da_i = -\frac{1}{32\,\pi} \langle \dot{h}^{TT\alpha\beta} \hat{x}^i h^{TT}_{\alpha\beta,i} \rangle\, r^2 \sin\theta\, d\theta d\phi \qquad (8.106)$$

(where the minus sign comes from raising the temporal index using the Minkowski metric). Since, ultimately, we will be using $h_{\alpha\beta}$ from (8.95) as our field (with transverse-traceless gauge enforced), we know how to compute the spatial derivatives, $\hat{x}^i h_{\alpha\beta,i} = \frac{\partial}{\partial r} h_{\alpha\beta}$, just the derivative with respect to r (think of the more traditional way of writing this dot product: $\frac{\mathbf{x}}{|\mathbf{x}|} \cdot \nabla f(r) = \frac{\partial f(r)}{\partial r}$). Using primes to denote derivatives with respect to r, our final form looks like:

$$t^{0i}\, da_i = -\frac{1}{32\,\pi} \langle \dot{h}^{TT\alpha\beta} h^{TT'}{}_{\alpha\beta} \rangle\, r^2 \sin\theta\, d\theta d\phi. \qquad (8.107)$$

Given that we are only interested in the portion of the stress tensor that exists as $r \longrightarrow \infty$, it is clear that we need only keep the $1/r$ terms in both $\dot{h}_{\alpha\beta}$ and $h'_{\alpha\beta}$ – those will cancel with the r^2 from the area element. But, if we are dropping terms of order $1/r^2$ and higher, then $h'_{\alpha\beta} = -\dot{h}_{\alpha\beta}$ since the relevant r-dependence is carried inside $t_- = t - r$, the retarded time used to relate h_{jk} to T_{jk}, the matter sources, as in (8.90).

We are almost ready to integrate over the surface of our sphere at infinity – the only remaining issue is the transverse-traceless projection of the field $h_{\alpha\beta}$. In general, if we are observing the wave at a location defined by unit vector $\hat{x}^i$, then the tensor we observe will have only spatial components, and these are related to the spatial components of (8.95) via:

$$h_{ij}^{TT} = \left[\left(\delta_i^k - \hat{x}^k\,\hat{x}_i \right) \left(\delta_j^\ell - \hat{x}^\ell\,\hat{x}_j \right) - \frac{1}{2} \left(\delta_{ij} - \hat{x}_i\,\hat{x}_j \right) \left(\delta^{k\ell} - \hat{x}^k\,\hat{x}^\ell \right) \right] h_{k\ell}, \quad (8.108)$$

and from this, we learn that (using the fact that (8.95) is traceless):

$$\dot{h}^{TTij}\,\dot{h}_{ij}^{TT} = \dot{h}^{ij}\,\dot{h}_{ij} - 2\,\hat{x}^i\,\hat{x}^j\,\dot{h}_{ik}\,\dot{h}_j^k + \frac{1}{2}\,\hat{x}^i\,\hat{x}^j\,\hat{x}^k\,\hat{x}^\ell\,\dot{h}_{ij}\,\dot{h}_{k\ell}. \quad (8.109)$$

All of the spatial integration dependence (over θ and ϕ) is stored in the products $\hat{x}^i\,\hat{x}^j$ and $\hat{x}^i\,\hat{x}^j\,\hat{x}^k\,\hat{x}^\ell$ – the field $h_{k\ell}$ itself is independent of these variables. So the spatial integration relies only on knowing how products of unit vectors average over a sphere – think of the three components written in terms of angles: $\hat{x}^1 = \sin\theta\,\cos\phi, \hat{x}^2 = \sin\theta\,\sin\phi$, and $\hat{x}^3 = \cos\theta$ – any product of two distinct elements will have $\cos\phi$ or $\sin\phi$ by itself, or $\cos\phi\,\sin\phi$, all of which integrate to zero when we perform the ϕ integration. The only nonzero components, post-integration, will be the $\left(\hat{x}^1\right)^2, \left(\hat{x}^2\right)^2$, and $\left(\hat{x}^3\right)^2$ diagonal elements of $\hat{x}^i\,\hat{x}^j$. From symmetry, we expect all three of these products to yield the same value when integrated, so we need only compute one of them – take $I = \int \hat{x}^3\,\hat{x}^3\,\sin\theta\,d\theta$, then:

$$\begin{aligned} I &= \int_0^{2\pi} \int_0^{\pi} \hat{x}^3\,\hat{x}^3\,\cos^2\theta\,\sin\theta\,d\theta\,d\phi \\ &= \int_0^{2\pi} \frac{2}{3}\,d\phi \\ &= \frac{4\pi}{3}, \end{aligned} \qquad (8.110)$$

and we conclude that:

$$\int \hat{x}^i\,\hat{x}^j\,\sin\theta\,d\theta d\phi = \frac{4\pi}{3}\,\delta^{ij}. \quad (8.111)$$

Similarly, we can reduce the product $\hat{x}^i\,\hat{x}^j\,\hat{x}^k\,\hat{x}^\ell$, under integration, to a spray of deltas – there are three ways to pair up the four unit vectors, so we expect:

$$\int \hat{x}^i\,\hat{x}^j\,\hat{x}^k\,\hat{x}^\ell\,\sin\theta\,d\theta d\phi = \alpha\left(\delta^{ij}\,\delta^{k\ell} + \delta^{ik}\,\delta^{j\ell} + \delta^{i\ell}\,\delta^{jk} \right) \quad (8.112)$$

and to set α, we just need one component – the easiest is $\left(\hat{x}^3\right)^4$, then:

$$\begin{aligned} I &= \int_0^{2\pi} \int_0^{\pi} \cos^4\theta\,\sin\theta\,d\theta d\phi \\ &= \int_0^{2\pi} \frac{2}{5}\,d\phi \\ &= \frac{4\pi}{5}. \end{aligned} \qquad (8.113)$$

We conclude that $3\alpha = \frac{4\pi}{5}$, and we can write:

$$\int \hat{x}^i \hat{x}^j \hat{x}^k \hat{x}^\ell \sin\theta \, d\theta d\phi = \frac{4\pi}{15} \left(\delta^{ij} \delta^{k\ell} + \delta^{ik} \delta^{j\ell} + \delta^{i\ell} \delta^{jk} \right). \tag{8.114}$$

Putting it all together, we have:

$$\oint_{\partial\Omega} t^{0i} \, da_i = \frac{r^2}{32\pi} \left(\int_0^{2\pi} \int_0^\pi \dot{h}^{TT\alpha\beta} \dot{h}^{TT}_{\alpha\beta} \sin\theta \, d\theta d\phi \right)$$

$$= \frac{r^2}{32\pi} \left\langle \dot{h}^{ij} \dot{h}_{ij} (4\pi) - 2\dot{h}^{ij} \dot{h}_{ij} \left(\frac{4\pi}{3} \right) + \frac{1}{2} \dot{h}^{ij} \dot{h}_{ij} \left(2\frac{4\pi}{15} \right) \right\rangle$$

$$= \frac{r^2}{20} \langle \dot{h}^{ij} \dot{h}_{ij} \rangle, \tag{8.115}$$

where we have again anticipated the fact that $\dot{h}^i_i = 0$. Finally, we are ready to input our field expression from (8.95). Inserting h^{ij}, we have:

$$\oint_{\partial\Omega} t^{0i} \, da_i = \frac{1}{10} 8 \left(4 R^2 m \omega^2 \right)^2 \omega^2 = \frac{64}{5} m^2 R^4 \omega^6, \tag{8.116}$$

where all the time-dependence goes away automatically, so there is no need to perform the temporal averaging.

The original (8.100) now reads:

$$\frac{dE}{dt} = -\frac{64}{5} m^2 R^4 \omega^6. \tag{8.117}$$

Energy loss for a Newtonian orbit implies a decrease in orbital radius, and increase in orbital speed (hence a decrease in the period of the orbit). In 1993, Hulse and Taylor won the Nobel prize for observing the shift associated with gravitational wave energy loss (observed over $\approx$ 20 years) in a binary system. This is the first indirect evidence for gravitational radiation. Current experiments like LIGO [22] and the space-based LISA (see [23]) are attempting direct observation through a separation distance measurement.

Problem 8.1

Plot the "cross" polarization pattern for two pairs of test masses using (8.37) – set the masses on the 45° lines between the x and y-axes, and take an initial separation of 2 for both x and y distances. For the polarization, use $P_{xy} = 0.5$ – this is "large" by the standards of gravitational radiation, but we want to be able to see the pattern. In addition, set $\omega = 1$.

Problem 8.2

Take the metric perturbation $h_{\mu\nu}$ in transverse-traceless gauge, i.e. $g_{\mu\nu} = \eta_{\mu\nu} + \epsilon h_{\mu\nu}$ with $h_{\mu\nu}$ given by $h_{\mu\nu} = P_{\mu\nu} e^{i k_\mu x^\mu}$, and the polarization tensor in the form (8.29). What is the geodesic equation for a particle of mass m to first order in ϵ, assuming the particle moves slowly (so that the terms quadratic in velocities can be ignored, as in (7.36)). Can you see why particle *separations* are necessary for gravitational wave detection?

Problem 8.3

The Green's function for a (linear) differential operator $\mathcal{D}$ is defined to be the solution to $\mathcal{D}G(\mathbf{x}) = -\delta(\mathbf{x})$ (think of $\delta(\mathbf{x})$ as a point source at the origin). Find the (spherically symmetric) Green's function for the following operators (we are only interested in the solutions that die at infinity, except for ∇^2 in $D = 2$, there the interesting solution is infinite as $r \longrightarrow \infty$) – all dimensions are spatial with Euclidean metric, and $r \equiv |\mathbf{x}|$:

(a) $D = 2, \mathcal{D} = \nabla^2$.
(b) $D = 2, \mathcal{D} = \nabla^2 - \mu^2$ for $\mu \in \mathbb{R}$.
(c) $D = 3, \mathcal{D} = \nabla^2$.
(d) $D = 3, \mathcal{D} = \nabla^2 - \mu^2$ for $\mu \in \mathbb{R}$.

Problem 8.4

Use the Lorentz gauge requirement written for the Fourier transform of A^μ:

$$-\frac{i\,\omega}{c}\,\tilde{A}^0 + \partial_j\,\tilde{A}^j = 0 \tag{8.118}$$

together with the spatial solution for $\tilde{A}^j$ in (8.80) to find $V(t, \mathbf{x})$ for the dipole source (i.e. the potential V associated with the vector potential $\mathbf{A}$ in (8.84)).

Problem 8.5

Use the Lorentz gauge requirement in $\mathbf{x}$–t space, $\partial_\mu A^\mu = 0$, directly with:

$$\mathbf{A} = \frac{\mu_0\,\dot{\mathbf{p}}(t - |\mathbf{x}|/c)}{4\,\pi\,|\mathbf{x}|} \tag{8.119}$$

to find V (this is the same as the previous problem, but without the use of the Fourier transform).

Problem 8.6

We have the spatial components of the metric for a mass quadrupole:

$$h^{ij}(t, \mathbf{x}) = \frac{2}{3\,|\mathbf{x}|}\,\ddot{q}^{ij}(t_-) \tag{8.120}$$

in the far-field approximation, with $t_- = t - |\mathbf{x}|/c$. Use the Lorentz gauge condition to find h^{0i}.

*Problem 8.7

We know, from the discussion in Section 8.1, that it should be possible to perform additional gauge fixing in the context of electrodynamics, even once Lorentz gauge has been introduced. For the dipole radiation field, we have:

$$V = \frac{1}{4\pi\epsilon_0}\left[\frac{\mathbf{x}\cdot\mathbf{p}(t_-)}{|\mathbf{x}|^3} + \frac{\mathbf{x}\cdot\dot{\mathbf{p}}(t_-)}{c\,|\mathbf{x}|^2}\right]$$

$$A = \frac{\mu_0\,\dot{\mathbf{p}}(t_-)}{4\pi\,|\mathbf{x}|} \tag{8.121}$$

with $t_- = t - |\mathbf{x}|/c$. By introducing $\mathbf{q}(t_-)$ such that $\dot{\mathbf{q}}(t_-) = \mathbf{p}(t_-)$, find a function Ψ with $\Box^2\Psi = 0$, and $A'_\mu = A_\mu + \Psi_{,\mu}$ with $A'_0 = 0$, i.e. $V' = 0$. Show that the terms of the new potential that dominate far away (in the radiation zone) can be written as:

$$\mathbf{A}' = \frac{\mu_0}{4\pi\,|\mathbf{x}|}\left(\dot{\mathbf{p}}(t_-) - \hat{\mathbf{x}}\,(\hat{\mathbf{x}}\cdot\dot{\mathbf{p}}(t_-))\right), \tag{8.122}$$

where $\hat{\mathbf{x}} = \frac{\mathbf{x}}{|\mathbf{x}|}$.

*Problem 8.8

We can work out the final gauge fixing in E&M from the previous problem in a specific case. Consider a neutral dipole $\mathbf{p} = p_0\cos(\omega t)\hat{\mathbf{z}}$. Show that the potentials:

$$V' = 0$$

$$\mathbf{A}' = \frac{p_0}{4\pi\epsilon_0 c^2}\left[-2\cos\theta\left(\frac{c\cos(\omega t_-)}{r^2} + \frac{c^2\sin(\omega t_-)}{\omega r^3}\right)\hat{\mathbf{r}}\right.$$
$$\left.-\sin\theta\left(\frac{c\cos(\omega t_-)}{r^2} + \frac{c^2\sin(\omega t_-)}{\omega r^3} - \frac{\omega\sin(\omega t_-)}{r}\right)\hat{\boldsymbol{\theta}}\right] \tag{8.123}$$

lead to the same electric and magnetic fields as:

$$V = \frac{p_0\cos\theta}{4\pi\epsilon_0}\left(\frac{\cos(\omega t_-)}{r^2} - \frac{\omega\sin(\omega t_-)}{cr}\right)$$

$$A = -\frac{\mu_0}{4\pi r}p_0\omega\sin(\omega t_-)\left(\cos\theta\,\hat{\mathbf{r}} - \sin\theta\,\hat{\boldsymbol{\theta}}\right) \tag{8.124}$$

by showing that the two sets are related by a gauge transformation (use your results from Problem 8.7, or work from scratch), i.e. there is a Ψ such that:

$$V' = V - \frac{\partial\Psi}{\partial t} \qquad A' = A + \nabla\Psi. \tag{8.125}$$

Note that in both sets, $t_- = t - \frac{r}{c}$, the retarded time.

Problem 8.9

There do exist exact plane-wave solutions to Einstein's equation. We'll find these by making an astute guess of the form, and solving Einstein's equation in vacuum.

(a) Write the line element ds^2 motivated by the form of the polarization tensor (8.29) with $P_{xy} = 0$ (so we are focusing on + polarization):

$$ds^2 = -dt^2 + (1 - P_{yy}(z,t))\,dx^2 + (1 + P_{yy}(z,t))\,dy^2 + dz^2 \tag{8.126}$$

and transform to the right and left-traveling coordinates: $u = t - z$, $v = t + z$ (setting $c = 1$). We are taking the transverse-traceless form from linearized theory, but allowing freedom in $P_{yy}(z, t)$.

(b) To obtain a simple Ricci tensor, it is easiest to parametrize using $p(u)$ and $q(u)$ defined via:

$$p(u)^2 = 1 - P_{yy}(z, t) \quad q(u)^2 = 1 + P_{yy}(z, t). \tag{8.127}$$

Put these into your metric from part a (the resulting form is due to Rosen) and find the one nonzero component of the Einstein tensor (use the `Mathematica` package) – this should be an ODE relating $p(u)$ to $q(u)$ (and their derivatives).

(c) Take, for example, $q(u) = q_0 \, e^{i\,k\,(z-t)}$, and solve for $p(u)$ in this case. Rewrite the original line element (8.126). Is the resulting space-time flat?

Problem 8.10
Given the form for the linearized gravitational wave's stress-energy tensor:

$$t_{\mu\nu} = \frac{c^4}{32\,\pi\,G} \langle \partial_\mu h^{\alpha\beta} \, \partial_\nu h_{\alpha\beta} \rangle \tag{8.128}$$

(for $h_{\alpha\beta}$ in transverse-traceless gauge), find the t^{00}, t^{0i}, and t^{ij} components appropriate for a gravitational wave propagating in the z-direction:

$$h_{\mu\nu} \doteq \begin{pmatrix} 0 & 0 & 0 & 0 \\ 0 & -P_{yy} & P_{xy} & 0 \\ 0 & P_{xy} & P_{yy} & 0 \\ 0 & 0 & 0 & 0 \end{pmatrix} \cos(k\,(z - c\,t)). \tag{8.129}$$

Take the temporal average over one full cycle for the angle brackets: $\langle f \rangle \equiv \frac{1}{T} \int_0^T f \, dt$.

Problem 8.11
The upshot of our gauge fixing in E&M is that given the magnetic vector potential:

$$\mathbf{A} = \frac{\mu_0 \, \dot{\mathbf{p}}(t_-)}{4\,\pi\,|\mathbf{x}|} \tag{8.130}$$

we can transform via gauge choice so that $\mathbf{A}'$ is manifestly transverse to any direction $\hat{x}_i$ (a unit vector) in the radiation zone. Referring to the solution of Problem 8.7, we have:

$$A_i' = \underbrace{\left(\delta_i^j - \hat{x}_i \, \hat{x}^j \right)}_{\equiv p_i^j(\hat{x})} A_j \tag{8.131}$$

where $\hat{x}_i$ is the unit vector pointing from the origin to the observation point. We can use this to compute the magnetic vector potential at any location given the original form $\mathbf{A}$. The same basic procedure holds for h_{ij} (in the radiation zone) calculated via (8.90).

(a) Introduce two copies of the projection, one for each index, so that $\bar{h}_{ij}$ is defined to be:

$$\bar{h}_{ij} = p_i^k(\hat{x}) \, p_j^{\ell}(\hat{x}) h_{k\ell}. \tag{8.132}$$

Show that the perturbation $\bar{h}_{ij}$ is perpendicular to $\hat{x}_i$ (so $\bar{h}_{ij}$ is transverse).

(b) In addition to transverse, enforced above, we want to make the metric perturbation-traceless. Show that the projection:

$$h_{ij}^{TT} \equiv \left(p_i^k(\hat{x}) \, p_j^{\ell}(\hat{x}) - \frac{1}{2} p_{ij}(\hat{x}) \, p^{k\ell}(\hat{x}) \right) h_{k\ell} \tag{8.133}$$

is traceless, and (still) transverse. This projection gives us a way to put a perturbation, computed from (8.90), into transverse-traceless gauge, as we are free to do.

(c) Use your projector to take the h_{jk} we get for the circular case, of the form:

$$h_{jk} \doteq \begin{pmatrix} h_{11} & h_{12} & 0 \\ h_{12} & -h_{11} & 0 \\ 0 & 0 & 0 \end{pmatrix}, \tag{8.134}$$

and transform to the transverse-traceless expression that corresponds to measuring the perturbation along the $\hat{x}$-axis (i.e. with $\hat{x}_i = \delta_{i1}$).

Problem 8.12

Write the transverse-traceless product $h^{TTij} h_{ij}^{TT}$ in terms of h_{ij} as in (8.109) (use the projectors from the previous problem to simplify the calculation) – assume that $h_i^i = 0$.

Problem 8.13

We have:

$$\frac{dE}{dt} = -\frac{64}{5} m^2 R^4 \omega^6, \tag{8.135}$$

with $G = c = 1$.

(a) By requiring that the left-hand side be in Joules per second, find the combination of G and c that must be introduced on the right.

(b) Write the classical energy $E = \frac{1}{2} m v^2 - \frac{GMm}{R}$ for the circular orbit in terms of T, the period of the orbital motion ($T = \frac{2\pi R}{v}$) and constants (m, M, G, for example).

(c) By taking the time-derivative of your expression for E in terms of T, solve for $\frac{dT}{dt}$ in terms of $\frac{dE}{dt}$.

(d) Rewrite your unit-corrected expression for the right-hand side of (8.135) in terms of T and constants, and insert this for $\frac{dE}{dt}$ from your answer in part c. This final expression gives an ordinary differential equation relating $\frac{dT}{dt}$ to T.

*Problem 8.14

It is possible to solve Einstein's equation in matter itself – our radiation examples are still *vacuum* solutions, meaning that they are valid away from the source. So the source (and our observing location) set the geometry and strength of the radiation. In this problem, we will set up and find the equations governing the metric *inside* a distribution of matter (similar to solving for the electric field, say, inside a sphere with constant charge density).

(a) Write the stress tensor $T_{\mu\nu}$ for an infinite region filled with uniform mass density $\rho(t)$ (we will *find* the time-dependence of $\rho(t)$, all we know for now is that the density is the same everywhere, we do not know how it changes with time).

(b) For such uniform distributions of mass, it is possible to solve Einstein's equation using a single undetermined metric function that is a time-dependent factor sitting in front of the otherwise flat spatial portion:

$$ds^2 = -dt^2 + A(t)^2 \left(dx^2 + dy^2 + dz^2\right). \tag{8.136}$$

Solve Einstein's equation $G_{\mu\nu} = 8\pi\, T_{\mu\nu}$ for $A(t)$ (use the Mathematica package to compute $G_{\mu\nu}$). Note that the time-dependence of $\rho(t)$ is then fixed – what must $\rho(t)$ be in order for Einstein's equation to hold?

(c) This metric represents the simplest possible "cosmological model". It is an example of the Robertson–Walker cosmological model for a dust-filled universe. Is your solution a disguised flat space-time? Compute the curvature of the metric.

9

Additional topics

We close with some additional topics of interest both to general relativity, and to the program of "advanced classical mechanics". First, because it is of the most immediate astrophysical interest, we discuss qualitatively the Kerr solution for the exterior of a spinning, spherically symmetric mass. The Kerr metric is stated, and its linearization compared to the closest electromagnetic analogue: a spinning sphere of charge. Using this comparison, we can interpret the two parameters found in the Kerr metric (when written in Boyer–Lindquist coordinates) as the mass and angular momentum (per unit mass) of the source.

The Kerr metric is nontrivial to derive using our Weyl method, so we are content to verify that it is a solution to Einstein's equation in vacuum. Some of the physical implications of the Kerr solution are available in linearized form, but the more interesting and exotic particle motions associated with the geodesics are easier to explore numerically (see [6] for an exhaustive analytical treatment). For this reason, we include a brief discussion of numerical solutions to the geodesic ODEs that arise in the context of the Kerr space-time. In some ways, a hands-on approach to these trajectories can provide physical insight, or at the very least, predictions of observations that are interesting and accessible.

Finally, given the work we have done on variational methods and geometry, we are in good position to understand extremization in the context of physical area minimization. This is the famous "minimal surface" problem that shows up in a variety of physical configurations. If we move the physical area problem to Minkowski space-time, then we are minimizing a generalized area, and the natural physical object is a relativistic string. We close by studying the equations of motion for a classical (non-quantum) relativistic string, making contact with the now familiar notions of action variation, gauge choice, and boundary conditions.

9.1 Linearized Kerr

We know that Newtonian gravity, as a theory, requires both source-field modification:

$$\nabla^2 \phi = 4 \pi \rho, \tag{9.1}$$

has no time-dependence, and hence, cannot enforce information propagation velocity at c, and field-particle modification:

$$m \ddot{x} = -m \nabla \phi \tag{9.2}$$

allows particles to travel faster than c in finite time (for ϕ the potential of a spherically symmetric source, for example).

We have seen in Section 7.2 how GR, in the weak-field limit, can produce gravitomagnetic forces, and why these forces are necessary from a special-relativistic point of view. What we will do now is use our ideas about spinning charged spheres from electricity and magnetism to predict, qualitatively, the form of the metric associated with a spinning massive sphere, and show that the weak-field Kerr metric takes just such a form. This association allows us to interpret the parameters found in the Kerr metric in terms of the central body's mass and spin.

We will exploit the correspondence between E&M and Newtonian gravity to start the process, extending the analogy to magnetostatics.[1]

9.1.1 Spinning charged spheres

A sphere of radius R and total charge Q spins about the $\hat{z}$-axis with uniform angular velocity $\boldsymbol{\omega} = \omega\,\hat{z}$, as shown in Figure 9.1.

This sphere has electrostatic potential given by:

$$V = \frac{Q}{4 \pi \epsilon_0 r} \tag{9.3}$$

and magnetic vector potential:

$$\mathbf{A} = \frac{1}{2} \frac{\mu_0 \, Q \, \ell \, \sin\theta}{4 \pi \, r^2} \hat{\phi} \tag{9.4}$$

for $\ell = \frac{I\omega}{M}$ the angular momentum per unit mass, with moment of inertia $I = \frac{2}{5} M R^2$ for a uniform sphere (see [17] for these potentials).

Consider the motion of a test particle of mass m moving under the influence of these potentials. The Lagrangian governing test particle motion for nontrivial V and $\mathbf{A}$ is the usual:

$$L = \frac{1}{2} m v^2 - q V + q \mathbf{v} \cdot \mathbf{A}, \tag{9.5}$$

---

[1] Some of this material can be found in [15].

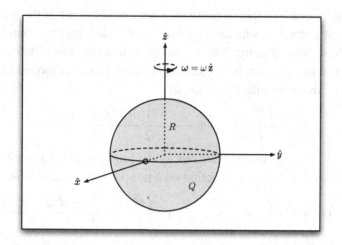

Figure 9.1   A sphere of radius R carrying total charge Q spins with constant angular speed ω about the $\hat{\mathbf{z}}$-axis.

leading to the Lorentz force law when varied. If we input our expressions for this physical configuration, and work in spherical coordinates, the Lagrangian can be written explicitly, prior to variation, as:

$$L = \frac{1}{2} m \left(\dot{r}^2 + r^2 \dot{\theta}^2 + r^2 \sin^2 \theta \, \dot{\phi}^2 \right) - \frac{q \, Q}{4 \pi \, \epsilon_0 \, r} + \frac{1}{2} \frac{q \, \mu_0 \, Q \, \ell \, \sin^2 \theta}{4 \pi \, r} \dot{\phi}, \quad (9.6)$$

where we have used the relevant expression for the $\hat{\boldsymbol{\phi}}$ component of the test particle velocity, $v_\phi = r \sin \theta \, \dot{\phi}$, to construct the $\mathbf{v} \cdot \mathbf{A}$ term.

9.1.2 Static analogy

We know that for Newtonian gravity, the potential energy associated with a spherically symmetric central body is:

$$m \, \phi = -\frac{m \, G \, M}{r} \quad (9.7)$$

for a test particle with mass m. Comparing with the electrostatic potential, we see that structurally, the solutions are similar, with both satisfying Laplace's equation away from the source. So the fundamental difference, aside from the physical forces involved, is one of units. If we take the electrostatic result, we can map to the Newtonian result by sending:

$$\frac{q \, Q}{4 \pi \, \epsilon_0} \longrightarrow -m \, M \, G. \quad (9.8)$$

Now, since we know that a gravitomagnetic term is *possible*, given only special relativistic considerations as seen in Section 2.6.2, and given the similarity between the electrostatic and Newtonian result, it is reasonable to ask if the magnetostatic force on a test particle has an analogue in an extended, special relativistic modification of Newtonian gravity. If there was such an analogous source and force, it would give rise to a term in the Lagrangian (we suspect) that looked like the $\mathbf{v} \cdot \mathbf{A}$ term in (9.6). Given our map (9.8), we send:

$$\frac{1}{2} \frac{q \, \mu_0 \, Q \, \ell \, \sin^2 \theta}{4 \, \pi \, r} \dot{\phi} \longrightarrow -\frac{1}{2} \frac{m \, M \, G \, \ell \, \sin^2 \theta}{r \, c^2} \dot{\phi}, \tag{9.9}$$

noting that $\frac{1}{\mu_0 \, \epsilon_0} = c^2$ as usual. Then our model Lagrangian for a spinning *massive* sphere of radius R with mass M acting on a test particle of mass m is:

$$L_G = \frac{1}{2} m \left(\dot{r}^2 + r^2 \, \dot{\theta}^2 + r^2 \, \sin^2 \theta \, \dot{\phi}^2 \right) + \frac{m \, M \, G}{r} - \frac{1}{2} \frac{m \, M \, G \, \ell \, \sin^2 \theta}{r \, c^2} \dot{\phi}. \tag{9.10}$$

At this stage, all we can say is that if we saw a Lagrangian of this form, we would know how to interpret it by analogy with the E&M configuration. Given our discussion of weak, linearized gravity in Section 7.2, and our discussion of the need for source currents as a purely special relativistic issue from Section 2.6, perhaps the existence of such a Lagrangian is reasonable.

9.1.3 General relativity and test particles

We know the basic idea: Einstein's equation tells us the space-time metric associated with a distribution of energy, just as Maxwell's equations tell us the electric and magnetic fields (or potentials) associated with a distribution of charge. Once we have a metric that solves Einstein's equation, in this case, in vacuum (so, outside the distribution), we can solve for test-particle trajectories – those are geodesics (minimal length trajectories) of the space-time defined by the metric.

The geodesic action is, in arbitrary parametrization:

$$S = -m \, c \int \sqrt{-\dot{x}^\mu \, g_{\mu\nu} \, \dot{x}^\nu} \, d\tau, \tag{9.11}$$

where τ is just a curve parameter, with no physical significance attached to it. Suppose the metric can naturally be broken up into a temporal coordinate and three spatial ones, then we can imagine the partitioning:

$$g_{\mu\nu} \doteq \begin{pmatrix} g_{00} & g_{0i} \\ g_{0i} & g_{ij} \end{pmatrix} \tag{9.12}$$

for coordinates $dx^\mu \doteq \begin{pmatrix} c \, dt & dx & dy & dz \end{pmatrix}^T$. In terms of this, the geodesic action reads:

$$S = -m \, c \int \sqrt{-c^2 \, g_{00} \, \dot{t}^2 - 2 \, g_{0i} \, c \, \dot{x}^i - g_{ij} \, \dot{x}^i \, \dot{x}^j} \, d\tau, \tag{9.13}$$

where i, j run from $1 \longrightarrow 3$ and represent the spatial coordinates. The action, as written above, is reparametrization-invariant as we have seen before. That is, if we take $s(\tau)$ to be some new parameter, then $\frac{dx^\mu}{d\tau} = \frac{dx^\mu}{ds}\frac{ds}{d\tau}$ and the action, in this new s parametrization, is:

$$S = -m\,c \int \sqrt{-\frac{dx^\mu}{ds} g_{\mu\nu} \frac{dx^\nu}{ds} \frac{ds}{d\tau} \frac{d\tau}{ds}}\,ds$$

$$= -m\,c \int \sqrt{-\frac{dx^\mu}{ds} g_{\mu\nu} \frac{dx^\nu}{ds}}\,ds. \tag{9.14}$$

So we can take the parameter to be anything we like without changing either the interpretation of the variation (length extremization) or the form of the action. In particular, for comparison with the Lagrangian (9.10), a classical Lagrangian with time as the parameter of the curve, we choose $\tau = t$ in (9.13). Then our geodesic action, in temporal parametrization, is:

$$\boxed{S = -m\,c \int \sqrt{-c^2\,g_{00} - 2\,g_{0i}\,c\,v^i - g_{ij}\,v^i\,v^j}\,dt,} \tag{9.15}$$

with $v^i = \frac{dx^i}{dt}$, the "usual" time-parametrized tangent to the curve defined by $x^i(t)$, i.e. velocity.

9.1.4 Kerr and the weak field

The Kerr metric is an exact solution to Einstein's equations in vacuum. It is axisymmetric in the spherical-coordinates sense – that is, it is independent of ϕ. As we shall see, the interpretation of the Kerr metric as governing the space-time outside of a spinning massive source is natural. For now, we will only state the form of the metric in a particular coordinate system. Issues of derivation are explored in [6].

In coordinates labeled by $dx^\mu \doteq \begin{pmatrix} c\,dt & dr & d\theta & d\phi \end{pmatrix}^T$, the Kerr metric takes the form (in Boyer–Lindquist coordinates):

$$g_{\mu\nu} \doteq \begin{pmatrix} -\left(1 - \frac{2MGr}{c^2 \rho^2}\right) & 0 & 0 & -\frac{2aMGr\sin^2\theta}{c^3 \rho^2} \\ 0 & \frac{\rho^2}{\Delta} & 0 & 0 \\ 0 & 0 & \rho^2 & 0 \\ -\frac{2aMGr\sin^2\theta}{c^3 \rho^2} & 0 & 0 & \left(\left(r^2 + \left(\frac{a}{c}\right)^2\right) + \frac{2a^2 MGr\sin^2\theta}{c^4 \rho^2}\right)\sin^2\theta \end{pmatrix}$$

$$\rho^2 \equiv r^2 + \left(\frac{a}{c}\right)^2 \cos^2\theta$$

$$\Delta \equiv \left(\frac{a}{c}\right)^2 - 2\frac{MGr}{c^2} + r^2. \tag{9.16}$$

All we know, for now, is that $R_{\mu\nu} = 0$ (check it). There are two "parameters" in the above, M and a, and it is our goal to understand these parameters and the physical source configuration to which they refer.

Let's look at the limiting behavior of the Kerr metric – if we set $a = 0$, we recover Schwarzschild space-time, in Schwarzschild coordinates. That is the motivation for the parameter label M in the Kerr metric. When whatever a refers to is not there, we have the metric appropriate to the exterior of a static, spherically symmetric body of mass M. If we set $M = 0$, we get back a flat space-time, i.e. Minkowski in disguised coordinates (how do we know this is flat?), and that tells us very little about the physics of a.

There are, just going from units, two natural lengths defined by the above. We have our usual $\frac{MG}{c^2}$, but in addition, there is $\frac{a}{c}$. In order to work in the weak-field limit, where we can interpret the metric as flat space-time with a perturbation that causes an effective force (one we could compare with (9.10)), we must be far away w.r.t. both of these intrinsic lengths. So, expanding the Kerr metric with $r \gg \frac{MG}{c^2}$ and $r \gg \frac{a}{c}$, we get a flat space portion and a correction:

$$g_{\mu\nu} \approx \begin{pmatrix} -1 & 0 & 0 & 0 \\ 0 & 1 & 0 & 0 \\ 0 & 0 & r^2 & 0 \\ 0 & 0 & 0 & r^2 \sin^2\theta \end{pmatrix} + \begin{pmatrix} \frac{2MG}{c^2 r} & 0 & 0 & -\frac{2aMG\sin^2\theta}{c^3 r} \\ 0 & \frac{2MG}{c^2 r} & 0 & 0 \\ 0 & 0 & 0 & 0 \\ -\frac{2aMG\sin^2\theta}{c^3 r} & 0 & 0 & 0 \end{pmatrix}.$$

$$(9.17)$$

It is from this form that we can interpret the parameter a, and we do this by inserting the linearized metric into (9.15).

Linearized Kerr effective potential

We are ready to use (9.15) with (9.17) – if we factor out a c in the action to make the orders in small velocity clear, we can write the associated Lagrangian (just the integrand) as:

$$L = -m c^2 \sqrt{-g_{00} - 2 g_{0i} \frac{v^i}{c} - g_{ij} \frac{v^i}{c} \frac{v^j}{c}}. \tag{9.18}$$

The final approximation we will make is that the motion of the test particle is small relative to c – and then we can input the linearized Kerr metric and expand in powers of $\frac{v}{c}$. If we do this, we end up with:

$$L \sim -m c^2 \left(1 - \frac{MG}{c^2 r} + \frac{2aMG\sin^2\theta}{c^4 r} \dot{\phi} - \frac{1}{2} \frac{v^2}{c^2} \right)$$

$$= -m c^2 + \frac{1}{2} m v^2 + \frac{mMG}{r} - \frac{2amMG\sin^2\theta}{c^2 r} \dot{\phi}. \tag{9.19}$$

Now in this final form, we are close to making the required association. As a finishing move, we take v^2 above at face value, as the flat-space speed (squared) of the test particle. This is part of our approximation – technically there is an additional term coming from the g_{rr} perturbative component of the metric, but that term goes like $\frac{M G}{c^2 r} \frac{v_r}{c} \frac{v_r}{c}$ for radial speed v_r – this expression is cubic in small terms, one from the metric linearization and two factors of v/c, so we drop it.

Our final expression for the slow-moving, weak-field limit of motion, under the influence of the Kerr metric, is:

$$
\begin{aligned}
L_{eff} &= -m c^2 + \frac{1}{2} m \left(\dot{r}^2 + r^2 \dot{\theta}^2 + r^2 \sin^2 \theta \, \dot{\phi}^2 \right) \\
&\quad + \frac{m M G}{r} - \frac{2 a m M G \sin^2 \theta}{c^2 r} \dot{\phi},
\end{aligned}
\tag{9.20}
$$

and this is directly comparable to (9.10). The additive constant $-m c^2$ doesn't change equations of motion or dynamics, just sets a background energy scale (rest energy). We see that the correct interpretation of the constant a in the Kerr metric is $a \sim \ell$, so that a is related to the flat-space angular momentum per unit mass from the spinning sphere example on the E&M side. Notice that there is also a factor of four difference between the linearized term above and the E&M prediction for corrected Newtonian gravity. This famous counting factor tells us, effectively, that we have a linearized second-rank tensor theory standing in for a first-rank tensor theory, and was predicted in our original weak-field expansion from Section 7.2 (where $h_{0i} = 4 A_i$, as in (7.31)).

9.1.5 Physics of Kerr

Any physical prediction you would make about test particle motion under the influence of the electromagnetic spinning charged sphere holds in qualitative approximation for a spinning massive sphere. In particular, for E&M we know that if a test body is itself spinning, and placed in the exterior dipole field of a spinning central body, we will observe precession of the test body spin. The same should hold for a spinning massive test body placed in the weak-field exterior of spinning central bodies – in this context, the precession is known as Lense–Thirring precession, and the Gravity Probe B (see [21]) experiment is currently in orbit attempting a direct measurement of this effect.

Most importantly, we have a simple physical picture for the source of the Kerr metric. The actual orbital motion of test bodies in the strong field regime is exotic, and can only be described completely numerically, but we now have a rough, qualitative way to predict behavior far away from central sources. The Kerr metric is most relevant to astrophysics since most bodies are roughly spherically symmetric, and spinning.

9.2 Kerr geodesics

If we think about the geodesic equations of motion for the Kerr metric, it is clear that only very special solutions are accessible in closed form. In order to find specific orbits, we can use numerical methods to solve the geodesic equations, so we will discuss, in simplified form, a numerical approach to the problem of particle motion in the Kerr space-time.

9.2.1 Geodesic Lagrangian

In Boyer–Lindquist coordinates, the Kerr geodesic Lagrangian, $L = \frac{1}{2} \dot{x}^\mu g_{\mu\nu} \dot{x}^\nu$, reads (setting $G = c = 1$):

$$L = \frac{1}{2} \left(-1 + \frac{2 M r}{r^2 + a^2 \cos^2 \theta} \right) \dot{t}^2$$

$$+ \frac{1}{2} \left(\frac{r^2 + a^2 \cos^2 \theta}{a^2 - 2 M r + r^2} \right) \dot{r}^2 + \frac{1}{2} \left(r^2 + a^2 \cos^2 \theta \right) \dot{\theta}^2$$

$$+ \frac{1}{2} \left((r^2 + a^2) + \frac{2 M a^2 r \sin^2 \theta}{r^2 + a^2 \cos^2 \theta} \right) \dot{\phi}^2 - \left(\frac{2 M a r \sin^2 \theta}{r^2 + a^2 \cos^2 \theta} \right) \dot{t} \dot{\phi}, \quad (9.21)$$

where t, r, θ, and ϕ describe the location of a massive test particle, parametrized by τ, its proper time (so all four coordinates are functions of τ, and $\dot{r} = \frac{dr}{d\tau}$, for example).

We can get rid of both the $\dot{t}$ and $\dot{\phi}$ terms by noting that:

$$\frac{d}{d\tau} \left(\frac{\partial L}{\partial \dot{t}} \right) = 0 \qquad \frac{d}{d\tau} \left(\frac{\partial L}{\partial \dot{\phi}} \right) = 0 \qquad (9.22)$$

from the Euler–Lagrange equations – these two equations correspond to energy and angular momentum conservation. If we set:

$$-E = \frac{\partial L}{\partial \dot{t}} \quad \text{and} \quad J_z = \frac{\partial L}{\partial \dot{\phi}}, \qquad (9.23)$$

then the Lagrangian, in terms of $\dot{r}, \dot{\theta}, E$, and J_z, becomes:

$$L = \frac{1}{2} \left(\frac{r^2 + a^2 \cos^2 \theta}{a^2 - 2 M r + r^2} \right) \dot{r}^2 + \frac{1}{2} \left(r^2 + a^2 \cos^2 \theta \right) \dot{\theta}^2$$

$$- \left(\frac{a^2 E^2 \cos(2\theta) - 2 J_z^2 \csc^2 \theta + \frac{a^4 E^2 + 2 a^2 J_z^2 + E r \left(2 M a (a E - 4 J_z) + 3 a^2 E r + 2 E r^3 \right)}{a^2 - 2 M r + r^2}}{4 \left(r^2 + a^2 \cos^2 \theta \right)} \right).$$

$$(9.24)$$

In the interests of getting to a manageable equation of motion, we will now specialize our trajectories – consider only those geodesics that lie in the equatorial
plane: $\theta = \frac{1}{2}\pi$ with $\dot\theta = 0$.[2] This assumption is not necessary – there is another
constant of the motion that can be used to eliminate $\dot\theta$ – this is Carter's constant.
But Carter's constant comes from a second-rank Killing tensor (i.e. it comes from a
symmetric tensor $f_{\mu\nu}$ satisfying Killing's equation $f_{(\mu\nu;\gamma)} = 0$), and the associated
infinitesimal transformation does not have a direct geometrical interpretation. Its
existence does render the equations of motion integrable, and there are a variety of
bound, nonplanar solutions one can characterize using the full set of constants of
the motion (see [6] for an exhaustive treatment).

Our current goal is to generate ordinary differential equations that we can solve
numerically, and the restriction to the equatorial plane allows us to focus on that
program – in the end, what we are about to do can be carried out directly from the
geodesic equations of motion for Kerr ((3.100) with $g_{\mu\nu}$ given by (9.16), and the
appropriate metric connection), or from the Euler–Lagrange equations associated
with the full Lagrangian (9.24). We'll take a slightly shorter route by simplifying
the Lagrangian, and treating it as a constant of the motion. The Lagrangian, under
our planar assumption, becomes:

$$L = \frac{-2\,M\,(J_z - a\,E)^2 + \left(J_z^2 - a^2\,E^2\right)r + r^3\left(\dot r^2 - E\right)}{2r\left(a^2 - 2\,M\,r + r^2\right)}. \tag{9.26}$$

Using the fact that $L = -\frac{1}{2}$, numerically, we can solve for $\dot r^2$ entirely in terms of r
and constants:

$$\dot r^2 = \frac{1}{r^3}\left(2\,M\,(J_z - a\,E)^2 + \left(a^2\,(E^2 - 1) - J_z^2\right)r + 2\,M\,r^2 + (E^2 - 1)r^3\right). \tag{9.27}$$

In general, we avoid solving differential equations that contain square roots – in
this case, the sign of the r-velocity changes depending on whether the test particle
is moving toward or away from the central body. While we could keep track of
that sign separately, it is easier to take the derivative of the above, solve for $\ddot r$, and
use this ordinary differential equation as the basis of a full solution. The sign of $\dot r$
is then correctly, and automatically, tracked by our numerical routine. Taking the
derivative of both sides of (9.27) and solving for $\ddot r$ gives:

$$\ddot r = \frac{1}{r^4}\left(-3\,M\,(a\,E - J_z)^2 + \left(a^2\,(1 - E^2) + J_z^2\right)r - M\,r^2\right). \tag{9.28}$$

---

[2] That there exist such orbits, ones that remain in the plane of motion, is by no means obvious – we can establish
that for this choice, $\ddot\theta = 0$, for example, only from the full geodesic equations, obtainable by taking:

$$\frac{d}{d\tau}\left(\frac{\partial L}{\partial \dot r}\right) - \frac{\partial L}{\partial r} = 0 \qquad \frac{d}{d\tau}\left(\frac{\partial L}{\partial \dot\theta}\right) - \frac{\partial L}{\partial \theta} = 0, \tag{9.25}$$

and solving for $\ddot r$ and $\ddot\theta$. Such planar trajectories do exist, as the reader may verify.

In addition, it is useful, when producing output, to keep track of ϕ as well as r – in our planar setting, $\dot{\phi}$ is given by the simple expression (obtained from the solution of (9.23) for $\dot{\phi}$):

$$\dot{\phi} = \frac{2\,M\,a\,E - 2\,M\,J_z + J_z\,r}{r\,(a^2 - 2\,M\,r + r^2)}. \tag{9.29}$$

We can turn the pair, $\dot{r}$ and $\dot{\phi}$, into a set of three, first-order ODEs (this should remind you of the Hamiltonian formulation for geodesics, another good starting point for this investigation of Kerr geodesics):

$$\frac{d}{d\tau}\begin{pmatrix} r \\ \dot{r} \\ \phi \end{pmatrix} = \begin{pmatrix} \dot{r} \\ \frac{1}{r^4}\left(-3\,M\,(J_z - a\,E)^2 + \left(a^2\,(1 - E^2) + J_z^2\right)r - M\,r^2\right) \\ \frac{2\,M\,a\,E - 2\,M\,J_z + J_z\,r}{r(a^2 - 2\,M\,r + r^2)} \end{pmatrix}. \tag{9.30}$$

So, we have reduced the problem of finding geodesic, planar, trajectories to one of solving three first-order ordinary differential equations.

Problem 9.1

From the form of the Kerr metric in (9.16), it should be clear that for $a = 0$, we recover Schwarzschild. Characterize, as completely as you can, the space-time in the case that $M = 0$, $a \neq 0$.

Problem 9.2

Show that the Kerr metric (9.16) satisfies Einstein's equation in vacuum.

Problem 9.3

Find the radii of the circular, equatorial light-like orbits in the Kerr space-time – leave your answer in terms of E and J_z, the energy and angular momentum of the orbit (there are two equations for r that define the circular orbits – one allows us to solve r in terms of E and J_z, and the other constrains the allowed relationship between E and J_z).

Problem 9.4

We can cast the geodesics of Kerr in Hamiltonian form, allowing additional analysis and a complementary point of view.

(a) Using Mathematica, find the form of the contravariant metric associated with (9.16) (set $G = c = 1$).

(b) Construct the Hamiltonian $H = \frac{1}{2}\,p_\alpha\,g^{\alpha\beta}\,p_\beta$ specialized to the equatorial plane by setting $\theta = \frac{1}{2}\pi$ and $p_\theta = 0$. The Hamiltonian is itself a constant of the motion, set $H = -\frac{1}{2}$ and solve for p_r.

(c) Returning to the definition of L in (9.24), find p_r in terms of $\dot{r}$ and verify that your Hamiltonian-generated equation for p_r matches the Lagrangian-generated equation for $\dot{r}$ in (9.27) (keep in mind that $p_0 = -E$ and $p_\phi = J_z$).

*Problem 9.5

As we noted in the Schwarzschild case (see Problem 7.13), true singularities must be uncovered by computing scalars. Since $R = 0$ for the Kerr metric, and $R_{\mu\nu} = 0$, the next most natural scalar to construct is the square of the Riemann tensor: $R^{\alpha\beta\gamma\delta} R_{\alpha\beta\gamma\delta}$. Form this scalar, and find the singularity of the Kerr metric (there is only one) – it corresponds to a choice of both r and θ.

9.2.2 Runge–Kutta

We now turn to a numerical method that allows approximate solution to equations of the form (9.30). Consider the one-dimensional ordinary differential equation for $p(\tau)$:

$$\dot{p} = Q(\tau, p), \tag{9.31}$$

where $Q(\tau, p)$ is some specified function, and we must also provide the value of p at some point along the trajectory ($p(0) = p_0$, for example). If we can numerically solve this problem, then we can numerically solve the vector form (where p and Q are three-dimensional, to make contact with (9.30)).

Our first step is to introduce a discrete grid of values for τ, on which we will approximately solve for $p(\tau)$ – take $\tau_n = n\,\Delta\tau$ for some fixed $\Delta\tau$. Then we can use the definition of the derivative to approximate $\dot{p}(\tau)$ at the nth grid point:

$$\dot{p}(\tau_n) \approx \frac{p(\tau_{n+1}) - p(\tau_n)}{\Delta\tau}. \tag{9.32}$$

So our *finite difference* equation is:

$$\frac{p(\tau_{n+1}) - p(\tau_n)}{\Delta\tau} = Q(\tau_n, p(\tau_n)). \tag{9.33}$$

Let's take a concrete, but simple, form for the right-hand side: $Q(\tau, p) = \tau$. We know the solution in this case, $p(\tau) = \frac{1}{2}\tau^2 + p_0$, where p_0 is the value of p at $\tau = 0$ (this must be specified). We can solve the difference equation (9.33) for $p(\tau_{n+1})$ to compare the numerical solution with the exact solution:

$$p(\tau_{n+1}) = p(\tau_n) + \Delta\tau\, Q(\tau_n, p(\tau_n)) \quad p(0) = p_0. \tag{9.34}$$

This method of update is called Euler's method, and it provides a recursion that gives us the value of p at τ_{n+1} from the value of p at τ_n. Using our $Q(\tau, p) = \tau$, we can solve the recursion relation explicitly – from $p(\tau_{n+1}) = p(\tau_n) + n \, \Delta\tau^2$ (since $Q(\tau_n, p(\tau_n)) = \tau_n = n \, \Delta\tau$), we get:

$$p(\tau_n) = p_0 + \sum_{j=0}^{n-1} j \, \Delta\tau^2 = p_0 + \frac{1}{2} n (n-1) \, \Delta\tau^2$$

$$= p_0 + \frac{1}{2} \tau_n^2 - \frac{1}{2} n \, \Delta\tau^2. \tag{9.35}$$

Our approximate numerical solution matches the exact solution up to a factor of $\tau_n \, \Delta\tau$. So there is "numerical" error associated with taking the finite difference approximation for the derivative defined in (9.32). How might we modify our approach so as to cut down on this error?

One way to improve the *accuracy* of an integration method is to modify the evaluation of $Q(\tau_n, p(\tau_n))$. In our current example, Q depends only on τ, so suppose we try introducing shifted evaluation points on the right-hand side of (9.33):

$$\frac{p(\tau_{n+1}) - p(\tau_n)}{\Delta\tau} = Q(\tau_n + \beta, p(\tau_n) + \gamma) \tag{9.36}$$

for arbitrary β and γ (to be determined). Then our recursion relation reads, for $Q(\tau_n + \beta, p(\tau_n) + \gamma) = \tau_n + \beta$:

$$p(\tau_{n+1}) = p(\tau_n) + \Delta\tau \, (n \, \Delta\tau + \beta) \qquad p(0) = p_0. \tag{9.37}$$

Now our recursion is solved by:

$$p(\tau_n) = p_0 + \sum_{j=0}^{n-1} (j \, \Delta\tau^2 + \beta \, \Delta\tau) = p_0 + \frac{1}{2} \tau_n^2 - \frac{1}{2} n \, \Delta\tau^2 + n \, \beta \, \Delta\tau. \tag{9.38}$$

In order to kill the offending term, we take $\beta = \frac{1}{2} \Delta\tau$. Our updated method, in this case, relied on shifting the evaluation point for the right-hand side of the finite difference equation (9.33) – we now have:

$$p(\tau_{n+1}) = p(\tau_n) + \Delta\tau \, Q(\tau_{n+\frac{1}{2}}, p(\tau_n)). \tag{9.39}$$

Runge–Kutta methods are characterized by shifts in both the evaluation of the first argument in Q (the β in (9.36)), and in the evaluation of the second argument (γ in (9.36)). The goal is improved accuracy, and we generally choose β and γ based on a Taylor expansion. In our example, which has a polynomial right-hand side, we obtain perfect correspondence with the exact solution (from a Taylor series point of view, this exactness comes from the fact that $\frac{\partial^2 Q}{\partial \tau^2} = 0$). An example of a Runge–Kutta method (called "midpoint Runge–Kutta") that provides an update accurate to order $\Delta\tau^3$ at each step for generic $Q(\tau, p(\tau))$ reads:

$$
\begin{aligned}
k_1 &= \Delta\tau\, Q(\tau_n, p(\tau_n)) \\[4pt]
k_2 &= \Delta\tau\, Q\!\left(\tau_n + \frac{1}{2}\,\Delta\tau,\, p(\tau_n) + \frac{1}{2}k_1\right) \\[4pt]
p(\tau_{n+1}) &= p(\tau_n) + k_2.
\end{aligned}
\tag{9.40}
$$

One can verify, through Taylor expansion of Q in both its arguments, that the desired accuracy is achieved.

9.2.3 Equatorial geodesic example

Moving a method like (9.40) to two or three dimensions is as easy as providing three-dimensional vectors for $Q(\tau_n, p(\tau_n))$ and $p(\tau_n)$. For Q, take the right-hand side of (9.30), and for p, take the left. All we need to do now is provide the relevant constants that start the procedure off – in this case, that amounts to specifying $r(0)$, $\dot{r}(0)$, and $\phi(0)$. If we start at a turning point of the motion (a specific choice of $r(0)$), we know that $\dot{r}(0) = 0$. Our starting angle can be taken to be zero, so we set $\phi(0) = 0$. For an extreme Kerr black hole, with $M = 1$, $a = 0.99$ (a has a maximum value of M), we'll consider a trajectory that has $E \approx 0.9733232697$, $J_z \approx 2.3608795160$, and (that gives us the) starting point $r(0) \approx 1.8421052632$ with $\dot{r}(0) = 0$. These numbers have been chosen based on the orbital turning points they imply. Those turning points can be used to find E and J_z directly from the roots of (9.27) (turning points have no radial velocity). This particular choice corresponds to an "extreme" orbit, one that would not be predicted by, for example, our linearized consideration of the Kerr metric.

We solve (9.30) using the midpoint Runge–Kutta method (9.40) using the above parameters, and the result for a few cycles is shown in Figure 9.2. Note that our curve is parametrized by the proper time of the test particle, but we could easily switch to the (more relevant) coordinate time by dividing the right-hand side of (9.30) by $\dot{t}$.

The trajectory shown in Figure 9.2 is a so-called "zoom-whirl" orbit. The test particle starts out close to what must be a black hole (the test body is just outside the event horizon in this case), wraps around a few times, and is then kicked out to its maximum turning point. This type of orbit can be thought of as an extreme case of the precession we worked out for the Schwarzschild space-time in Chapter 7.

There is much left to explore for Kerr geodesics – but that exploration requires no additional physical insight. Once you have the metric, you can define the ODE governing geodesics (of any flavor), and proceed to solve them (generally numerically). So we will leave Kerr space-time. Actions related to length are

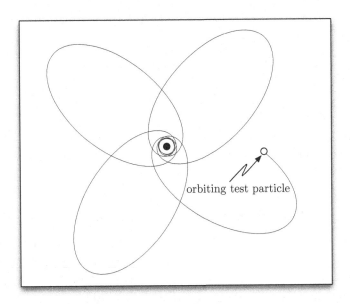

Figure 9.2   A massive test particle travels along a geodesic in the Kerr space-time – at the center of the orbit is a massive spinning body with $M = 1, a = 0.99$.

what produce the geodesic equations of motion, and so it is somewhat natural to ask if other geometrically significant constructs, used to define an action, lead to physically interesting equations of motion.

9.3 Area minimization

We have built up enough machinery to naturally address the issue of an extended body, relativistically. In particular, we'll start with the simplest extended body, a string. So we'll develop the relevant governing equations for a relativistic string. Just as we built the mechanics for relativistic point particles purely as an extremization of length, we will generate the equations of motion for a relativistic string as an extremization of area. The length we use for point particles is defined by the Minkowski metric, and similarly, the area we will define for strings relies on the Minkowski form. As a warm-up, let's look at area extremization for Euclidean space, then we just need to replace the Cartesian metric with the Minkowski one to go from physical surface minimization to the string equations of motion.

9.3.1 Surfaces

For a point particle, the ultimate goal of an action formulation is a description of the equations of motion that will allow us to solve for $\mathbf{x}(t)$, the position of the particle as a function of time. In the relativistic development of the action, we saw that a more natural parameter was τ, the proper time of the particle. The action itself was reparametrization-invariant, though, so we could go back and forth between the coordinate t and the proper time τ. In addition, we were led to the correct relativistic action by a minimization of path requirement (especially true in the curved GR setting).

Suppose we consider an extended object, that is, a surface with two parameters, defined by $\mathbf{x}(u, v)$. We would like to find equations of motion that allow us to solve for the surface – there is no *a priori* obvious action, and it is not even clear physically what we want out of a theory of surfaces. After all, there is no Newton's second law governing the time evolution of a surface, nor is there any obvious physics to be found in a free surface. So we proceed by analogy – if the point particle action is developed as a length (of trajectory) minimization, perhaps the surface action can be usefully thought of as an area minimization.

For a small change du and dv in the parameters, we span a parallelogram on the surface – this is clear from the relation between $\mathbf{x}(u + du, v)$ and $\mathbf{x}(u, v + dv)$ to $\mathbf{x}(u, v)$:

$$d\mathbf{x}_u = \frac{\partial \mathbf{x}}{\partial u} du \quad d\mathbf{x}_v = \frac{\partial \mathbf{x}}{\partial v} dv, \tag{9.41}$$

defining two vectors that share an origin. This is really just a gradient, so it is no surprise that we get these two vector contributions. The area (squared) spanned by these vectors is:

$$dA^2 = (d\mathbf{x}_u \times d\mathbf{x}_v) \cdot (d\mathbf{x}_u \times d\mathbf{x}_v). \tag{9.42}$$

If we play our usual indexing games, we can make the above clear:

$$\begin{aligned} dA^2 &= (\epsilon_{ijk} dx_u^j dx_v^k)(\epsilon_{i\ell m} dx_u^\ell dx_v^m) \\ &= (\delta_{j\ell} \delta_{km} - \delta_{jm} \delta_{k\ell}) dx_u^j dx_v^k dx_u^\ell dx_v^m \\ &= (d\mathbf{x}_u \cdot d\mathbf{x}_u)(d\mathbf{x}_v \cdot d\mathbf{x}_v) - (d\mathbf{x}_u \cdot d\mathbf{x}_v)^2. \end{aligned} \tag{9.43}$$

By now, it should come as no surprise that this area element can be written as a determinant – consider the matrix:

$$g_{ij} \doteq \begin{pmatrix} \frac{\partial \mathbf{x}}{\partial u} \cdot \frac{\partial \mathbf{x}}{\partial u} & \frac{\partial \mathbf{x}}{\partial u} \cdot \frac{\partial \mathbf{x}}{\partial v} \\ \frac{\partial \mathbf{x}}{\partial v} \cdot \frac{\partial \mathbf{x}}{\partial u} & \frac{\partial \mathbf{x}}{\partial v} \cdot \frac{\partial \mathbf{x}}{\partial v} \end{pmatrix}, \tag{9.44}$$

then we have:

$$dA = \sqrt{g} \, du \, dv. \tag{9.45}$$

That's interesting, and g_{ij} here is called an induced metric (hence the suggestive letter), it tells us how areas transform under the map from (u, v) space to

$\mathbf{x} = x(u, v)\,\hat{\mathbf{x}} + y(u, v)\,\hat{\mathbf{y}} + z(u, v)\,\hat{\mathbf{z}}$. The matrix g_{ij} is certainly symmetric, and invertible (by assumption), so it is a candidate for metric interpretation. The clincher: it transforms like a second-rank tensor under $u \longrightarrow \bar{u}$, $v \longrightarrow \bar{v}$. Then we know, from Section 5.4, that the action:

$$S = \int dA = \int \sqrt{g}\, du\, dv \qquad (9.46)$$

is a scalar.

9.3.2 Surface variation

Now we ask the usual question: what surface $\mathbf{x}(u, v)$ minimizes the area action? Let $\mathbf{X}' \equiv \frac{\partial \mathbf{x}}{\partial u}$ and $\dot{\mathbf{X}} \equiv \frac{\partial \mathbf{x}}{\partial v}$ (we use capital $\mathbf{X}$ to denote the surface solution rather than the coordinates $\mathbf{x}$). Then explicitly, we have:

$$S = \int \sqrt{\dot{\mathbf{X}}^2\, \mathbf{X}'^2 - (\dot{\mathbf{X}} \cdot \mathbf{X}')^2}\, du\, dv. \qquad (9.47)$$

The equations of motion are given, as always (note that there is no $\mathbf{X}$-dependence) by:

$$\frac{\partial}{\partial u}\left(\frac{\partial \mathcal{L}}{\partial \mathbf{X}'}\right) + \frac{\partial}{\partial v}\left(\frac{\partial \mathcal{L}}{\partial \dot{\mathbf{X}}}\right) = 0, \qquad (9.48)$$

which is just:

$$\frac{\partial}{\partial u}\left(\frac{\dot{\mathbf{X}}^2\, \mathbf{X}' - (\mathbf{X}' \cdot \dot{\mathbf{X}})\, \dot{\mathbf{X}}}{\sqrt{\dot{\mathbf{X}}^2\, \mathbf{X}'^2 - (\dot{\mathbf{X}} \cdot \mathbf{X}')^2}}\right) + \frac{\partial}{\partial v}\left(\frac{\mathbf{X}'^2\, \dot{\mathbf{X}} - (\mathbf{X}' \cdot \dot{\mathbf{X}})\, \mathbf{X}'}{\sqrt{\dot{\mathbf{X}}^2\, \mathbf{X}'^2 - (\dot{\mathbf{X}} \cdot \mathbf{X}')^2}}\right) = 0. \qquad (9.49)$$

An unenlightening result if there ever was one – but we display it here to emphasize the difficulty, given a generic pair $\{u, v\}$, in writing down, much less solving, the "equations of motion" for the surface area-minimizing solution.

The point, as we saw in the case of the relativistic point particle action, is that we can use reparametrization invariance *prior* to varying to simplify the result.

Example – minimal surface in cylindrical geometry

Suppose we want to find the area-minimizing two-dimensional surface that connects two rings as shown in Figure 9.3.

Using cylindrical coordinates for $\mathbf{X}$: $d\mathbf{X} \doteq (ds(u, v), d\phi(u, v), dz(u, v))^{\mathrm{T}}$, the natural parametrization here is to set $u = \phi$, $v = z$. We can do this with impunity, they are just labels. The dot product takes its usual form for cylindrical coordinates with metric given by $d\ell^2 = ds^2 + s^2\, d\phi^2 + dz^2$, and this must be respected in the action. Finally, we do not expect the cylindrical radial coordinate to depend on ϕ – that's a

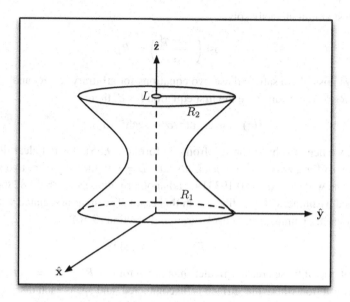

Figure 9.3 Two rings, of radius R_1 and R_2, sitting at $z = 0$ and $z = L$. We want to find the surface connecting the rings that minimizes the total area.

symmetry ansatz we are adding. It should be supported by the solution to the full equations of motion (i.e. leaving in the full $\{u, v\}$-dependence), and as usual, we must be careful about making this assumption before we vary. Forging ahead, we can write the action (noting that $\frac{\partial \phi(u,v)}{\partial u} = 1$, $\frac{\partial \phi(u,v)}{\partial v} = 0$, since $(u = \phi, v = z)$, and $\frac{\partial z(u,v)}{\partial u} = 0$, $\frac{\partial z(u,v)}{\partial v} = 1$ for the same reason), assuming $s(u, v) = s(v)$:

$$S = \int_0^{2\pi} \int_0^L \underbrace{s(z) \sqrt{1 + s'(z)^2}}_{\equiv \mathcal{L}} \, ds \, d\phi. \tag{9.50}$$

The variation is straightforward, and tells us that:

$$\frac{d}{dz} \frac{\partial \mathcal{L}}{\partial s'} - \frac{\partial \mathcal{L}}{\partial s} = 0 = \frac{s \, s'' - s'^2 - 1}{(1 + s'^2)^{3/2}}, \tag{9.51}$$

the solution to which is:

$$s(z) = \alpha \, \cosh\left(\frac{z - \beta}{\alpha}\right). \tag{9.52}$$

In order to set boundary conditions, we need to solve for α and β – this is not as easy as it appears. The $z = 0$ boundary condition gives:

$$\alpha \, \cosh\left(\frac{\beta}{\alpha}\right) = R_1, \tag{9.53}$$

and we must simultaneously solve:

$$\alpha \cosh\left(\frac{L - \beta}{\alpha}\right) = R_2. \tag{9.54}$$

Is it always possible to satisfy these two equations for arbitrary R_1, R_2 and L? If we set $R_1 = R_2 = 1$, we can solve the first equation for β, then:

$$s(z) = \alpha \cosh(z/\alpha - \text{sech}^{-1}(\alpha)). \tag{9.55}$$

To find α, we need to choose an L – from the form of (9.55), it is not clear that solutions exist for all α (they do not). Let's take $L = 0.9$, then there are two solutions to (9.55), one with $\alpha \equiv \alpha_1 \approx 0.193419$, and another with $\alpha \equiv \alpha_2 \approx 0.882804$ (determined by numerical root-finding, in this case). The areas associated with these two solutions are (running them back through the action (9.50)):

$$A_1 \approx 6.771 \qquad A_2 \approx 5.448. \tag{9.56}$$

Notice that one of these areas is greater than 2π – for our $R_1 = R_2 = 1$ case, there is a spurious solution where the surface is disconnected, and spans each disk separately (think of a soap film on each of the rings separately) – that solution has area $\pi R_1^2 + \pi R_2^2 = 2\pi$ here, and so the actual minimal surface that joins the two rings is given by the second value of α. The two solutions, for α_1 and α_2, are shown in Figure 9.4.

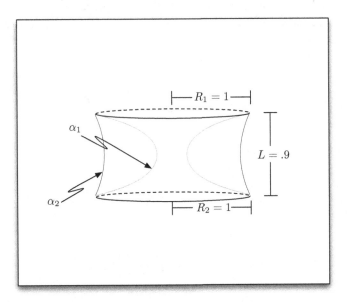

Figure 9.4   The solutions for the minimal surface problem with rings of equal size ($R = 1$) separated by a distance $L = 0.9$.

Problem 9.6

We should be able to solve for the cylindrical minimal surface directly from (9.49). Set:

$$\mathbf{X} \doteq \begin{pmatrix} s(z) \\ \phi \\ z \end{pmatrix} \tag{9.57}$$

and use the correct (cylindrical) interpretation for the dot product to show that you obtain a single equation governing $s(z)$ that has the same content as the one we got from variation (9.51).

Problem 9.7

Find the solution to:

$$s(z)\,s''(z) - s'(z)^2 - 1 = 0, \tag{9.58}$$

"by hand".

Problem 9.8

Show that the method (9.40) is accurate as claimed – Taylor expand both sides of $p(\tau_n + \Delta\tau) = p(\tau_n) + k_2$ and show that they match up to errors of order $\Delta\tau^3$. Note that when expanding $p(\tau_n + \Delta\tau)$, you will need to make use of the fact, true from the ODE itself, that:

$$\dot{p} = Q \quad \ddot{p} = \frac{\partial Q}{\partial \tau} + \frac{\partial Q}{\partial p}\,\dot{p} = \frac{\partial Q}{\partial \tau} + \frac{\partial Q}{\partial p}\,Q, \tag{9.59}$$

where the second relation follows by differentiating both sides of the ODE with respect to τ (in fact, you will also need the analogous expression for $\dddot{p}$).

9.3.3 Relativistic string

Very little changes structurally if we consider a string rather than a spatial surface. We are still minimizing an "area", but this time, our target space is not just three-dimensional Euclidean space with its natural dot product. Instead, we are interested in Minkowski space-time, which also has a dot product. If we make the formal replacement:

$$\boxed{\mathbf{X} \cdot \mathbf{X} \longrightarrow X^\mu\,\eta_{\mu\nu}\,X^\nu \equiv X \cdot X} \tag{9.60}$$

in (9.44), then almost everything we have done so far carries over without change. We can also relabel our parameters – these are typically called τ and σ rather than u and v, and this just reminds us to think of τ as a time-like parameter (going from

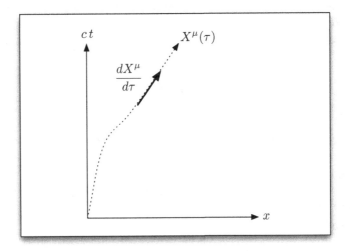

Figure 9.5   A particle moves along a world-line as shown. At any point, the tangent to the curve is given by $\frac{dX^\mu}{d\tau}$, and this must be a time-like vector, defining, as it does, the purely temporal direction of motion.

negative to positive infinity), and σ as a curve parameter that takes us along a string (so it is finite or closed).

The only other modification we must make to our Euclidean area action is to change the sign under the square root, this is a familiar shift, but comes about here in a more subtle way (remember that g_{ij} from (9.44) is an induced metric). We want a string to "be relativistic", but what does this mean for the derivatives of $X^\mu(\tau, \sigma)$ w.r.t. τ and σ?

Well, if we think of a point particle, parametrized by τ, then we know that the world-line of the point particle has, at any point, a time-like vector that points in the direction of increasing τ. This vector must be time-like, as it defines the local temporal axis for the particle at rest (and hence, near the point, any change $d\tau$ results in a change only in the temporal component of the particle's motion), as shown in Figure 9.5.

When we think of a string, we imagine a world-sheet (the natural extension of a world-line), a surface mapped out in the (ct, x) plane shown in Figure 9.5 (and more generally, in the full four-dimensional space-time). Our first requirement is that points along the string should behave like point particles, i.e. there should exist, for any σ, a time-like direction. The space-like direction is along the σ-parametrized direction of the curve, and just represents a tangent vector in that direction. We do not allow degenerate cases, i.e. we cannot have two space-like vectors, else a portion of the string is moving faster than light (or at the speed of light – in fact, the endpoints of a string can move at this speed), and two time-like directions implies that we have a point particle with no spatial extent, a situation we already know something about.

Once we have two linearly independent directions, one time-like, one spacelike, it is relatively easy to show that the quantity:

$$(\dot{X} \cdot \dot{X})(X' \cdot X') - (\dot{X} \cdot X')^2 < 0 \tag{9.61}$$

(with $\dot{X} \doteq \frac{dX^\mu}{d\tau}$, $X' \doteq \frac{dX^\mu}{d\sigma}$). This combination is g^2, so to get a non-imaginary area, we need to put a minus sign under the square root in the action:

$$\boxed{S_{NG} \sim \int \sqrt{-g}\, d\tau \, d\sigma.} \tag{9.62}$$

Aside from normalization, this is the Nambu–Goto classical string action. The Nambu–Goto action is the starting point for classical string theory, and is covered extensively in [19, 33, 42].

9.4 A relativistic string solution

Here we vary the Nambu–Goto action, and fix all gauge choices in the equations of motion to find the physical content of our relativistic, extended body theory. Finally, we consider the simplest possible consistent solution, a spinning string. The gauge choices must be made prior to physical interpretation, and we have a couple of natural choices. We'll fix, throughout, "static gauge", where we let the temporal parameter τ be coordinate time. Then we can use arc-length parametrization for the spatial parameter σ, that provides an interesting connection to relativistic point particles. To recover the wave equation (of motion) governing strings, we use a different σ-parametrization, and it is in this setting that we will actually solve for our rotating string. Finally, we'll return to the equations of motion that arise in arc-length parametrization and compare with the classical case.

It is interesting that if we tried to construct a rotating string solution, *a priori*, we would probably guess the correct answer – for a rigid body of length L spinning about its center with constant angular velocity:

$$\mathbf{X}(s, t) = s \left(\cos\left(\frac{2vt}{L}\right) \hat{\mathbf{x}} + \sin\left(\frac{2vt}{L}\right) \hat{\mathbf{y}} \right), \tag{9.63}$$

for $s \in [-L/2, L/2]$. This seems pretty reasonable – as we shall see, this is only correct provided $v = c$, so that the string endpoints move with a . . . specific speed.

9.4.1 Nambu–Goto variation

The Nambu–Goto action from the previous section, now with some units in it, is:

$$S_{NG} = -\frac{T_0}{c} \int \sqrt{-g}\, d\sigma\, d\tau = -\frac{T_0}{c} \int \sqrt{(\dot{X} \cdot X')^2 - (\dot{X})^2\, (X')^2}\, d\sigma\, d\tau,$$

(9.64)

with $\dot{X} \equiv \frac{\partial X^\mu}{\partial \tau}$, $X' \equiv \frac{\partial X^\mu}{\partial \sigma}$, and the dot product is understood in the Minkowski sense, $A \cdot B = A^\alpha B^\beta \eta_{\alpha\beta}$. The arbitrary constant T_0 has units of force.

Defining $\mathcal{L} = -\frac{T_0}{c}\sqrt{-g}$, we can denote the momenta conjugate to $\dot{X}$ and X' via $\Pi_\alpha^\tau \equiv \frac{\partial \mathcal{L}}{\partial \dot{X}^\alpha}$, and $\Pi_\alpha^\sigma \equiv \frac{\partial \mathcal{L}}{\partial X'^\alpha}$. The variation of the above, i.e. the introduction of arbitrary $\delta X^\alpha(\tau, \sigma)$, gives the full set:

$$\delta S_{NG} = \int_{\sigma_0}^{\sigma_f} \int_{\tau_0}^{\tau_f} d\sigma\, d\tau \left[\left(\frac{\partial}{\partial \tau} (\delta X^\mu\, \Pi_\mu^\tau) \right) + \left(\frac{\partial}{\partial \sigma} (\delta X^\mu\, \Pi_\mu^\sigma) \right) \right.$$
$$\left. - \left(\frac{\partial \Pi_\mu^\tau}{\partial \tau} + \frac{\partial \Pi_\mu^\sigma}{\partial \sigma} \right) \delta X^\mu \right]$$
$$= \int_{\sigma_0}^{\sigma_f} d\sigma \left(\delta X^\mu\, \Pi_\mu^\tau \right) \Big|_{\tau=\tau_0}^{\tau_f} + \int_{\tau_0}^{\tau_f} d\tau \left(\delta X^\mu\, \Pi_\mu^\tau \right) \Big|_{\sigma=\sigma_0}^{\sigma_f}$$
$$+ \int_{\sigma_0}^{\sigma_f} \int_{\tau_0}^{\tau_f} d\sigma\, d\tau\, \delta X^\mu \left(\frac{\partial \Pi_\mu^\tau}{\partial \tau} + \frac{\partial \Pi_\mu^\sigma}{\partial \sigma} \right),$$

(9.65)

and physically, we assume that the string's initial and final configuration is given, so that $\delta X^\mu(\tau_f, \sigma) = \delta X^\mu(\tau_0, \sigma) = 0$. This kills the first boundary term in the usual way. The second boundary term, the one at the string endpoints σ_0 and σ_f, we treat more carefully. There are two options – first, we can leave the spatial endpoints fixed at that boundary. Physically, this means that the spatial portion of X^μ is time-independent at $\sigma = \sigma_0$ and $\sigma = \sigma_f$:[3]

$$\mathbf{X}(\tau, \sigma_0) = \text{const.} \longrightarrow \frac{d\mathbf{X}(\tau, \sigma_0)}{dt} = 0$$

(9.66)

and the same statement holds for $\mathbf{X}(\tau, \sigma_f)$. Now, the temporal coordinate $X^0(\sigma, \tau)$ depends on τ in some manner (remember that τ is a proper time of sorts, and so we expect $\frac{dt}{d\tau} \neq 0$, so the derivative of X^0 with respect to τ will similarly be nonzero). Then from a change of variables, we find that a string fixed at both ends will have:

$$\frac{d\mathbf{X}(\tau, \sigma_0)}{d\tau} = \frac{dt}{d\tau} \frac{d\mathbf{X}(\tau, \sigma_0)}{dt} = 0,$$

(9.67)

---

[3] We will use bold-faced variable names to refer to the spatial portion of a four-vector.

and since $\frac{dt}{d\tau} \neq 0$, we have:

$$\boxed{\frac{d\mathbf{X}(\tau, \sigma_0)}{d\tau} = \frac{d\mathbf{X}(\tau, \sigma_f)}{d\tau} = 0.}$$ (9.68)

This requirement applies only to the spatial part of our solution, and is called a "Dirichlet" boundary condition. Because this is a spatial condition, we must also have $\Pi_0^\sigma(\tau, \sigma_0) = \Pi_0^\sigma(\tau, \sigma_f) = 0$ since we are guaranteed that $\frac{dX^0(\tau,\sigma_0)}{d\tau} \neq 0$, nor is the derivative at σ_f zero.

Our second option is to set:

$$\boxed{\Pi_\mu^\sigma(\tau, \sigma_0) = \Pi_\mu^\sigma(\tau, \sigma_f) = 0,}$$ (9.69)

called a "free endpoint" boundary condition – no constraint is placed on the variation. We will use the free endpoint condition for what follows, the relativistic rotating string has endpoints that are free to move.

The variation here follows the same pattern as the multi-dimensional variation we did in Section 5.2. I have included the boundary conditions only to highlight the choices, although these have always existed implicitly. When we actually solve the equations of motion, we need to specify the boundary conditions we will use. In addition, as with the special relativistic point Lagrangian, we need to provide an interpretation for τ, the temporal parameter (see Section 2.2.2). For our string, we also need to specify σ – the string action is reparametrization-invariant in both of these parameters separately. The choices for τ and σ are referred to as "gauge choices", and to the extent that a solution to the equations of motion exists without any specific interpretation attached to τ or σ, they do not reflect fundamental physics, but are, rather, structural placeholders. We know that in order to physically interpret a solution, we must have an appropriate understanding of the parameters that appear. In the case of a relativistic point particle, that means we must fix a single parameter, τ. For extended relativistic objects, like the string, we have to also place restrictions on σ, a spatial variable.

For our string, we will take the time-like parameter to be t itself, and then work on the spatial σ parameter. Remember that imposing, for example, arc-length parametrization amounted to fixing the magnitude of the derivative of position with respect to that parameter (as in (4.39)). In the current setting, we will impose our gauge choice by fixing the derivative of $\mathbf{X}$ (the spatial part of our solution) with respect to σ. We have additional freedom here, basically, we are allowed to choose the magnitude of $\mathbf{X}'$ and in addition, its relation to $\dot{\mathbf{X}}$, the vector tangent to curves of constant σ.[4] Once the temporal parametrization is fixed, we will explore the spatial options.

9.4.2 Temporal string parametrization

In terms of interpretation, setting $\tau = t$, the coordinate time, is useful, just as it was for the point particle. Our two natural four-vectors, $\frac{\partial X^\mu}{\partial \sigma}$ and $\frac{\partial X^\mu}{\partial \tau}$ become, in this gauge:

$$\frac{\partial X^\mu(t, \sigma)}{\partial \sigma} \doteq \begin{pmatrix} 0 \\ \frac{\partial \mathbf{X}}{\partial \sigma} \end{pmatrix} \qquad \frac{\partial X^\mu(t, \sigma)}{\partial t} \doteq \begin{pmatrix} c \\ \frac{\partial \mathbf{X}}{\partial t} \end{pmatrix}, \tag{9.71}$$

where we have chosen Cartesian Minkowski space-time: $dx^\mu \doteq (c\,dt, dx, dy, dz)^T$. We are again using the bold vectors to refer to the spatial components of X^μ. Choosing one of the coordinates as the parameter is similar to our area minimization example from Section 9.3.2. In terms of physics, we have, at any time t, an instantaneously at rest string parametrized by σ, hence the name given to this choice, "static gauge".

In this gauge, the Nambu–Goto action becomes:

$$S'_{NG} = -\frac{T_0}{c} \int \sqrt{(\mathbf{X}' \cdot \dot{\mathbf{X}})^2 - (-c^2 + \dot{\mathbf{X}} \cdot \dot{\mathbf{X}})(\mathbf{X}' \cdot \mathbf{X}')} \, dt \, d\sigma. \tag{9.72}$$

From here, we can verify the sign under the square root – for a piece of string that is not moving, so that $\dot{\mathbf{X}} = 0$, we have a positive area functional under the square root – if we have a string at rest (for all points along the string), we expect the area of the world-sheet to be $c\,(t_f - t_0)(\sigma_f - \sigma_0)$.

9.4.3 A σ string parametrization (arc-length)

Given a time t, we have a one-dimensional curve, $\mathbf{X}(t, \sigma)$, and we know how to form a unit tangent vector to the curve: choose arc-length parametrization. If we let $\sigma = s$ (denoting the arc-length parameter as s), then arc-length parametrization immediately gives $|\mathbf{X}'| = 1$. We can use this to define the transverse component of the string velocity:

$$\mathbf{v}_\perp = \dot{\mathbf{X}} - (\dot{\mathbf{X}} \cdot \mathbf{X}') \mathbf{X}' \tag{9.73}$$

---

[4] Think of the dot product of $\dot{\mathbf{X}}$ and $\mathbf{X}'$:

$$\dot{\mathbf{X}} \cdot \mathbf{X}' = |\dot{\mathbf{X}}| \, |\mathbf{X}'| \cos \theta \tag{9.70}$$

for θ the angle between $\dot{\mathbf{X}}$ and $\mathbf{X}'$ – we are specifying $|\mathbf{X}'|$, the magnitude of $\mathbf{X}'$ and θ as our gauge choice. The end result for arc-length parametrization will be to define σ so that $|\mathbf{X}'| = 1$ and to take $\theta = \frac{1}{2}\pi$.

where now, finally, $\dot{\mathbf{X}} = \frac{\partial \mathbf{X}}{\partial t}$ (since we set $\tau = t$) and $\mathbf{X}' = \frac{\partial \mathbf{X}}{\partial s}$, since we chose $\sigma = s$. Then (9.73) is just the full spatial velocity with the component tangent to the string (the longitudinal component) projected out. The *magnitude* of the transverse velocity, then, takes the form:

$$v_\perp^2 = \dot{\mathbf{X}} \cdot \dot{\mathbf{X}} - (\dot{\mathbf{X}} \cdot \mathbf{X}')^2, \tag{9.74}$$

suggesting that we can write the action in terms of $v_\perp^2$ itself. In this gauge, we have (keeping the constraint $\mathbf{X}' \cdot \mathbf{X}' = 1$ in mind):

$$S_{NG}' = -\frac{T_0}{c} \int dt\, ds\, \sqrt{-v_\perp^2 + c^2} = -T_0 \int dt\, ds\, \sqrt{1 - \frac{v_\perp^2}{c^2}}. \tag{9.75}$$

Notice how close the associated Lagrangian is to the relativistic point Lagrangian (2.60). By fixing both parameters, we have recovered the usual form of the point Lagrangian, the only difference is that the speed here is not the particle speed, but the perpendicular string speed.

Boundary conditions in arc-length parametrization

Consider the boundary condition (9.69) in this parametrization – the free endpoint boundary condition (9.69), together with our pre-gauge fixed expression for $\frac{\partial \mathcal{L}}{\partial X'^\sigma}$ will give four total constraints, and these tell us physically relevant facts about the ends of an open string.

Recall:

$$
\begin{aligned}
\Pi_\mu^\sigma &= \frac{\partial \mathcal{L}}{\partial X'^\mu} = -\frac{T_0}{c} \frac{(\dot{\mathbf{X}} \cdot \mathbf{X}') \dot{X}_\mu - (\dot{\mathbf{X}})^2 X_\mu'}{\sqrt{(\dot{\mathbf{X}} \cdot \mathbf{X}')^2 - \dot{X}^2 (X')^2}} \\
&= -\frac{T_0}{c} \frac{(\mathbf{X}' \cdot \dot{\mathbf{X}}) \dot{X}_\mu - (\dot{\mathbf{X}} \cdot \dot{\mathbf{X}} - c^2) X_\mu'}{\sqrt{(\mathbf{X}' \cdot \dot{\mathbf{X}})^2 - (-c^2 + \dot{\mathbf{X}} \cdot \dot{\mathbf{X}})(\mathbf{X}' \cdot \mathbf{X}')}}
\end{aligned}
\tag{9.76}
$$

and this becomes, in our current gauge:

$$\Pi_\mu^\sigma = -\frac{T_0}{c^2} \frac{(\mathbf{X}' \cdot \dot{\mathbf{X}}) \dot{X}_\mu - (\dot{\mathbf{X}} \cdot \dot{\mathbf{X}} - c^2) X_\mu'}{\sqrt{1 - \frac{v_\perp^2}{c^2}}}. \tag{9.77}$$

The vanishing of this "momentum" at the endpoints provides a set of one "scalar" and one "vector" (in the spatial sense) constraints. Take $\mu = 0$, for which $\dot{X}_0 = -c$ and $X_0' = 0$, then we learn that $\mathbf{X}' \cdot \dot{\mathbf{X}} = 0$, which here tells us that *at the endpoints,* the string motion is perpendicular to the string itself – i.e. the motion of the endpoints is necessarily transverse.

The vector portion, for $\mu = 1, 2, 3$, gives (this time only the second term in the numerator contributes, the first term is zero by the above):

$$-\frac{T_0}{c} \frac{c^2 - v^2}{\sqrt{c^2 - v_\perp^2}} \mathbf{X}' = -\frac{T_0}{c} \sqrt{c^2 - v^2} \mathbf{X}', \tag{9.78}$$

where, again from above, $v_\perp^2 = v^2$ at the endpoint. Now this must be zero, and $\mathbf{X}' \neq 0$, since it is a unit vector, hence we must have $v = c$, so *the string endpoint moves at the speed of light*.

9.4.4 Equations of motion

One has to be careful with gauge conditions, as we have stressed over and over (and over). They cannot necessarily be applied before variation, and our current example is no different. They can *always* be applied to equations of motion (i.e. after variation), and there is a natural σ parametrization (that is not arc-length) of the string equations of motion (from whence it gets its name – after all, we have yet to see a string-like equation) that makes their structure familiar and their form easily solvable.

The equation of motion for an arbitrarily parametrized string reads (compare with (9.48)):

$$\boxed{\frac{\partial}{\partial \sigma} \Pi_\mu^\sigma + \frac{\partial}{\partial \tau} \Pi_\mu^\tau = 0.} \tag{9.79}$$

Again, for arbitrary τ and σ, we can write the "momenta":

$$\Pi_\mu^\tau = -\frac{T_0}{c} \frac{(\dot{X} \cdot X') X_\mu' - (X')^2 \dot{X}_\mu}{\sqrt{(\dot{X} \cdot X')^2 - (\dot{X})^2 (X')^2}}$$

$$\Pi_\mu^\sigma = -\frac{T_0}{c} \frac{(\dot{X} \cdot X') \dot{X}_\mu - (\dot{X})^2 X_\mu'}{\sqrt{(\dot{X} \cdot X')^2 - (\dot{X})^2 (X')^2}}. \tag{9.80}$$

Now suppose we take $\tau = t$, imposing static gauge. Assume, further, that it is possible to develop a σ parameter that forces the spatial vectors $\mathbf{X}' \cdot \dot{\mathbf{X}} = 0$.[5] Then the four-vector dot products are:

$$\dot{X} \cdot \dot{X} = (-c^2 + \dot{\mathbf{X}} \cdot \dot{\mathbf{X}}) \quad X' \cdot X' = \mathbf{X}' \cdot \mathbf{X}' \quad \dot{X} \cdot X' = \dot{\mathbf{X}} \cdot \mathbf{X}' = 0. \tag{9.81}$$

---

[5] This is always possible – it has nothing to do with any intrinsic σ, but rather an identification we make on a two-dimensional grid. The choice amounts to fixing the angle in the Euclidean dot product: $\mathbf{X}' \cdot \dot{\mathbf{X}} = X' \dot{X} \cos\theta$.

Notice that we have *not* imposed arc-length parametrization at this point, the magnitude of $\mathbf{X}'$ is as yet unspecified.

Our equations of motion simplify through their momenta – with our current choices, we have:

$$\Pi_\mu^\tau = \frac{T_0}{c} \frac{(\mathbf{X}' \cdot \mathbf{X}') \dot{X}_\mu}{\sqrt{(c^2 - v^2)\mathbf{X}' \cdot \mathbf{X}'}}$$

$$\Pi_\mu^\sigma = -\frac{T_0}{c} \frac{(c^2 - v^2) X_\mu'}{\sqrt{(c^2 - v^2)\mathbf{X}' \cdot \mathbf{X}'}},$$
(9.82)

using $v^2 = \dot{\mathbf{X}} \cdot \dot{\mathbf{X}}$.

As usual, this is a set of four equations. For the $\mu = 0$ case, we know that $X_0' = 0$, so the equation of motion here is:

$$\frac{\partial}{\partial \tau}\left(\frac{T_0 (\mathbf{X}' \cdot \mathbf{X}')}{\sqrt{(c^2 - v^2)\mathbf{X}' \cdot \mathbf{X}'}}\right) = 0,$$
(9.83)

telling us that the expression in parentheses is τ-constant, i.e. equal to some arbitrary $f(\sigma)$. We can use this to *define* the σ-parametrization. One choice, as we just saw, is arc-length, then $\mathbf{X}' \cdot \mathbf{X}' = 1$. But in general, we have the full class:

$$\sqrt{\mathbf{X}' \cdot \mathbf{X}'} = \frac{f(\sigma) \sqrt{c^2 - v^2}}{T_0}.$$
(9.84)

We can input this into the spatial part of (9.79):

$$\frac{\partial}{\partial \tau}\left(\frac{1}{c} f(\sigma) \dot{\mathbf{X}}\right) - \frac{\partial}{\partial \sigma}\left(\frac{T_0^2}{c} \frac{\mathbf{X}'}{f(\sigma)}\right) = 0.$$
(9.85)

Remember, $f(\sigma)$ is up to us – we can take *any* function of σ only. Here's a good choice: $f(\sigma) = f_0$, a constant. Then we have a familiar reduction, the wave equation:

$$\boxed{-\frac{f_0^2}{T_0^2} \ddot{\mathbf{X}} + \mathbf{X}'' = 0.}$$
(9.86)

To complete the traditional picture, let's set the constant $f_0 = \frac{T_0}{c}$, then we recover the wave equation with propagation speed c.

We have a few side-constraints, now, so let's tabulate our results:

$$\boxed{-\frac{1}{c^2} \ddot{\mathbf{X}} + \mathbf{X}'' = 0 \quad \dot{\mathbf{X}} \cdot \mathbf{X}' = 0 \quad \mathbf{X}' \cdot \mathbf{X}' = 1 - \frac{1}{c^2} \dot{\mathbf{X}} \cdot \dot{\mathbf{X}}.}$$
(9.87)

Here we can see that a choice for σ has been made (fixed by $f(\sigma)$) – whatever you want to call that choice, we are clearly *not* using arc-length parametrization at this stage. The wave equation for the relativistic string does not appear in the above form in arc-length parametrization, so to make this wave association, we had to use σ such that $\mathbf{X}' \cdot \mathbf{X}' = 1 - \frac{v^2}{c^2}$.

Take free endpoint boundary conditions (the string is not fixed). We are not in arc-length parametrization, and the spatial constraint that $\Pi^\sigma_\mu = 0$ defining this boundary condition (see (9.69)) automatically has $\Pi^\sigma_0 = 0$ (because $\dot{\mathbf{X}} \cdot \mathbf{X}' = 0$), leaving:

$$\mathbf{\Pi}^\sigma = -\frac{T_0}{c^2}\frac{(c^2 - v^2)\,\mathbf{X}'}{\sqrt{1 - \frac{v^2}{c^2}}} = 0 \tag{9.88}$$

so that at the endpoints, we must have $\mathbf{X}'(\sigma = \sigma_0) = \mathbf{X}'(\sigma = \sigma_f) = 0$.[6] There are now sufficient boundary and initial conditions to actually solve the string equations of motion. Since we have fixed all of the available gauge freedom, our solution should have direct physical interpretation. Rather than discuss the general case, let's work our specific rotating string example, and see how the above equations constrain the solutions.

9.4.5 A rotating string

Consider a string that rotates with some constant angular velocity – we can make left and right-traveling ansätze in the usual way:

$$\mathbf{X}(t, \sigma) = \delta\,(A\,\cos(\kappa\,(\sigma - c\,t)) + B\,\cos(\kappa\,(\sigma + c\,t)))\,\hat{\mathbf{x}}$$
$$+ \delta\,(F\,\sin(\kappa\,(\sigma - c\,t)) + G\,\sin(\kappa\,(\sigma + c\,t)))\,\hat{\mathbf{y}} \tag{9.89}$$

for constants $\{A, B, F, G\}$. We have automatically solved the wave equation with our choice of left and right-traveling waves. Now we have the boundary condition $\mathbf{X}'(t, \sigma = 0, \sigma_f) = 0$. Taking the $\sigma = 0$ case, we learn that $B = A$ and $G = -F$. Then imposing the boundary condition at σ_f puts a constraint on κ:

$$\kappa\,\sigma_f = n\,\pi \longrightarrow \kappa = \frac{\pi}{\sigma_f}, \tag{9.90}$$

where we have chosen to set $n = 1$ (so that as $\sigma = 0 \longrightarrow \sigma_f$, we go across the string once, our choice). Our current solution, then, is:

$$\mathbf{X}(t, \sigma) = 2\,\delta\,\cos(\pi\,\sigma/\sigma_f)\,(A\,\cos(\pi\,c\,t/\sigma_f)\,\hat{\mathbf{x}} - F\,\sin(\pi\,c\,t/\sigma_f)\,\hat{\mathbf{y}}). \tag{9.91}$$

---

[6] Remember that we saw this same sort of equation (9.78), in arc-length parametrization. There, we know that $\mathbf{X}' \neq 0$, and we concluded that $v = c$ at the ends of the string. In the current setting, where we know that $\mathbf{X}'$ is perpendicular to $\dot{\mathbf{X}}$, we cannot conclude that $\mathbf{X}' \neq 0$, nor do we know that $v = c$ at the ends.

Now setting the condition $\mathbf{X}' \cdot \dot{\mathbf{X}} = 0$ gives the relation $A = \pm F$, and taking the negative sign to adjust the phase, we have $F = -A$ above. Finally, we use the last of the relations in (9.87) to fix the magnitude of A:

$$\mathbf{X}' \cdot \mathbf{X}' + \frac{1}{c^2} \dot{\mathbf{X}} \cdot \dot{\mathbf{X}} = 1 \longrightarrow A = \pm \frac{\sigma_f}{2\delta\pi}, \tag{9.92}$$

and our final solution reads:

$$\boxed{\mathbf{X}(t, \sigma) = \frac{\sigma_f}{\pi} \cos\left(\frac{\pi\sigma}{\sigma_f}\right) \left(\cos\left(\frac{\pi c t}{\sigma_f}\right) \hat{\mathbf{x}} + \sin\left(\frac{\pi c t}{\sigma_f}\right) \hat{\mathbf{y}}\right).} \tag{9.93}$$

From this solution, we can obtain the perpendicular component of velocity – since the motion is transverse to the string already, the velocity is perpendicular to the string everywhere, and has value:

$$v_\perp = \sqrt{\dot{\mathbf{X}} \cdot \dot{\mathbf{X}}} = c \cos\left(\frac{\pi\sigma}{\sigma_f}\right). \tag{9.94}$$

As expected, then, the ends of the string travel at the speed of light. Rather than continue in this parametrization, we'll now develop the same solution in arc-length parametrization, this provides a way to interpret the string equations of motion classically, in addition to highlighting the interpretation of the same solution in different gauges.

9.4.6 Arc-length parametrization for the rotating string

We'll start with $\mathbf{X}(t, \sigma)$ from (9.93) and impose arc-length parametrization using its definition and a change of variables. We could also go back to (9.84) and develop/solve the equations of motion in this parametrization (we'll do that to make contact with a classical string).

Arc-length parametrization is defined by $s(\sigma)$ such that $\frac{\partial \mathbf{X}}{\partial s}$ has unit magnitude. Given that we know the magnitude in σ-parametrization from (9.87), we can perform the change of variables (let $\mathbf{X}' = \frac{\partial \mathbf{X}}{\partial \sigma}$):

$$\mathbf{X}' \cdot \mathbf{X}' = \underbrace{\frac{\partial \mathbf{X}}{\partial s} \cdot \frac{\partial \mathbf{X}}{\partial s}}_{=1} \left(\frac{\partial s}{\partial \sigma}\right)^2 = 1 - \frac{1}{c^2} \dot{\mathbf{X}} \cdot \dot{\mathbf{X}} \longrightarrow \frac{\partial s}{\partial \sigma} = \sin\left(\frac{\pi\sigma}{\sigma_f}\right), \tag{9.95}$$

where we have simplified the requirement $\frac{\partial s}{\partial \sigma} = \sqrt{1 - \frac{\dot{\mathbf{X}} \cdot \dot{\mathbf{X}}}{c^2}}$ by using our solution (9.93). The ODE above is easily solved – we require $s(0) = -\frac{L}{2}$ and $s(\sigma_f) = \frac{L}{2}$

to get $s \in [-\frac{L}{2}, \frac{L}{2}]$, and then the solution is:

$$s(\sigma) = -\frac{L}{2} \cos\left(\frac{\pi \sigma}{\sigma_f}\right) \tag{9.96}$$

with $\sigma_f = \frac{\pi L}{2}$.

In this s-parametrization, the string solution is, from (9.93):

$$\boxed{\mathbf{X}(t, s) = -s \left(\cos\left(\frac{2ct}{L}\right) \hat{\mathbf{x}} + \sin\left(\frac{2ct}{L}\right) \hat{\mathbf{y}}\right).} \tag{9.97}$$

From this form, it is clear that the magnitudes of the velocity and spatial derivatives are:

$$\dot{\mathbf{X}} \cdot \dot{\mathbf{X}} = \frac{4c^2 s^2}{L^2} \qquad \mathbf{X}' \cdot \mathbf{X}' = 1 \tag{9.98}$$

(primes now refer to s-derivatives). Once again, the endpoints travel at the speed of light – in this case, we have a constant (unit) magnitude s derivative along the curve, so the speed of the endpoints *must* be c.

Finally, we can return to the canonical momenta Π_μ^τ and Π_μ^σ – it is easy to verify that (9.79) is satisfied, and our immediate concern is the energy density of this configuration. By analogy with the point particle, the energy of points along the string should be related to the t-canonical momentum's zero component. We can calculate the full tensor:

$$\Pi_\mu^\tau \doteq \begin{pmatrix} -\dfrac{T_0}{c\sqrt{1-\frac{4s^2}{L^2}}} \\[2ex] \dfrac{T_0\,\dot{\mathbf{X}}}{c^2\sqrt{1-\frac{4s^2}{L^2}}} \end{pmatrix}. \tag{9.99}$$

For particles, the analogous tensor would be p_μ, and we know that $p_0 = -\frac{\mathcal{E}}{c}$ for $\mathcal{E}$ the energy density – then in this case, we should associate:

$$\mathcal{E} = \frac{T_0}{\sqrt{1 - \frac{4s^2}{L^2}}} \tag{9.100}$$

with the energy per unit length of the string (shown in Figure 9.6). Notice that it is a function of s – the endpoints carry infinite energy, apparently, but the total energy, given by:

$$E = \int_{-L/2}^{L/2} \mathcal{E}\, ds = \frac{\pi L T_0}{2}, \tag{9.101}$$

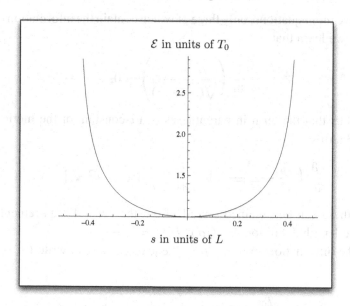

$\mathcal{E}$ in units of T_0

s in units of L

Figure 9.6   The energy density of the rotating string (9.100).

depends only on the "tension" and length, and is finite (its units are correct, as well, with T_0 having units of force).

9.4.7 Classical correspondence

To understand what this relativistic string corresponds to classically, we will redevelop the string equations of motion in the static (temporal) gauge with σ taken to be the arc-length parameter. In a lab frame, we measure time as coordinate time (for the lab), so static gauge is appropriate. In addition, we typically define, say, transverse oscillation in terms of an x-coordinate – this amounts to arc-length parametrization in our current language.

So we want to write (9.79) in terms of t and s, where:

$$\mathbf{X}' \cdot \mathbf{X}' = 1 \tag{9.102}$$

for $\mathbf{X}' \equiv \frac{\partial \mathbf{X}}{\partial s}$. We still have $\mathbf{X}' \cdot \dot{\mathbf{X}} = 0$ here. For a background Minkowski spacetime written in Cartesian coordinates, the relevant canonical momenta are (returning to (9.80)):

$$\Pi_\mu^\tau = -\frac{T_0}{c} \frac{\left(-\dot{X}_\mu\right)}{\sqrt{(c^2 - v^2)}}$$

$$\Pi_\mu^\sigma = -\frac{T_0}{c} \frac{(c^2 - v^2) X_\mu'}{\sqrt{(c^2 - v^2)}}. \tag{9.103}$$

Now we have four equations, only three of which contain the information of interest. For $\mu = 0$, we learn that:

$$\frac{\partial}{\partial t}\left(\frac{T_0}{\sqrt{(c^2 - v^2)}}\right) = 0, \qquad (9.104)$$

which tells us that the term in parentheses is a t-constant of the motion. For the $\mu = 1, 2, 3$ terms:

$$\frac{\partial}{\partial t}\left(\frac{T_0}{c}\frac{\dot{\mathbf{X}}}{\sqrt{c^2 - v^2}}\right) - \frac{\partial}{\partial s}\left(\frac{T_0}{c}\sqrt{c^2 - v^2}\,\mathbf{X}'\right) = 0. \qquad (9.105)$$

Note that these can be obtained directly from the general requirement in (9.84) with the arc-length definition for $f(\sigma)$: $f(\sigma) = \frac{T_0}{\sqrt{c^2 - v^2}}$.

Using the information from the $\mu = 0$ equation, we can write (9.105) suggestively:

$$\frac{T_0}{c^2\sqrt{1 - \frac{v^2}{c^2}}}\ddot{\mathbf{X}} - \frac{\partial}{\partial s}\left(T_0\sqrt{1 - \frac{v^2}{c^2}}\,\mathbf{X}'\right) = 0. \qquad (9.106)$$

Referring to our discussion of field Lagrangians in Section 5.1.1 (and (5.14) in particular), the term sitting in front of $\ddot{\mathbf{X}}$ is what we would call the "effective mass density" of the classical string. The additional term in the $\frac{\partial}{\partial s}$ derivative is the effective tension:

$$\mu_{eff} = \frac{T_0}{c^2\sqrt{1 - \frac{v^2}{c^2}}} \qquad T_{eff} = T_0\sqrt{1 - \frac{v^2}{c^2}}. \qquad (9.107)$$

Thinking of our rotating string solution, we see that $v = v_\perp$, and the effective string tension vanishes at the endpoints, while the effective mass density becomes infinite there.

As a subject that ties together our work on extremization of an action, gauge choices, and the metric, classical strings are interesting theoretical constructs. We have only touched on the solution of the equations of motion, and this introduction is meant to provide only the highlights of these familiar elements. The idea of an extended, relativistic body is interesting, and the action is natural (as the relativistic analogue of a soap film). The equations of motion emphasize the importance of fixing gauges prior to physical interpretation, and we can familiarize ourselves with the content of the classical strings by comparing them to waves in a strange medium. To go beyond these observations, and in order to appreciate the role of string theory in modern physics, one must make the move to quantum mechanics, which is not our current subject.

Problem 9.9

Starting from (9.80), we'll use static gauge and arc-length parametrization directly to arrive at (9.97).

(a) Simplify Π^σ_μ and Π^τ_μ from (9.80) using the following gauge choice:

$$\dot{X}^\mu \doteq \begin{pmatrix} c \\ \dot{\mathbf{X}} \end{pmatrix} \qquad X'^\mu \doteq \begin{pmatrix} 0 \\ \mathbf{X}' \end{pmatrix}$$

$$\dot{\mathbf{X}} \cdot \mathbf{X}' = 0 \tag{9.108}$$

$$\mathbf{X}' \cdot \mathbf{X}' = 1.$$

(b) Show, from the $\mu = 0$ equation of $\frac{\partial}{\partial \tau} \Pi^\tau_\mu + \frac{\partial}{\partial \sigma} \Pi^\sigma_\mu = 0$, that the combination $\frac{T_0}{\sqrt{c^2 - v^2}}$ is independent of t (here, $v^2 = \dot{\mathbf{X}} \cdot \dot{\mathbf{X}}$), and hence v is independent of t.

(c) With $v = v(\sigma)$ a function of σ only, we can write:

$$\dot{\mathbf{X}} = v(\sigma) \left[\cos(f(\sigma, t)) \, \hat{\mathbf{x}} + \sin(f(\sigma, t)) \, \hat{\mathbf{y}} \right]. \tag{9.109}$$

We know that $\mathbf{X}'$ should be perpendicular to this vector and have unit magnitude. Take:

$$\mathbf{X}' = \left[-\sin(f(\sigma, t)) \, \hat{\mathbf{x}} + \cos(f(\sigma, t)) \, \hat{\mathbf{y}} \right] \tag{9.110}$$

to enforce these final two gauge requirements. Using the integrability condition, $\frac{\partial}{\partial \sigma} \dot{\mathbf{X}} = \frac{\partial}{\partial t} \mathbf{X}'$, find $f(\sigma, t)$ and $v(\sigma)$. Assume we require $v(\ell/2) = v(-\ell/2) = c$, appropriate for arc-length parametrization (see (9.78)). In the end, you should recover (9.97) (up to direction of rotation and starting point).

Problem 9.10

What we have, in (9.87), is a set of spatial functions of σ and τ that satisfy:

$$\Box^2 X^i = 0 \qquad i = 1, 2, 3, \tag{9.111}$$

where $\Box^2 = -\frac{\partial^2}{\partial \tau^2} + \frac{\partial^2}{\partial \sigma^2}$ in units where $c = 1$. Note that, from our choice of $\tau = t$, it is also the case that $\Box^2 X^0 = 0$, so each element of the four-vector X^μ satisfies the wave equation. We know how to write an action that leads to the wave equation for a scalar (Klein–Gordon), and in this problem, we'll start with four copies of that action, and recover the wave equation in τ–σ coordinates by introducing an auxiliary metric field.

(a) Using:

$$h_{ab} \doteq \begin{pmatrix} -1 & 0 \\ 0 & 1 \end{pmatrix} \tag{9.112}$$

and its inverse h^{ab} (to raise and lower indices referring to the two-dimensional τ–σ space, we'll use roman indices for this space), show that the action:

$$S = \int X^\mu_{,a} \, \eta_{\mu\nu} \, X^\nu_{,b} \, h^{ab} \, d\sigma \, d\tau \tag{9.113}$$

gives back $\Box^2 X^\mu = 0$ when varied with respect to X^ν – commas here refer to the derivatives of X^μ with respect to τ and σ (so that $X^\mu{}_{,0} = \frac{\partial X^\mu}{\partial \tau}$ and $X^\mu{}_{,1} = \frac{\partial X^\mu}{\partial \sigma}$). The metric $\eta_{\mu\nu}$ is the usual four-dimensional Minkowski metric in Cartesian coordinates.

(b) We can promote the metric h_{ab} to dynamical variable, and use it to recover the auxiliary conditions. Remember that the wave equation is not enough for a relativistic string, we have to enforce all of the gauge conditions in the solution. By making h_{ab} itself a field, we can return those gauge conditions as field equations (a setting reminiscent of Lagrange multipliers). The "Polyakov" action is:

$$S_P = \int \sqrt{-h}\, h^{ab}\, X^\mu{}_{,a}\, \eta_{\mu\nu}\, X^\nu{}_{,b}\, d\sigma\, d\tau, \qquad (9.114)$$

varying with respect to both X^μ and h^{ab} (don't forget our result $\frac{\partial h}{\partial h^{ab}} = -h\, h_{ab}$ for the derivative of a matrix determinant with respect to the matrix elements from Problem 3.13). This time, then, we are assuming that $h_{ab}(\sigma, \tau)$ is itself an undetermined function of σ and τ.

(c) We know, from Problem 4.7, that any two-dimensional metric can be written in conformally flat form – here, the statement is $h_{ab} = f(\sigma, \tau)\, \bar\eta_{ab}$ where $\bar\eta_{ab}$ is just the two-dimensional Minkowski metric of (9.112). Use this conformal flatness in your field equations to show that you recover the wave equation for each of the X^μ, and a set of two constraints representing gauge conditions: $\dot X^\mu \eta_{\mu\nu} X'^\nu = 0$ and $\left(\dot X^\mu \dot X^\nu + X'^\mu X'^\nu\right) \eta_{\mu\nu} = 0$. Show, finally, that the gauge conditions in (9.87) satisfy this set.

Bibliography

[1] Arnold, V. I. *Mathematical Methods of Classical Mechanics*. Springer-Verlag, 1989.

[2] Bergmann, P. G. *Introduction to the Theory of Relativity*. Prentice-Hall, 1947.

[3] Bondi, H. "Negative mass in general relativity", *Rev. Mod. Phys.* **29**(3): 423–8, 1957.

[4] Carroll, S. M. *Spacetime and Geometry: An Introduction to General Relativity*. Addison-Wesley, 2004.

[5] Carroll, S. M. *Lecture Notes on General Relativity*. gr-qc/9712019.

[6] Chandrasekhar, S. *The Mathematical Theory of Black Holes*. Oxford University Press, 1992.

[7] Deser, S. "Self-interaction and gauge invariance", *Gen. Rel. Grav.* **1**: gr-qc/0411023, 1970.

[8] Deser, S. *Lecture Notes on General Relativity* (unpublished). Orsay, 1972.

[9] D'Inverno, R. *Introducing Einstein's Relativity*. Oxford University Press, 1992.

[10] Dirac, P. A. M. *General Theory of Relativity*. John Wiley & Sons, 1975.

[11] Einstein, A., Lorentz, H. A., Weyl, H. and Minkowski, H. *The Principle of Relativity*. Dover Publications, Inc., 1952.

[12] Eisenhart, L. P. *Riemannian Geometry*. Princeton University Press, 1926.

[13] Flanders, H. *Differential Forms with Applications to the Physical Sciences*. Dover Publications, Inc., 1989.

[14] Fock, V. *The Theory of Space, Time and Gravitation*. Macmillan Company, 1964.

[15] Franklin, J. and Baker, P. T. "Linearized Kerr and spinning massive bodies: an electrodynamics analogy", *Am. J. Phys.* **75**(4): 336–42, 2007.

[16] Goldstein, H., Poole, C. and Safko, J. *Classical Mechanics*. Addison-Wesley, 2002.

[17] Griffiths, D. J. *Introduction to Electrodynamics*. Prentice-Hall, 1999.

[18] Hartle, J. B. *Gravity: An Introduction to Einstein's General Relativity*. Addison-Wesley, 2003.

[19] Hatfield, B. *Quantum Field Theory of Point Particles and Strings*. Westview Press, 1992.

[20] Hobson, M. P., Efstathiou, G. P. and Lasenby, A. N. *General Relativity: An Introduction for Physicists*. Cambridge University Press, 2006.

[21] http://einstein.stanford.edu/

[22] http://www.ligo.caltech.edu/

[23] http://lisa.nasa.gov/

[24] Jackson, J. D. *Classical Electrodynamics*. John Wiley & Sons, 1999.

[25] Kreyszig, E. *Differential Geometry*. Dover Publications, Inc., 1991.

[26] Lanczos, C. *The Variational Principles of Mechanics* (fourth edition). Dover Publications, Inc., 1986.

[27] Landau, L. D. and Lifshitz, E. M. *The Classical Theory of Fields.* Butterworth-Heinenann, 1975.

[28] Levi-Civita, T. *The Absolute Differential Calculus (Calculus of Tensors).* Dover, 1977.

[29] Lightman, A. P., Press, W. H., Price, R. H. and Teukolsky, S. A. *Problem Book in Relativity and Gravitation.* Princeton University Press, 1975.

[30] Milne-Thomson, L. M. *Theoretical Hydrodynamics.* Dover, 1996.

[31] Misner, C. W., Thorne, K. S. and Wheeler, J. A. *Gravitation.* W. H. Freeman and Company, 1973.

[32] Panofsky, W. K. H. and Phillips, M. N. *Classical Electricity & Magnetism.* Dover, 2005.

[33] Polchinksi, J. *String Theory, Vol. 1.* Cambridge University Press, 2005.

[34] Rubakov, V. (transl. by S. Wilson). *Classical Theory of Gauge Fields.* Princeton University Press, 2002.

[35] Synge, J. L. and Schild, A. *Tensor Calculus.* Dover Publications, Inc., 1949.

[36] Schutz, B. *A First Course in General Relativity.* Cambridge University Press, 2009.

[37] Schwinger, J. *Particles, Sources, and Fields.* Addison-Wesley, 1970.

[38] Thirring, W. E. *Classical Mathematical Physics: Dynamics Systems and Fields.* Springer, 1997.

[39] Wald, R. M. *General Relativity.* The University of Chicago Press, 1984.

[40] Weinberg, S. *Gravitation and Cosmology: Principles and Applications of the General Theory of Relativity.* John Wiley & Sons, 1972.

[41] Wheeler, N. *Analytical Dynamics of Fields.* Reed College, 1999.

[42] Zwiebach, B. *A First Course in String Theory.* Cambridge University Press, 2004.

Index

Printed in the United States
By Bookmasters